Environmental Science and Engineering

Series Editors

Ulrich Förstner, Buchholz, Germany

Wim H. Rulkens, Department of Environmental Technology, Wageningen
The Netherlands

The ultimate goal of this series is to contribute to the protection of our environment, which calls for both profound research and the ongoing development of solutions and measurements by experts in the field. Accordingly, the series promotes not only a deeper understanding of environmental processes and the evaluation of management strategies, but also design and technology aimed at improving environmental quality. Books focusing on the former are published in the subseries Environmental Science, those focusing on the latter in the subseries Environmental Engineering.

Haoqing Xu
Editor

The 6th International Symposium on Water Resource and Environmental Management

Water-Energy-Environment-Governance from Interdisciplinary Perspectives

Springer

Editor
Haoqing Xu
Jiangsu University of Science
and Technology
Zhenjiang, China

ISSN 1863-5520 ISSN 1863-5539 (electronic)
Environmental Science and Engineering
ISBN 978-3-031-55988-4 ISBN 978-3-031-55989-1 (eBook)
https://doi.org/10.1007/978-3-031-55989-1

This Springer imprint is published by the registered company Springer Nature Switzerland AG
The registered company address is: Gewerbestrasse 11, 6330 Cham, Switzerland

Paper in this product is recyclable.

Preface

This book serves as the inaugural volume in the new Sustainable Development of Water Resource and Environmental Management Series, offering a comprehensive introduction to the critical themes within. Delving into the intricate realm of sustainable development and management in the water sector, it addresses the profound challenges facing water resources amidst current global changes.

In the wake of these transformations, the availability and quality of water resources confront imminent threats. Notably, sustainable development emerges as a pivotal concern across all sectors related to water resources management, carrying profound implications for both present and future generations.

Curated from the noteworthy contributions of the 6th International Symposium on Water Resource and Environmental Management (WREM 2023), this book presents a collection of carefully selected papers that explore key themes in water resource and environmental management. Readers will find a wealth of information encompassing the fundamental principles of sustainable water resources management, recent advances, insights into future research directions, and the formulation of policies aimed at fostering sustainable water resources management.

Designed to cater to a diverse audience, this book is a valuable resource for beginners, researchers, and professionals engaged in the fields of water pollution and treatment, water engineering, and engineering structures. Whether you are initiating your journey into this field or are an established expert, the content within promises to enhance your understanding and contribute to the ongoing discourse in sustainable water resources management.

Haoqing Xu, Ph.D.
Master Supervisor, Associate Professor
Deputy Director of Department
of Geotechnical Engineering
School of Architecture and Civil
Engineering
Jiangsu University of Science
and Technology
Zhenjiang, Jiangsu, China

Contents

Chapter 1
Assessing the Drinking Water Quality Indexes of Borehole and Surface Water Close to a Mining Dump in Welkom, South Africa

Silent Ruzvidzo and Saheed Oke

Abstract Groundwater is an important source of drinking water in most parts of Southern Africa where access to tap water is limited. In most cases, groundwater is used untreated as it is considered safe to drink than surface water. However, anthropogenic activities such as agriculture, landfills, and mining have resulted in the contamination of groundwater sources. This study examines the drinking water quality standards of 5 boreholes and a stream of water which are located near a mining dump in Welkom, Free State. In the study, water samples were periodically collected between the Autumn and Summer seasons of 2022. The concentrations of the physical, chemical, and microbiological determinants; Electrical Conductivity (EC), Total Dissolved Solids (TDS), calcium (Ca), magnesium (Mg), sodium (Na), potassium (K), fluoride (F), chloride (Cl), nitrate (NO_3), sulphate (SO_4), calcium hardness ($CaCO_3$), phosphate (PO_4), Magnesium hardness, Total coliform, Faecal coliforms, and *Escherichia Coli* were assessed against the SANS 241 (Guidelines for drinking water quality, 2015) and the WHO (Guidelines for drinking-water quality. World Health Organization, Geneva, 2017) guidelines for water quality standards. The study results indicated that none of the 5 boreholes had safe drinking water as their Water Quality Indices were poor. Abnormally high concentrations of Total Coliforms and *Escherichia Coli* coupled with seasonal variations were also noticed for all boreholes. Recommendations based on these findings are that the Welkom area residents should avoid drinking any borehole water before purification and treatment. Furthermore, the local authorities should be urgently alerted to remedy the situation and discuss solutions on the safe disposal of mining waste with the mining company.

S. Ruzvidzo
Department of Life Sciences, Central University of Technology, Bloemfontein 9300, South Africa

S. Oke (✉)
Department of Civil Engineering, Central University of Technology, Bloemfontein 9300, South Africa
e-mail: soke@cut.ac.za

H. Xu (ed.), *The 6th International Symposium on Water Resource and Environmental Management*, Environmental Science and Engineering,
https://doi.org/10.1007/978-3-031-55989-1_1

Keywords Groundwater · Drinking · Water quality · Contamination · Mining dump · Human Health

1.1 Introduction

The access to safe drinking water is a fundamental human right (WHO 2008, 2017). The pollution of groundwater sources has become a global problem with adverse effects on human health and the environment (Satheeskumar et al. 2020). In areas close to mining operations or mining tailings, the pollution of ground and surface water sources need to be investigated as the implications of water contamination might have serious health consequences for humans as well as animals (Ding et al. 2021).

To protect humans' health, different countries and international organisations have developed statutory regulations which stipulate the standards of the quality of drinking water for their populations. In South Africa, the South African National Standard 241 (SANS 241), provides the drinking water quality guidelines for all the portable water. Moreover, the World Health Organisation (WHO) also provides water quality guidelines to safeguard the general world population in areas where the drinking water standards are not explicitly detailed. In all, the drinking water quality standards highlight the physical, chemical, and microbiological determinants and concentrations that determine the suitability of drinking water in terms of the public health regulations.

To assess the drinking water quality standards encompassing a variety of determinants, the Water Quality Index (WQI) was developed (Zhou et al. 2021). Initially proposed by Horton (1965) and Brown et al. (1970) in 1972, the WQI has been modified and developed into different varieties so that it can adapt to different environments (Malan et al. 2003; Wertz and Shank 2019). Most of the WQI methods utilise physical and chemical parameters to calculate the index in different ways (Lumb et al. 2011). The WQI is used to measure the quality of the drinking water as per the regulatory standards. It is a rating that reflects the compound effect of various water quality determinants on the total quality of water. In essence, the WQI is a mathematical tool used to convert large amounts of water quality information to a single number, which provides the tools and basis for water quality managers to determine the quality and potential uses of a given body of water. The inclusion of different water quality parameters in the Water Quality Index depends on the context, purpose, and approach of the investigation (WRC 2017).

This study was undertaken to investigate the drinking water quality of the groundwater from 5 boreholes in Welkom including an open water stream close to a gold mine dump which is the suspected contaminant source. The investigation involved the use of the SANS 241 (SANS 2015) as well as the WHO (2017) water quality guidelines to establish the water quality from the sources.

1.2 Materials and Methods

Groundwater from the Central University of Technology campus boreholes, located in Welkom, was periodically sampled, and analysed in the Autumn and Summer seasons of 2022. To provide snapshots of the water quality at different times of the seasons, the grab method was used to collect the samples from 6 sampling sites at least 3 times per each season. 5 of the sampling sites were boreholes, and the 6th one was an open water stream located at no less than 100 m from a nearby mining dump. The accessibility of the boreholes, as well as their proximity to the mining dump, made them ideal sampling sites for the investigation. Figure 1.1 shows the location of the sampling points 1–6 across the CUT campus in Welkom and the coordinates are represented in Table 1.1.

Fig. 1.1 Sampling points (1–6) along the Central University of Technology Welkom campus and an open stream close to the mining dump (Google Maps 2023)

Table 1.1 Geographical coordinates of the sampling points

Sampling point	Description	Longitude	Latitude	Elevation (m)
1	Borehole 1	27° 56′ 57.9″ S	26° 47′ 05.7″ E	1,396
2	Borehole 2	27° 56′ 58.6″ S	26° 47′ 05.5″ E	1,391
3	Borehole 3	27° 56′ 52.4″ S	26° 47′ 08.6″ E	1,386
4	Borehole 4	27° 56′ 57.4″ S	26° 46′ 58.4″ E	1,397
5	Borehole 5	27° 56′ 57.4″ S	26° 46′ 55.3″ E	1,399
6	Open stream	27° 56′ 52.7″ S	26° 46′ 41.1″ E	1,403

The collected samples were analysed for Physical, Chemical, and Microbiological elements. The physical water quality parameters such as electrical conductivity (EC), Total Dissolved Solids (TDS), temperature, and pH, were analysed onsite using the HANS instruments whilst the chemical and microbiological determinants were analysed at the Institute of Ground Water Services laboratories (IGS) at the University of the Free State (UFS). The chemical and microbiological elements analysed were, pH, calcium (Ca), magnesium (Mg), sodium (Na), potassium (K), fluoride (F), chloride (Cl), nitrate (NO_3), sulphate (SO_4), calcium hardness ($CaCO_3$), phosphate (PO_4), Magnesium hardness, Total coliform, Faecal coliforms, and *Escherichia Coli.* For laboratory chemical analyses, sterile 500 ml bottles were used, while for microbiological analyses, sterile 100 ml bottles were used to collect water samples. All samples destined for the laboratory were labelled appropriately and placed in an icebox before being transported to the laboratory where they were analysed within a 12-h period.

The results from the physical, chemical, and micro-biological analysis were compared with the South African National Standards (SANS) 241 of 2015 as well as the World Health Organisation (WHO) 2017 standards to establish the percentage compliance for each borehole sample as well as for each parameter. The water quality for each borehole and the stream water was also calculated using the weighted arithmetic index method. Additionally, the microbiological parameters from the water samples were compared with the SANS 241 (2015) standards.

1.2.1 Methodology for Calculating the Water Quality Index (WQI)

As the water samples were collected at different time intervals of the season and at varying frequencies, the Weighted Arithmetic Index (WAI) method was selected as the most suitable method for calculating the water quality. Furthermore, the WAI utilises less complicated calculations that are easy to apply as compared to the other water quality index calculation methods.

The WAI method consisted of the following steps:

Step 1: Data collection of the physico-chemical water quality parameters including pH, electrical conductivity, calcium (Ca), magnesium (Mg), sodium (Na), potassium (K), fluoride (F), chloride (Cl), nitrate (NO_3), and sulphate (SO_4).

Step 2: The proportionality constant "K" value was calculated using the formula:

$$\mathrm{k} = (1/(1/\sum si)) \tag{1.1}$$

where

"si" are the standard permissible for the *n*th parameter.

Table 1.2 Water class parameters for the weighted arithmetic index method

Water quality index	Water quality class
0–25	Excellent
26–50	Good
51–75	Poor
76–100	Very poor
Above 100	Not suitable for drinking purposes

Rickwood and Carr (2007), Ratikane (2013)

Step 3: The quality rating for the *n*th parameter (qn) was calculated for n parameters using the formula:

$$qn = 100\{(v_n - vi_o)/(s_n - vi_o)\} \tag{1.2}$$

where

v_n Estimated value of the *n*th parameter of the given sampling station.
vi_o Ideal value of *n*th parameter in pure water, and
S_n Standard permissible value of the *n*th parameter.

Step 4: Calculation for the unit weight for the *n*th parameters using formula:

$$Wn = (k/Sn) \tag{1.3}$$

Step 5: Calculation of the Water Quality Index (WQI) using formula:

$$WQI = ((\Sigma Wn \times qn)/\Sigma\, Wn) \tag{1.4}$$

The WQI ranges and their classes were defined as shown in Table 1.2.

1.3 Results

The comparison of the average physical and chemical determinants with SANS 241 (2015) and WHO drinking water quality (2017) guidelines revealed that the water quality parameters in the boreholes and mine stream exceeded the recommended limits by varying margins. Table 1.3 indicates the percentage compliance of each borehole (B1–B5) as well as the stream water (S6) against the SANS 241 and WHO (2017) water quality guidelines.

The samples B3 and S6 from borehole 3 and stream water had the lowest compliance rate out of all the 6 samples assessed. While sample B3 exceeded the physical

Table 1.3 Comparison of the physical and chemical determinants with the SANS 241 and WHO, 2017 limit standards

Parameter	Limit standard (SANS 241)	Limit standard (WHO)	Risk	B1	B2	B3	B4	B5	Mean for boreholes	S6 (open stream)	Minimum	Maximum
EC (Electrical Conductivity)	(170 mS/m)	-	Acute Health	84.62	<14.7	202.31	62.75	138.89	122.14	2051.42	<14.7	2051.42
TDS (Total Dissolved Solids)	(1200 mg/l)	1000	Aesthetic	579.77	85.97	1234.44	433.16	891.73	645.01	17531.61	85.97	17531.61
pH	(5-9.7)	8.5	Operational	7.04	7.55	6.80	7.64	6.81	7.17	7.17	6.81	7.55
Calcium (Ca)	(150mg/L)	75	Aesthetic	68.60	10.82	79.63	55.61	93.06	61.54	896.04	10.82	896.04
Magnesium (Mg)	(70mg/L)	30	Aesthetic	28.69	3.21	41.37	22.98	36.53	26.56	856.49	3.21	41.37
Sodium (Na)	(200mg/L)	200	Aesthetic	73.92	13.22	307.61	49.72	146.79	118.25	3170.74	13.22	3170.74
Potassium (K)	(50mg/L)	10	Aesthetic	12.67	4.30	29.20	8.25	25.47	15.98	304.65	4.30	29.20
Fluoride (F)	(1.5mg/L)	1.5*	Chronic Health	0.37	<0.10	0.39	0.30	0.35	0.35	1.89	<0.10	1.89
Chloride (Cl)	(300mg/L)	250*	Aesthetic	85.25	18.89	317.05	63.82	194.54	135.91	6197	18.89	6197
Nitrate (NO_3)	(11mg/L)	50*	Acute Health	1.76	0.36	0.29	2.00	4.16	1.71	0.75	0.36	4.16
Sulphate ($S0_4$)	(500mg/L)	250	Acute Health	134.85	1.11	146.51	99.40	89.36	94.25	3075.46	1.11	3075.46
Total Hardness ($CaCO_3$)	(100mg/L)	-	Aesthetic	289.65	40.25	369.21	233.49	382.81	263.1	5764.44	40.25	5764.44
% Compliance				92	100	58	92	92	100	17		

Key		
	Sample exceeds the SANS (2015) limit standard	(orange)
	WHO Guidelines for drinking water quality (2017); standards of health concern	*

parameters limits of EC, and TDS, by approximately 19% and 3% margins respectively, the samples from the stream water exceeded the same parameters by more than 200% and 14% respectively.

For the chemical parameters, the borehole samples B1, B3, B4, and B5 from boreholes 1, 3, 4, and 5 all exceeded the total hardness due to $CaCO_3$ as per the SANS 241 and WHO (2017) recommended guidelines. Additionally, sample B3 had excess Sodium and Chloride concentrations.

The stream water sample S6 exceeded all the chemical water quality guidelines with the exception of pH and Nitrate. The S6 sample also had the lowest compliance percentage of 17% with dangerous levels of Fluoride and Sulphates which are highlighted as the causes of some chronic and acute diseases in humans.

Some of the important water quality parameters in terms of aesthetics, and health were plotted against the SANS 241 limit standards. Figure 1.2 shows a plot of the EC recorded for all the samples against the SANS 241 limit standard.

The orange line in Fig. 1.2 represents the SANS limit standard of 170 mS/m whilst the blue line represents the average recorded EC values for each sample.

Figure 1.3 shows the limit standard for Nitrate values plotted against the SANS limit standard of 11 mg/L.

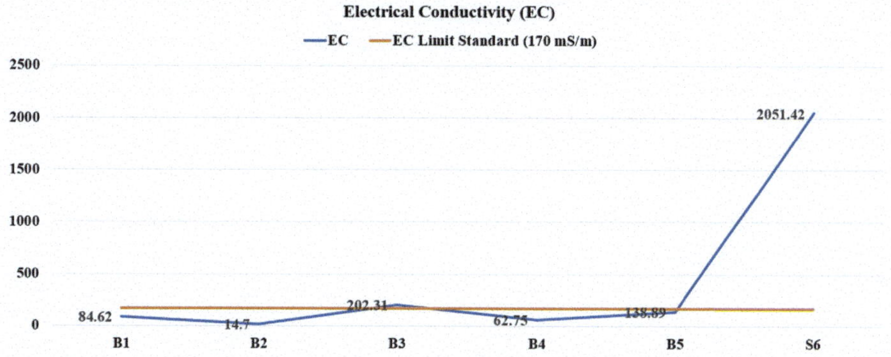

Fig. 1.2 Electrical Conductivity (EC) plotted against their SANS 241 limit standard

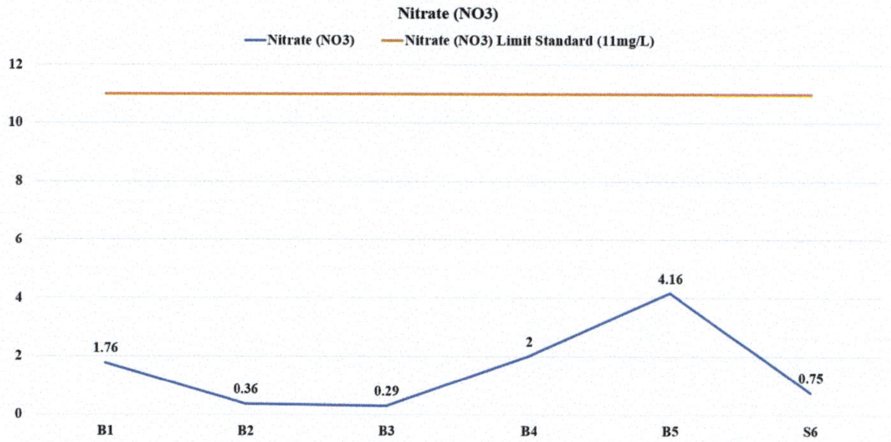

Fig. 1.3 Nitrates concentrations plotted against the SANS 241 limit standard

All the plotted nitrate values were below the SANS 241 limit standard. Figure 1.4 shows the plot of the average Fluoride concentrations per sample against the SANS 241 limit standard of 1.5 mg/L.

All samples except S6 had Fluoride concentration levels below the SANS limit standard.

1.3.1 Comparison of Microbiological Parameters

The analysis of the microbial parameters involved the assessment of Total Coliforms, Faecal Coliforms, and *Escherichia Coli* in all the borehole and stream samples in the Summer and Automn seasons of 2022 against the SANS 241 (2015) standards.

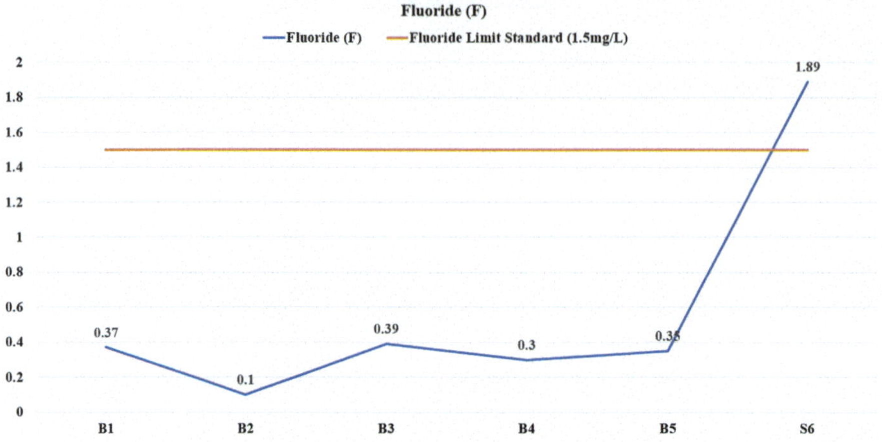

Fig. 1.4 Fluoride concentrations plotted against the SANS 241 limit standards

Samples B2, B3, and B5 had excess numbers of Total and Faecal Coliforms in the summer seasons for the boreholes whilst sample S6 had exceeded all the SANS 241 limits for all the parameters under investigation. Table 1.4 shows the percentage compliance of the microbiological parameters against the SANS 241 standards for the summer season.

For the Autumn season, all the 5 boreholes together with the stream sample exceeded the Total coliforms SANS 241 standard. None of the samples exceeded Faecal Coliforms SANS 241 standards and all samples but B3 exceeded the number of *Escherichia Coli* SANS standard as indicated in Table 1.5.

Table 1.4 Comparison of microbiological parameters-summer season

Parameter	SANS 241 Standard	Risk	B1	B2	B3	B4	B5	S6
Total coliforms	(<10 counts/100ml)	Operational	2	>2420	>2420	<1	127	>2420
Faecal Coliform	(Not detected in 100ml)	Acute Health	<1	67	4	<1	<1	613
Escherichia Coli	(Not detected in 100ml)	Acute Health	<1	<1	<1	<1	<1	1
Compliance %				33	33	100	67	0

Key	Sample exceeds the SANS (2015) limit standard.	

Table 1.5 Comparison of microbiological parameters-autumn season

Parameter	SANS 241 Standard		B1	B2	B3	B4	B5	S6
Total coliforms	(<10 counts/100ml)	Operational	>2420	>2420	10	2420	>1000	>2420
Faecal Coliform	(Not detected in 100ml)	Acute Health	326	1986	1	112	816	613
Escherichia Coli	(Not detected in 100ml)	Acute Health	5	>2420	<1	1	6	1
Compliance %				33	66	33	33	33

Key	Sample exceeds the SANS (2015) limit standard.	

Table 1.6 Water quality indexes

Borehole sample	Water quality index	water quality class
1	64.26	Poor
2	44.79	Good
3	63.06	Poor
4	31.78	Good
5	51.06	Poor
Mean	68.16	Poor
Stream water	544.31	Not suitable for drinking purposes

Table 1.6 shows the Water Quality Indexes for all the 6 samples together with their associated Water Quality Classes. All the Water Quality Indexes were calculated through the WAI method using the SANS 241 and WHO (2017) water quality guidelines. 3 out of the 5 boreholes had poor water quality whilst the other 2 boreholes had good water quality. The mean WQI calculated from all the 5 boreholes using the physical and chemical determinants indicated that the overall water quality was poor.

The stream WQI of 544.31 was 5 times higher than the minimum guideline requirements of 100 for very poor water quality making the water not suitable for drinking purposes.

1.4 Discussion

The results from the study indicate that the groundwater quality from the boreholes located at the CUT campus cannot be considered safe for drinking purposes. Despite the water quality indexes indicating that the water can be used for drinking purposes, the water is still not fit for human consumption as the microbiological water quality determinants for all samples exceed the threshold for drinking water quality guidelines as set by the SANS 241 and WHO (2017) standards.

The high levels of determinants found in the stream close to the mining dump can be possibly linked to the heavy leaching of contaminants from the mining dump. Similarly, the above average contamination of borehole 3 can also be linked to its close proximity to the contaminated stream water. No consistent direct correlations were found between the concentration of determinants in any sample and the number of microbial organisms found in both seasons. Further WQI assessments involving heavy metals might be necessary to explore further correlations with the microbial organisms.

The microbial contamination of the groundwater varied greatly for the Autumn and Summer seasons respectively with Total coliforms and *E. coli* being more dominant in the Autumn season whilst Total coliforms and Faecal coliforms are more dominant in the summer season. The variations in the seasonal microbial compositions can be attributed to the different weather patterns found in the Autumn and Summer seasons contributing variably to the microbial growths. The seasonal variations in microbial growth need to be closely traced as it can pose unexpected seasonal health risks to the Welkom population.

1.5 Conclusion

The groundwater in the CUT Welkom area is heavily contaminated by the physical, chemical, and micro-biological determinants from the nearby mine dump. The contamination deteriorates the water quality hence making it unsafe for drinking purposes. Despite some of the borehole water samples meeting some of the SANS 214 and WHO 2017, water quality guidelines for some determinants, none of the water samples collected from the boreholes were fully compliant in all the parameters. The use of the groundwater for drinking purposes might hence lead to some health complications in humans. Further testing is required to assess the presence of any heavy and or trace metals in the groundwater in order to establish vulnerability of the groundwater from the mine dump.

References

Brown RM, McClelland NI, Deininger RA, Tozer RG (1970) A water quality index: Do we dare? water sew works 117:339–343

Ding L, Yang Q, Yang Y, Ma H, Martin JD (2021) Potential risk assessment of groundwater to address the agricultural and domestic challenges in Ordos Basin. Environ Geochem Health 43:717–732

Horton RK (1965) An index number system for rating water quality. J Water Pollut Control Federation 37:300–306

https://www.google.com/maps/@28.0044172,26.6182962,11z/data=!3m1!4b1!4m3!11m2!2sv_Qv9I4yRSi57QBv7oNNBg!3e3?entry=ttu. Last accessed 20 July 2023

Lumb A, Sharma TC, Bibeault J (2011) A review of genesis and evolution of water quality index (WQI) and some future directions. Water Qual Expo Health 3(1):11–24

Malan H, Bath A, Day A, Joubert A (2003) A simple flow-concentration modelling method for integrating water quality and water quantity in rivers. Water SA 29(3):305–312

Ratikane M (2013) Quality of drinking water sources in the Bloemfontein area of the Mangaung Metropolitan Municipality. Master's dissertation, Central University of Technology, Bloemfontein

Rickwood C, Carr GM (2007) Global drinking water quality index development and sensitivity analysis report. United Nations Environment Programme Global Environment Monitoring System (GEMS) Water Programme, Burlington, Ontario

Satheeskumar V, Subramani T, Lakshumanan C, Roy PD, Karunanidhi D (2020) Groundwater chemistry and demarcation of seawater intrusion zones in the Thamirabarani delta of south India based on geochemical signatures. Environ Geochem Health 43:757–770

South African National Standards 241 (2015) Guidelines for drinking water quality, 2nd edn

Wertz TA, Shank MK (2019) Land use from water quality: development of a water quality index across Pennsylvania streams. Ecosphere 10(11):1–20

World Health Organization (WHO) (2008) Guidelines for drinking-water quality: incorporating the first and second addenda, 3rd edn. World Health Organization, Geneva

World Health Organization (WHO) (2017) Guidelines for drinking-water quality, 4th edn. World Health Organization, Geneva

WRC (2017) Groundwater sampling manual. Water Research Commission (WRC), Pretoria, South Africa

Zhou Y, Li P, Chen M, Dong Z, Lu C (2021) Groundwater quality for potable and irrigation uses and associated health risk in southern part of Gu'an County. North China Plain. Environ Geochem Health 43:813–835

Chapter 2
Small Watersheds Based on Game Theory Comprehensive Weight Method Flash Flood Disaster Risk Assessment

Qinxin He, Yunchu Ren, Yuyu Liu, and Bo Huang

Abstract In order to accurately assess the risk of flash flood disasters in small-scale river basins, this paper takes the small watershed of Beidasha River in Shandong Province as an example, selects 10 influencing indicators in view of the single method of flash flood disaster risk assessment in the past and the one-sided shortcomings of determining the weight coefficient of indicators, synthesizes the analytic hierarchy method and entropy method based on the principle of game theory, determines the more scientific index weight value, introduces the fuzzy comprehensive analysis method, constructs a flash flood disaster risk assessment model in the small watershed, and uses ArcGIS to divide the risk area. Combined with the township zoning and administrative village point layers, the raster was used as the evaluation unit to analyze the risk of flash flood disaster in the whole river basin. The results show that the annual rainstorm days are the most weighted among the indicators, followed by the topography index, and the vegetation index has the lowest weight. The area with high risk level accounted for 40.52% of the whole river basin, which was mainly distributed in the lower reaches of the river basin, and the evaluation results were consistent with the distribution of historical disasters, which could provide data support for flash flood disaster early warning.

Keywords Flash floods · Risk assessment · Game theory comprehensive weight method · Flood prevention · Disaster mitigation

Q. He · Y. Liu · B. Huang (✉)
School of Water Conservancy and Environment, University of Jinan, Jinan 250022, China
e-mail: stu_huangb@ujn.edu.cn

Y. Ren
Shandong Yellow River Survey, Design and Research Institute Co. Ltd, Jinan, China

H. Xu (ed.), *The 6th International Symposium on Water Resource and Environmental Management*, Environmental Science and Engineering,
https://doi.org/10.1007/978-3-031-55989-1_2

2.1 Introduction

Flash flood disasters are a series of disasters induced by heavy rainfall in mountainous areas, which directly or indirectly cause loss of life and property. It is sudden and destructive. Many cities in hilly areas in China occupy an area far exceeding the world average level, and are one of the countries most affected by disasters. Therefore, mountain torrent disaster prevention and comprehensive management of small watersheds in hilly areas have received more and more attention from the state and relevant departments. Risk assessment plays an important role in mountain torrent disaster prevention (Jiang et al. 2020; Wei et al. 2022). At present, researchers at home and abroad have studied many methods of mountain torrent disaster risk assessment, such as principal component analysis, analytic hierarchy process, entropy value method, fuzzy comprehensive analysis method, random forest method, artificial neural network method, multi-factor comprehensive index However, each evaluation method has its own advantages and disadvantages. The key is to determine the weight value of the impact index. Compared with the single method to determine the weight of the index, the comprehensive weight method of game theory can combine the advantages of each method. objectivity and accuracy (Yu et al. 2013; He et al. 2022; Liu et al. 2023).

In this paper, the Beidasha River Basin in Shandong Province is taken as the research area, and the 34 sub-watersheds divided based on ArcSWAT software are used as the evaluation objects, and the 30×30 m grid is used as the evaluation unit. The fuzzy comprehensive analysis method is used to determine the degree of membership of the hazard level, and the risk analysis of mountain torrent disasters in the Beidasha River Basin is transformed into a quantitative analysis. This research can provide more intuitive and reliable data information for regional flood control and disaster reduction, and can also provide further ideas for the risk assessment of flash flood disasters in similar regions.

The Beidasha River is a first-class tributary of the Yellow River, located in the east of Changqing District, Jinan City, with a total basin area of about 576 km^2, including 6 sub-districts and 282 administrative villages. The topography of the basin varies greatly, with the lowest elevation being 6 m and the highest elevation being 976 m. The hilly area accounts for about 81.3% of the total area. The upper reaches of the basin are low mountains and hills, the middle is the piedmont plain, and the lower reaches is the lowland along the Yellow River. The study area belongs to the warm temperate continental monsoon climate zone, with an average annual precipitation of about 644 mm. The precipitation is concentrated during the flood season from June to September. Extreme rainfall is the main cause of mountain torrent disasters in this basin, especially the catastrophic flood in Changqing District in 2007. Heavy rain, the maximum rainfall in the district reached more than 200 mm, and the direct economic loss reached 44.4 million yuan. There are 6 rainfall stations and Gushan hydrological stations in the basin, including Jieshou, Shihutong, Guanmachang, Wande, Jijiayu and Changqing.

2.2 Risk Assessment Methods of Flash Flood Disasters

2.2.1 *Determine the Comprehensive Weight of Evaluation Indicators Based on Game Theory*

Subjective and objective weighting methods. The judgment of the importance of the indicators by the analytic hierarchy process is often based on expert scoring or experience to determine the scale, which is greatly influenced by human subjectivity, while the entropy method uses the degree of dispersion of each indicator to determine the weight of each indicator by using information entropy. The method does not consider that the hierarchical relationship is too absolute (Liu et al. 2017; Yuan et al. 2018). Therefore, this paper neutralizes these two methods based on the principle of game theory, and takes their optimal solution as the comprehensive weight of the evaluation index.

The AHP is to divide complex problems from the top layer into target layer, criterion layer, and index layer to establish a corresponding hierarchical analysis system. The upper layer is influenced by the lower layer, and the indicators in the layer are independent of each other (Duan et al. 2021). The importance of each indicator is quantified by establishing a judgment matrix, and the saaty scaling method is often used for judgment.

The entropy value method is to normalize the judgment matrix according to the definition of entropy to obtain the entropy of the index, and determine the entropy weight by calculation w_i^* namely:

The judgment matrix $A_{i \times m}$ represents an evaluation index m area, $A_{i \times j} = (v_{ij})$ (i = 1, 2, …, n; j = 1, 2, …, m).

Matrix $A_{i \times j}$ is normalized to obtain matrix $B_{i \times j} = (b_{ij})$

$$b_{ij} = \begin{cases} \frac{v_{ij} - \min v_j}{\max v_i - \min v_j}, \text{ Positive indicators} \\ \frac{\max v_i - v_{ij}}{\max v_i - \min v_j}, \text{ Negative indicators} \end{cases} \tag{2.1}$$

In the formula, $\max v_i$ and $\min v_j$ represent the maximum value and minimum value in different regions under the same indicator, respectively.

$$H_j = -\frac{1}{\ln n} \sum_{i=1}^{n} p_{ij} \ln p_{ij} \text{ and } H_j \geq 0 \tag{2.2}$$

$$w_i^* = \frac{1 - H_j}{n - \sum_{j=1}^{n} H_j} \text{ and } \sum_{i=1}^{n} w_i^* = 1 \tag{2.3}$$

In the formula, P_{ij} is the ratio of different regions under each index; H_j is the entropy value of each index in different regions; w_i^* is the weight of each index calculated by the entropy value, $w_i^* \in [0,1]$.

The comprehensive weight is determined based on game theory. Game theory is based on the coordination goal of NASH equilibrium to construct the optimal countermeasure model, which can find the optimal solution among multiple indicators. Combining weights can minimize the dispersion of each evaluation index, integrate the advantages of subjective weighting method and objective weighting method, and make the distribution results more representative (Zhu et al. 2013).

Assuming that the index weights determined by different evaluation methods are used to construct the base vector set as $\lambda = (\lambda_1, \lambda_2, \ldots, \lambda_L)$, and the linear equations are formulated according to the differential principle, and the optimal base vector set λ is obtained through linear algebra calculations. After normalization, the combination weight W^*.

2.2.2 *Fuzzy Comprehensive Evaluation Method*

Membership function is a mathematical tool used to characterize the "real degree" of elements in fuzzy sets. It can be determined by fuzzy statistics, assignment methods, expert experience methods, binary comparison and sorting methods, etc., but its determination is usually subjective and difficult. It directly affects the accuracy of the evaluation results. In order to evaluate the reliability of the results, this paper adopts the method of assigning trapezoidal or semi-trapezoidal distribution functions to determine the membership degree of the index (Jing et al. 2018). Let $u_{i,n}(x)$ be the membership degree of evaluation index i to level n in target layer u, then the membership function is as follows:

$$u_{i,1}(x) = \begin{cases} 1, & x < a_1 \\ \frac{a_2 - x}{a_2 - a_1}, & a_1 \le x \le a_2 \\ 0, & x > a_2 \end{cases} \tag{2.4}$$

$$u_{i,j}(x) = \begin{cases} \frac{x - a_{j-1}}{a_j - a_{j-1}} & a_{j-1} \le x \le a_j \\ 1, & a_j \le x \le a_{j+1} \\ \frac{a_{j+2} - x}{a_{j+2} - a_{j+1}} & a_{j+1} \le x \le a_{j+2} \\ 0, & x < a_{j-1}, x > a_{j+2} \end{cases} \tag{2.5}$$

$$u_{i,n}(x) = \begin{cases} 0, & x < a_{n-1} \\ \frac{x - a_{n-1}}{a_n - a_{n-1}} & a_{n-1} \le x \le a_n \\ 1, & x > a_n \end{cases} \tag{2.6}$$

In the formula, x is the actual value of the evaluation index; a_n is the critical value of the corresponding category interval. The final evaluation is made according to the principle of the maximum degree of membership.

2.3 Comprehensive Risk Assessment of the Study Area

2.3.1 Index System Construction

According to the theory of mountain torrent disaster system (Ren et al. 2023), the risk evaluation index system in this paper selects 10 evaluation indexes from the three criteria of the hazard of the hazard, the sensitivity of the disaster-forming environment, and the vulnerability of the carrier to construct the evaluation index system. Evaluation set $M = \{$very low risk, low risk, medium risk, high risk, very high risk$\}$ with 5 levels in total. The average annual maximum 24-h rainfall, annual average rainfall, and annual number of rainstorm days are selected as the risk indicators of watershed disaster-causing factors; due to the large fluctuations in the topography of small watersheds in hilly areas, the slope of the terrain, the hydrological characteristics of the watershed, and the distribution of vegetation It has a greater impact on floods, so the elevation, slope, topographic index, and vegetation index are selected as the sensitivity indicators of the evaluation object; the small watershed is smaller in scale than the large watershed, and the vulnerability of the carrier is mainly on both sides of the river. Therefore, the population density, building density and GDP density are selected as the evaluation indicators of this criterion (Ge and Luo 2018; Sun et al. 2021; Wu et al. 2018; Meng et al. 2022).

According to the constructed indicator system to collect the relevant indicators raw data and processed, data information sources as shown in Table 2.1.

The data collected in Table 2.1 were processed in the ArcGIS platform in 30 m × 30 m rasters to obtain the required indicator layer data in Table 2.2, in which the indicator layer of the disaster-causing factors (annual maximum 24-h rainfall, annual average rainfall, and the number of days of heavy rainfall) was obtained by spatially interpolating seven stations in the watershed using the inverse distance-weighting method; the positive value in the topographic index map corresponds to the mountain ridges, the negative value corresponds to the valleys, and the zero value is for the flat

Table 2.1 Indicator data sources and precision

Indicator data	Source	Accurate
Rainfall data	Hydrological Yearbook 2007–2019	Daily rainfall
Elevation	Geospatial data cloud	30 m
Vegetation Index	NASA National Aeronautics and Space Administration	1 km
Population density	Resource Environment Data Center, Chinese Academy of Sciences	1 km
GDP density	Resource Environment Data Center, Chinese Academy of Sciences	1 km
Digital orthophoto	Google Maps	8 m

Table 2.2 Critical value of flash flood disaster assessment indicators in the Beidasha River Basin

Indicator threshold	Annual maximum 24 h average rainfall/mm	Annual average rainfall/mm	Rainstorm days/d	Elevation/m	Slope/(°)
a_1	67.59	483.06	1.85	4	0
a_2	77.19	531.75	1.99	108	6.20
a_3	84.74	573.72	2.13	221	12.57
a_4	91.22	619.05	2.29	348	20.31
a_5	96.17	655.99	2.49	513	29.70
a_6	101.54	697.12	2.70	976	64.33
Indicator threshold	Terrain index	Vegetation index	Population/(person km^{-2})	Architecture/(room km^{-2})	GDP/(million yuan km^{-2})
a_1	−31.67	0.22	358	22.13	1,796
a_2	−3.49	0.36	480	227.69	3,045
a_3	−0.95	0.41	610	358.12	4,397
a_4	0.90	0.45	770	504.38	6,776
a_5	3.20	0.50	955	702.02	10,666
a_6	27.22	0.61	1640	1,030.11	19,339

area; and the building density map was obtained by kernel density analysis of the data collected from the building points in the digital orthophotographs (Fig. 2.1).

Each evaluation index layer is divided into five interval categories using the natural discontinuity grading method, and the critical value of each interval is used as the discontinuity point of the membership function to determine the degree of membership. The specific values of each index are shown in Table 2.2.

2.3.2 *Comprehensive Risk Assessment*

The weight values of each index are calculated by different methods, as shown in Table 2.3, method 1 is the AHP subjective weight method, method 2 is the entropy weight method (objective assignment method), and method 3 is the integrated weight method based on game theory, which is the weight value obtained by combining the first two methods.

According to the constructed trapezoidal or semi-trapezoidal distribution function, the judgment matrix of the membership degree of 34 risk levels is obtained, and the risk evaluation matrix of 34 watersheds is obtained through the linear algebra calculation of the comprehensive weight matrix of each index and the judgment matrix, and the risk level is determined according to the principle of the maximum degree of membership, using ArcGIS to assign values to the small watershed boundary and

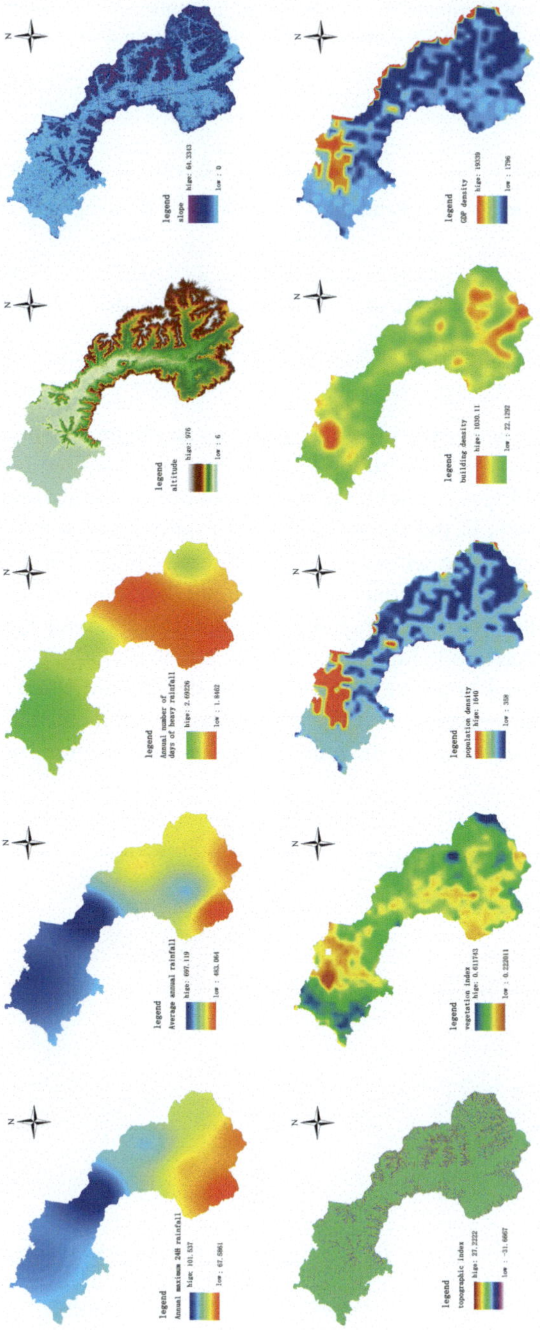

Fig. 2.1 Map of the spatial distribution of the evaluation indicators in the North Dasha River

Table 2.3 Index data source and accuracy

Impact indicators	Annual maximum 24 h average rainfall	Annual average rainfall	Number of rainstorm days per year	Elevation	Slope
Method 1	0.065	0.119	0.216	0.064	0.111
Method 2	0.077	0.080	0.133	0.104	0.098
Method 3	0.076	0.116	0.205	0.085	0.116

Impact indicators	Terrain index	Vegetation index	Population density	Building density	GDP density
Method 1	0.187	0.038	0.089	0.077	0.034
Method 2	0.062	0.074	0.173	0.049	0.1514
Method 3	0.158	0.055	0.129	0.074	0.158

township zoning layer, and perform zoning statistics to obtain the mountain torrent disaster risk zoning map of the Beidasha River Basin, as shown in Fig. 2.2.

It can be seen from Fig. 2.2 that the proportion of areas with extremely high and extremely low risk levels in the Beidasha River Basin is zero, and the overall risk of flash flood disasters in the whole region is relatively high. The high-risk areas of mountain torrent disasters are concentrated, mainly distributed in 10 small watersheds downstream of the watershed, including Guyunhu Street, Wenchang Street, and parts of Zhangxia Street, Guide Street, and Pingan Street, including 165 administrative villages; Medium-risk areas are evenly distributed in Wande Street and Zhangxia Street, including 47 administrative villages; lower-risk areas include 70 administrative villages. Through calculation, the areas of higher risk, medium risk, and lower

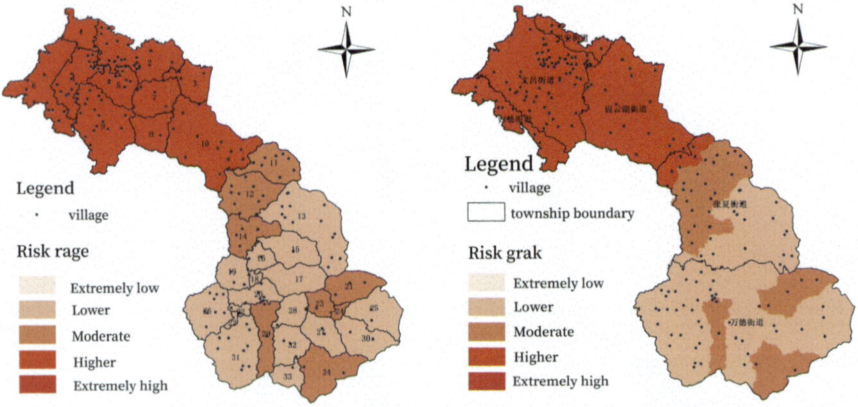

(a) Comprehensive evaluation map of mountain torrent disasters in 34 small watersheds.

(b) Comprehensive evaluation map of mountain torrent disasters in each township.

Fig. 2.2 Zoning map of mountain torrent disaster risk in the Beidasha River Basin

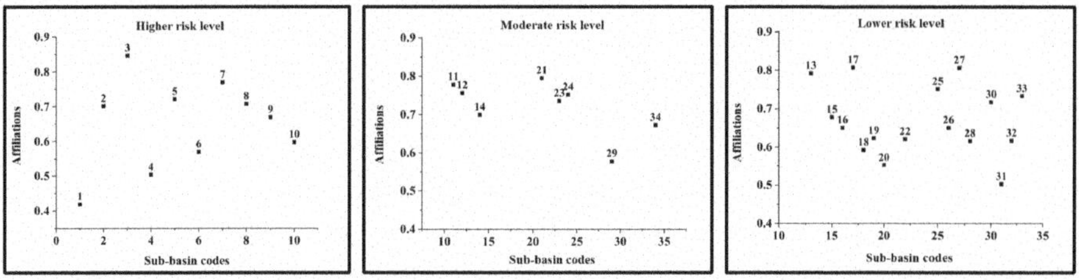

(a) higher risk level. (b) Moderate risk level. (c) lower risk level.

Fig. 2.3 The affiliation of the 34 Sub-basins of the Beidasha River under different risk levels

risk in the Beidasha River Basin are approximately 233.28, 112.91, and 229.58 km^2, accounting for 40.52, 19.61, and 39.87% of the entire basin area, respectively. Among them, the high-risk and medium-risk areas of Zhangxia Street accounted for 9.34 and 40.60% of the entire street area, respectively, and the medium-risk area of Wande Street accounted for 26.67% of the entire street area.

It can be seen from Fig. 2.3 that among the 10 small watersheds with higher risk levels, small watershed 3 has the highest degree of membership, and has a higher risk than other small watersheds. It is located in the northeast of Guyunhu Street and includes 4 administrative village; while watershed 1 has the lowest risk; in areas with medium risk levels, except for small watersheds 29 and 34 whose membership degrees are lower than 0.7, the risk level membership degrees of other small watersheds are all between 0.7 and 0.8; in areas with lower risk levels There are 9 small watersheds with low risk level and membership degree lower than 0.7.

Comparative analysis of the resultant map of watershed risk zoning and the historical flood data of Changqing District. It is obvious from Figs. 2.2 and 2.3a that watersheds 3, 5, and 2 with higher affiliation in the relatively high risk area are located in Ping'an Street and Wenchang Street, and that on July 18, 2007, heavy rainfall was widespread in Changqing District, and Ping'an Office and Wenchang Office were most seriously affected, which is in line with the actual situation; on July 22, 2016, heavy rainfall occurred in the southern part of Zhangxia Street, and some parts of the west bank of the Beidasha River that flowed through Zhangxia were villagers were besieged, and from the result map in Fig. 2.2b, it can be seen that the western part of Zhangxia Street has relatively high risk level and is prone to flooding; on July 12, 2022, Wande Street suffered from heavy rainfall, and Sunjiayu Village was severely affected on both sides of the river due to poor drainage downstream of the river, and the sub-watershed No. 29, which corresponds to Fig. 2.2a, belongs to the medium-risk area, which is in line with the situation. To summarize, the results are basically consistent with the historical flood disaster sites in Changqing District and have a certain degree of reliability.

2.4 Conclusion

The innovation is that the comprehensive weights of the indicators of the fuzzy comprehensive analysis method used in this paper are determined according to the principles of game theory, which not only solves the limitations of the subjective weighting method but also synthesizes the mathematical theoretical basis of the objective weighting method, and improves the reasonableness and scientificity of the determination of the indicator weight values. It provides further ideas for the risk analysis of the same type of flash flood disaster in this sub-basin, and has certain reference value, which can be further combined with the flood-prone area determined by the hydrological and hydraulic model to carry out the flood analysis in the study area (Aksoy et al. 2016).

The impact of each indicator of the watershed on flash flood disaster is different. According to the risk of the disaster-causing factor, the sensitivity of the disaster-forming environment, and the vulnerability of the carrier, the number of annual rainstorm days, the topographic index, and the population density are selected, respectively. Eventually, among the combined weights of the indicators, the number of storm days has the largest weight of 0.205, followed by the terrain index, and the vegetation index has the smallest weight, which indicates that heavy rainfall is the main cause of induced flooding, the terrain can contribute to the flooding, and the vegetation plays a role in weakening the flooding, which can be done by planting more vegetation in the high-risk places.

The high-risk areas in the North Dasha River Basin are mainly located in Mureyun Lake, Wenchang Street, and parts of Ping'an, Guide, and Zhangxia, and appropriate preventive measures should be taken in this area. The overall trend of risk level is increasing from south to north. Most of the areas in the whole basin have a high degree of risk, accounting for 40.52% of the whole basin, followed by lower risk areas, accounting for 39.87% of the whole basin, and the rest are medium risk areas. The results are basically consistent with the location of historical flooding sites in Changqing District.

Due to the complexity of the formation mechanism of flash floods and the large number of influencing factors, the risk analysis of flash floods is completely transformed into a quantitative analysis, and the determination of the weights of the indicators has a large impact on the evaluation results. The method is more advantageous in the determination of the comprehensive weight of the indicators, and the influence relationship between the indicators is considered comprehensively, but the influence of other indicators on the flash flood disaster in the basin needs to be further improved.

References

Aksoy H, Kirca VSO, Burgan HI, Kellecioglu D (2016) Hydrological and hydraulic models for determination of flood-prone and flood inundation areas. Proc IAHS 373:137–141

Duan YQ, Xie ZQ, Lin MN, Zou XQ, Hou ZQ (2021) Suitability evaluation of underground space development based on exponential scale AHP. J Underground Space Eng 17(S1):8–15

Ge X, Luo JX (2018) Research on the division method of mountain torrent disaster risk areas in small watersheds supported by GIS. China Rural Water Conserv Hydropower 10:170–176

He YF, Xiong JN, Cheng WM, Yang JW, He W, Yong ZY, Duan Y, Liu J, Yang G, Wang N (2022) Assessing the risk posed by flash floods to the transportation network in southwestern China. Geocarto Int 37(26):13210–13228

Jiang YT, Wang XL, Sun RF, Fan ZL, Sun FF (2020) Construction and application of flash flood disaster risk zoning division model. People's Yangtze River 51(04):7–13

Jing SY, Xin ML, Yu LH, Li WP, Ren B, Huang LZ (2018) Risk assessment of flash flood disasters in Datong City based on fuzzy comprehensive evaluation method. Water Conserv Hydropower Technol 49(05):84–89

Liu PX, Liu CL, Qiu SB, Lin W (2017) Risk assessment of flash flood disasters in Nanfeng County based on GIS and AHP. Hydropower Energy Sci 35(06):55–58

Liu JY, Wang Y, Tang JD, Luo JS, Wang PC (2023) Risk assessment of site groundwater environmental pollution based on the comprehensive weight method of game theory. Saf Environ Eng 30(01):221–230

Meng XQ, Tong J, Peng JH (2022) The fine mountain flood disaster risk zoning of Chengde city based on GIS. Desert Oasis Meteorol 16(06):118–123

Ren ZH, Sang YF, Yang MY, Wang YL, Shang L (2023) Research progress on early warning methods of rainstorm and mountain torrent disasters. Adv Geogr Sci 42(01):185–96

Sun Y, Sang GQ, Zhu LT, Wang JN, Zheng H (2021) Risk assessment of mountain torrent disasters in small watersheds based on DPSIR model. Hydropower Energy Sci 39(03):28–31

Wei YQ, Sheng D, Dong LY, Tan J, Qiu P (2022) Research status and development trend of mountain torrent disaster prevention and control. China Flood Control Drought Relief 32(07):30–35

Wu XJ, Fang XQ, Ren LL, Wu TY, Miao YQ (2018) Risk assessment of flash flood disasters based on random forest—taking Jiangxi Province as an example. Soil Water Conserv Res 25(03):142–149

Yu S, Wang MY, Liu J, Liu PB, Zhao YF (2013) Comprehensive evaluation method for ecological restoration of water-scarce rivers under the condition of artificial water replenishment. J Environ Sci 33(02):626–634

Yuan XM, Yang TT, Feng GN (2018) Application of fuzzy matter-element model based on entropy weight method in the evaluation of mountain torrent defense capability. Water Conserv Hydropower Technol 49(10):8–13

Zhu LL, Ai P, Mou P (2013) Application of game theory and fuzzy optimization model in the evaluation of water conservancy modernization. Hydropower Energy Sci 31(02):161–164

Chapter 3
Monitoring of Emerging Contaminants in Selected Wastewater Treatment Works Around Bloemfontein, South Africa: Initiative for Environmental Protection

Innocent Mugudamani, Saheed A. Oke, Thandi P. Gumede, and Samson Senbore

Abstract The study monitored the occurrence of emerging contaminants such as acetaminophen, carbamazepine, ibuprofen, triclosan, atrazine, metolachlor, simazine, terbuthylazine, 17-alpha-ethinyl-estadiol, progesterone, testosterone and estradiol in selected wastewater treatment works (influent and effluent) around Bloemfontein during autumn season. Samples collected from influent (n = 3) and effluent (n = 3) were analysed by high performance liquid chromatography. To assess the performance of wastewater treatment works in removing the targeted compounds the removal efficiency (%) method was used. The average concentration of acetaminophen, carbamazepine, ibuprofen, triclosan, atrazine, metolachlor, simazine, terbuthylazine, 17-alpha-ethinyl-estadiol, progesterone, testosterone and estradiol in influent samples were 28.15 μg/ml, 0.35 μg/ml, 7.95 μg/ml, 0.62 μg/ml, 0.04 μg/ml, 0.03 μg/ml, 0.15 μg/ml, 0.04 μg/ml, 37.53 μg/ml, <LOQ, 0.04 μg/ml, and 0.03 μg/ml while in effluent were as <LOQ, 0.63 μg/ml, 2.07 μg/ml, < LOQ, 0.05 μg/ml, 0.06 μg/ml, 0.24 μg/ml, 0.02 μg/ml, 41.4 μg/ml, <LOQ, and <LOQ respectively. In influent samples, 17-alpha-ethinyl-estradiol, acetaminophen, and simazine were the leading contaminants while in effluent it was 17-alpha-ethinyl-estradiol, ibuprofen, and simazine who recorded the highest mean concentrations. There was negative removal rates of compounds such as carbamazepine, atrazine, metolachlor, simazine and 17-alpha-ethinyl-estradiol. Considering the insufficient removal rates

I. Mugudamani (✉) · T. P. Gumede
Department of Life Sciences, Central University of Technology, Bloemfontein, Free State, South Africa
e-mail: imugudamani@gmail.com

T. P. Gumede
e-mail: TGumende@cut.ac.za

S. A. Oke · S. Senbore
Department of Civil Engineering, Centre for Sustainable Smart Cities, Central University of Technology, Bloemfontein, Free State, South Africa
e-mail: soke@cut.ac.za

H. Xu (ed.), *The 6th International Symposium on Water Resource and Environmental Management*, Environmental Science and Engineering,
https://doi.org/10.1007/978-3-031-55989-1_3

of these compounds, and the lack of established regulatory guidelines, it can be concluded that releasing effluent to nearby streams may pollute water and adversely affect aquatic organisms and human health. Therefore, in order to protect the environment, there is a need to develop and implement advance treatment techniques by the selected wastewater treatment works managers.

Keywords Emerging contaminants · Environmental protection · Removal efficiency · Wastewater treatment works · Water sources

3.1 Introduction

In the last decades, emerging contaminants such as disinfection by-products, endocrine disrupting compounds, flame retardants, personal care products, pesticides, pharmaceutical and micro-plastics among the few, are ever more released into the aquatic ecosystems thus causing environmental and public health concern (Necibi et al. 2021; Madikizela et al. 2020). The main cause of these compounds' presence in surface waters is the discharge of effluents from wastewater treatment works (Galindo-Miranda et al. 2019). According to Collivignarelli et al. (2023), hospital, domestic and industrial wastewater are disposed of into the municipal drains for treatment by a wastewater treatment works. However, numerous studies have found that most wastewater treatment processes are insufficiently effective at removing a significant number of emerging contaminants from wastewater (Necibi et al. 2021; Peña-Guzmán et al. 2019; Morin-Crini et al. 2022) as they are or were designed to decrease the high levels of nitrogen, phosphorus, and organic matter (Gosset et al. 2021). These emerging contaminants are consequently unceasingly discharged in lesser quantities (ng/l to μg/l) into adjacent streams as parent compounds, metabolites, or transformation products (Nannou et al. 2022).

Individually or in combination, emerging contaminants may trigger a serious danger to aquatic species and are harmful to human health either directly through exposure or indirectly through the food chain (Madikizela et al. 2020; Gosset et al. 2021; Nannou et al. 2022). At a very low levels, a number of these contaminants may effects chronic toxicity, endocrine disruption, as well as the growth of antibiotic resistant bacteria in aquatic species and human beings (Morin-Crini et al. 2022; Archer et al. 2022). Bearing in mind that the quantity of emerging contaminants is relentlessly fluctuating in relations to original compounds and their transformation or end products (Morin-Crini et al. 2022), inadequate removal efficiency (Peña-Guzmán et al. 2019), non-existence of established monitoring guidelines for the majority of these contaminants as well as their antagonistic risks on aquatic species and peoples health (Necibi et al. 2021; Archer et al. 2022), the ongoing assessments of loads of emerging contaminants in aquatic environments are needed in various global settings including Bloemfontein where there is no published data on emerging contaminants. This will help to indicate the level of their occurrence and potential health threats these contaminants pose. In this context, the aims of this study is to fill in such gap by monitoring the presence of emerging pollutants in three wastewater treatment works around Bloemfontein. This is the first scientific study targeting emerging contaminants in the city of Bloemfontein. Therefore, the outcomes of the study may be crucial for the development of advance treatment methods and ecological protection in the region.

3.2 Material and Methods

3.2.1 Study Area and Sample Collection

Bloemfontein is situated at 29°7′15.9″ S 26°12.84′ E at an elevation of 1 400 m above the sea level. The climate varies from warm and temperate with an annual rainfall of 1 020 mm in the east to semiarid with rainfall of only 380 mm in the far west. The water resources around the city include dams, rivers, wetlands, and groundwater resources which are utilised for a number of purposes in the area. Furthermore, households in the city have access to sanitation facilities and the municipality has several wastewater treatment works (Mugudamani et al. 2023). Three wastewater treatment plants were targeted during autumn season sampling campaign in this study. To collect both wastewater influent and effluent samples, a 750 ml hygienic glass bottles were used. The bottles used to collect samples were washed meticulously prior to sampling. All the samples were labelled and set aside in ice filled cool box during field work until laboratory analysis.

3.2.2 Laboratory Work

Sample preparation and extraction. The laboratory procedure was performed based on our previous studies (Mugudamani et al. 2023). Briefly, in the laboratory, the samples were filtered through glass fibre filters to eliminate particulate matter prior to being concentrated at a flow rate of 5 ml/minutes onto methanol conditioned C18-6 millilitre solid phase extraction cartridges (Strata, Phenomenex, Torrance, CA, USA). The bound sample was slowly eluted from the dried cartridges with 2 millilitre methanol and 2 ml ethyl acetate. The eluant was vacuum dried until almost dry (Thermo Scientific Savant Speedvac, Waltham, MA, USA) and reconstituted in 1 ml purified water.

Sample analysis. High-performance liquid chromatography (HPLC) linked to a QTRAP hybrid triple quadrupole ion trap mass spectrometer was used to analyse the wastewater samples. Analyst 1.5 (AB SCIEX, Framingham, MA, USA) software was used for all data acquisition and processing. Positive and negative ionisation modes were used to analyse the samples. During analysis, 20 μl of each extracted sample was separated on a C18 (150 mm × 4.6 mm, Gemini NX, Phenomenex) column at a flow rate of 300 μL/min using a 5 min gradient from 5% solvent A (H2O/0.1% formic acid) to 95% solvent B (MeOH/0.1% formic acid) with a total run time of 9 min and 10 min in positive and negative ionisation modes, respectively, to allow for column re-equilibration. Eluting analytes were electrospray ionised in the TurboV ion source with a heater temperature of 500 °C to evaporate the excess solvent, 40 psi nebuliser gas, 40 psi heater gas, and a curtain gas of 15 psi. In positive ionisation mode, the ion spray voltage was set at 5 500 V, while in the negative ionisation mode, it was set at – 4 500 V. Analysis of emerging contaminants in influents and effluents samples was carried out using multiple reaction monitoring transitions per analyte. The quantifier was the peak area on the chromatogram generated by the first and most sensitive transition, while the qualifier was the peak area generated by the second transition. The qualifier served as an additional level of confirmation for the analytes' presence (Mugudamani et al. 2023).

Quality assurance and control. For quality control and assurance, selectivity, linearity and limit of quantification were considered in this study. Samples were submitted in batches with solvent blank runs between each sample analysed and quality control samples of known concentration interspersed. For each analyte, a four-point calibration curve with a linear fit through the origin was generated, ranging in concentration from 0.001 ppm to 1 ppm. The selectivity was evaluated by comparing between the retention time of the extracted blank sample and the retention time of each spiked standard or IS in the extracted samples. If any interferences were observed, the signal of their peak area should be less than 20% of the peak area of limit of quantification, and less than 5% of the average peak area of IS in blank samples. The calibration values showed a concentration range of 5×10^{-5} to 1×10^{-1} mg per litre. There was no interference effect from the endogenous matrix at the retention time of all analytes and IS. The linear fit yielded a correlation coefficient (r) value exceeding 0.98. Furthermore, the quantification limits ranged from 0.0001 mg per litre to 0.1 mg per litre (Mugudamani et al. 2023).

3.2.3 *Removal Efficiency*

The treatment processes used in all wastewater treatment works selected in this study include three stages as shown in Fig. 3.1. The primary treatment entails the removal of suspended solids and floating debris. Additionally, clarified liquid stays above and is ready for processing while suspended solids gradually sink due to gravity to form "primary sludge." The biological remediation of soluble organic matter that has eluded primary treatment is the focus of the secondary treatment process. The dissolved organics are taken up by the bacteria, who then transform them into biomass and gases like carbon dioxide and nitrogen. The active process of activated sludge, which promotes microbial growth through mechanical mixing and aeration, is essential for secondary treatment. Before the discharge of effluent into nearby streams, the wastewater then goes to tertiary treatment which involved

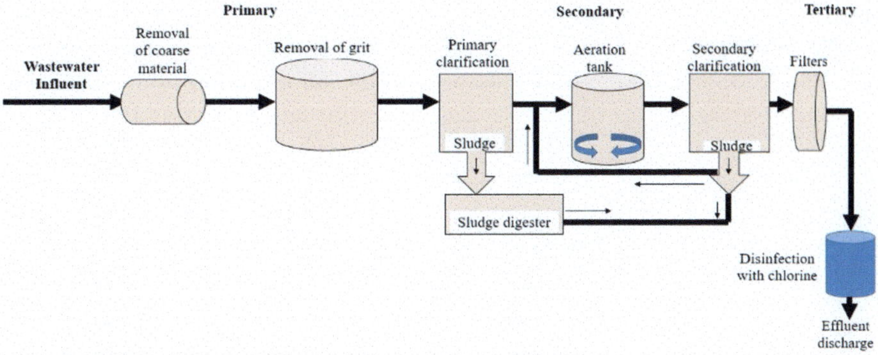

Fig. 3.1 Schematic of the selected wastewater treatment works

disinfection by a chlorine. However, in this study, all wastewater treatment works tertiary treatment was dysfunctional leading to effluent being discharged without being disinfected.

Therefore, in order to assess the rate of removing emerging contaminants from wastewater treatment works selected in this study, the removal efficiency method was adopted as suggested elsewhere (Kanama et al. 2018). The removal efficiency (%) expresses the decrease in a contaminants concentration throughout treatment and is based on comparison of the average influent and effluent concentrations (Nannou et al. 2022). The removal efficiencies of all the targeted analytes detected in selected wastewater treatment works were assessed using Eq. (3.1).

$$RF(\%) = \frac{influent\ concentration - effluent\ concentration}{influent\ concentration} \times 100 \quad (3.1)$$

3.3 Results and Discussion

All of the targeted wastewater treatment works receive domestic, industrial, and hospital wastewater. The treatment process used in all wastewater treatment works includes coarse grit, fine screen, grit chamber, primary sedimentation tanks, biological treatment, secondary sedimentation tank, maturation pond, sludge dewatering units, and disinfection units (which was not working in all wastewater treatment plants. Various groups of emerging contaminants namely herbicide, pharmaceutical, steroid hormone and personal care product were identified in selected wastewater treatment works as presented in Table 3.1.

Table 3.1 Levels of emerging contaminants in wastewater treatment works

Concentration (μg/ml)		
Compound	DF (n = 3)	Mean
Influent		
Acetaminophen	2	28.15
Carbamazepine	3	0.35
Ibuprofen	3	7.95
Triclosan	3	0.62
Atrazine	3	0.04
Metolachlor	3	0.03
Simazine	2	0.15
Terbuthylazine	3	0.04
17α ethinyl estradiol	3	37.53
Estradiol	0	<LOQ
Progesterone	2	0.04
Testosterone	2	0.03
Effluent		
Acetaminophen	1	<LOQ
Carbamazepine	3	0.63
Ibuprofen	3	2.07
Triclosan	0	<LOQ
Atrazine	3	0.05
Metolachlor	3	0.06
Simazine	3	0.24
Terbuthylazine	3	0.02
17α ethinyl estradiol	3	41.4
Estradiol	0	<LOQ
Progesterone	0	<LOQ
Testosterone	1	<LOQ

N = quantity of samples, DF = frequency of detection, LOQ = quantification limit

3.3.1 Presence of Emerging Contaminants in Influent Samples

Under the pharmaceutical and personal care products (PPCPs), acetaminophen, carbamazepine, ibuprofen and triclosan were detected at an average concentration of 28.15 μg/ml, 0.35 μg/ml, 7.95 μg/ml and 0.62 μg/ml correspondingly as presented in Table 3.1. Other studies also reported the occurrence of acetaminophen, and ibuprofen (Kanama et al. 2018), carbamazepine (Subedi et al. 2015) and triclosan

(Lin et al. 2020) in influent samples. Acetaminophen recorded the highest concentration among the PPCPs. Acetaminophen and ibuprofen are analgesics/non-steroid anti-inflammatory drugs commonly used alone or in combination with other formulations to treat a variety of ailments (Ekinci et al. 2020). Carbamazepine which showed lowest mean concentration among the pharmaceuticals belongs to the anti-convulsant group, and is a prescription drugs used to treat patients with seizures and bipolar disorder (Nannou et al. 2022). Triclosan as one of the personal care products detected in this study is broadly utilised in a diverse customer goods for its exceptional antimicrobial and antifungal properties. Most of the consumers' products such as soaps, disinfectants, toothpastes, body washes, and medical disinfectant, may have a concentration of triclosan ranging from 0.1 to 2% (Kanama et al. 2018). Therefore, the occurrence of this compounds may be attributed to the excretion from patients suffering from pain, fever, seizures, and bipolar disorder. Moreover, flushing of unused or expired drugs may introduce these contaminants in influent. The presence of triclosan reveals its constant use in household, industrials, or hospitals around Bloemfontein.

As shown in Table 3.1, among the herbicides, atrazine, metolachlor, simazine, and terbuthylazine showed an average concentration of 0.04 μg/ml, 0.03 μg/ml, 0.15 μg/ml and 0.04 μg/ml respectively. Simazine had the utmost concentration in influent samples which is opposite to the study in Baoding city of China which reported atrazine with the utmost concentration amongst the herbicide identified (Wang et al. 2022). In this study, wastewater treatment works, receive domestic wastewater from all residents, including fruit and vegetable cleaning, and other irregular uses. Therefore, the dominance of these herbicides in wastewater treatment works clearly shows their abundant use in the study city, households, agricultural area, or pesticide production factories.

The mean concentration of steroids hormones such as 17-alpha-ethinyl-estradiol, estradiol, progesterone and testosterone were as 37.53 μg/ml, <LOQ, 0.04 μg/ml, 0.03 μg/ml respectively (Table 3.1). An occurrence of 17-alpha-ethinyl-estradiol in influent samples with concentration lower than the current findings was also reported in other areas (Kanama et al. 2018). Manickum and John, (2014) also detected 17-alpha-ethinyl-estradiol, progesterone, and testosterone in influent samples which were lower than the current findings. Synthetic oestrogens such as 17-alpha-ethinyl-estradiol are commonly applied as the key ingredient in numerous oral contraceptives (Kanama et al. 2018). Progesterone is regarded as a female pregnancy hormones while testosterone is necessary for normal spermatogenesis, and decrease risk of osteoporosis (Hu et al. 2010; Prior 2019). Therefore, the presence of these hormones in wastewater influent may be attributed to humans' high excretion rates (Kanama et al. 2018). According to Forghani et al. (Forghani et al. 2018), even without the use of hormonal drugs, every human excretes oestrogens through their bodies. Women and men excrete approximately 2.3–259 and 1.6 g/d of estradiol hormones, respectively, which explains their presence in wastewater treatment works within the study area (Forghani et al. 2018).

3.3.2 Presence of Emerging Contaminants in Effluent Samples

Under the PPCPs, acetaminophen, carbamazepine, ibuprofen and triclosan were detected at an average concentration of <LOQ, 0.63 μg/ml, 2.07 μg/ml and <LOQ respectively as presented in Table 3.1. The presence of this compounds indicates that WWTWs are not constructed to eliminate these contaminants. The occurrence of ibuprofen and carbamazepine in effluent samples were also reported by other researchers (Kanama et al. 2018; Matongo et al. 2015). Among the herbicides, atrazine, metolachlor, simazine, and terbuthylazine were detected at a mean concentrations of 0.05 μg/ml, 0.06 μg/ml, 0.24 μg/ml, and 0.02 μg/ml correspondingly. Most notable in effluent samples was the level of simazine which was above its influent concentration. The detection of simazine and terbuthylazine in wastewater effluent was also reported by other author (Kunene 2019). The mean concentration of steroids hormones such as 17-alpha-ethinyl-estradiol, estradiol, progesterone and testosterone were as 41.4 μg/ml, <LOQ, <LOQ and <LOQ respectively. 17-alpha-ethynyl-estradiol had the highest mean concentration among the steroid hormones. A study conducted in China detected concentration of 17-alpha-ethynyl-estradiol in effluent samples (Lin et al. 2020) which were lower than the ones detected in this study.

Generally, in this study the concentrations of targeted pollutants in effluent samples can be described as either reduced, or higher than influent concentration. Contaminants such carbamazepine, atrazine, metolachlor, simazine and 17-alpha-ethinyl-estradiol had concentrations higher than their influent concentration. According to Kortesmäki et al. (2020), non-removal and the present of higher concentration of some analytes in effluent than influent may be attributed to factors such as wastewater composition, structure of analyte and conversion of the conjugated metabolites to the parent compound by enzymatic processes in the wastewater treatment works. The presence of these contaminants in effluent samples is concerning as most of the effluents are discharged in nearby streams. According to Nannou, et al. (2022) and Gosset et al. (2021), the occurrence of emerging contaminants in aquatic environment individually or in mixture constitute a serious hazard to aquatic environment and detrimentally to human health via direct contact or indirectly through the food chain. At very low concentrations, some can cause chronic toxicity, endocrine disruption in humans and aquatic organisms, and the development of antibiotic resistant bacteria (Morin-Crini et al. 2022; Archer et al. 2022).

3.3.3 Removal Efficiency of Emerging Contaminants in Wastewater Treatment Works

The wastewater treatment works revealed the removal efficiency of 100%, −80%, 74%, 100%, −25%, −100%, −60%, 50%, −10%, <LOQ, 100%, and 100% for acetaminophen, carbamazepine, ibuprofen, triclosan, atrazine, metolachlor, simazine, terbuthylazine, 17-alpha-ethinyl-estradiol, estradiol, progesterone and testosterone respectively. There was 100% removal rate of acetaminophen, triclosan, progesterone and testosterone while the removal rate of ibuprofen was 74% (Table 3.2). High removal rates of acetaminophen in wastewater treatment works was also reported by other researchers (Nannou et al. 2022; Kanama et al. 2018). Furthermore, there was also a negative removal rate of compounds such as carbamazepine, atrazine, metolachlor, simazine and 17-alpha-ethynyl-estradiol which are comparable to the findings of other scholars (Kock-Schulmeyer et al. 2012; Ofrydopoulou et al. 2021). Most notable in this study was the negative removal rate of pesticides which were also reported by other authors (Wang et al. 2022; Campos et al. 2016). The negative removal rates of emerging contaminants can be attributed to particulate matter desorption and hydrolysis, as well as emerging contaminants degradation and transformation during the biochemical process (Wang et al. 2022; Kock-Schulmeyer et al. 2012). According to Kortesmäki et al. (2020), non-removal and the present of higher concentration of some analytes in effluent than influent may be attributed to factors such as wastewater composition, structure of analyte and conversion of the conjugated metabolites to the parent compound by enzymatic processes in the wastewater treatment works.

The wastewater treatment works chosen in this study were not built with the intention of removing emerging contaminants received from domestic, industrial,

Table 3.2 Elimination of emerging contaminants in wastewater treatment plants

Analytes	Removal efficiency (%)
Acetaminophen	100
Carbamazepine	−80
Ibuprofen	74
Triclosan	100
Atrazine	−25
Metolachlor	−100
Simazine	−60
Terbuthylazine	50
17α-ethinyl estradiol	−10
Estradiol	<LOQ
Progesterone	100
Testosterone	100

and hospital wastewater. All of the wastewater treatment works use primary treatment which involves removal of suspended solids by flotation, settling, and screening mechanism and secondary treatment for removal of remaining organic matter via trickling filters utilizing bacteria-coated stones and bacterial activated sludge. Therefore, it can be said that these processes are unable to remove some of the emerging contaminants particularly pesticides and need improvement as wastewater effluent is typically discharged into nearby streams, introducing pollutants into the aqueous environment.

3.4 Conclusion

The study showed occurrence of emerging contaminants such as acetaminophen, carbamazepine, ibuprofen, triclosan, atrazine, metolachlor, simazine, terbuthylazine, 17-alpha-ethinyl-estadiol, progesterone, testosterone and estradiol in wastewater treatment works around Bloemfontein. Among these contaminants wastewater treatment works showed insufficient removal efficiency of contaminants such as carbamazepine, atrazine, metolachlor, simazine and 17-alpha-ethinyl-estradiol. Most notable was the incapability of wastewater treatment work to eliminate pesticides. In South Africa, wastewater effluent is typically discharged into nearby streams, introducing pollutants into the aqueous environment. Therefore, based on the findings of their removal efficiency it can be concluded that the release of wastewater effluent to nearby streams will cause pollution and adversely affect the aquatic species and public health eventually. It is suggested that advance treatment methods be added in wastewater treatment works in order to limit emerging contaminants pollution in water sources. Moreover, prioritizing these substances in contamination monitoring studies is critical not only in shielding aquatic life but in preserving and sustaining South Africa's water resources.

References

Archer E, Holtonb E, Fidal J, Kasprzyk-Hordern B, Carstens A, Brocker L, Kjeldsen TR, Wolfaardt GM (2022) Occurrence of contaminants of emerging concern in the Eerste River, South Africa: Towards the optimisation of an urban water profiling approach for public- and ecological health risk characterisation. Sci Total Environ 859(Pt 1):160254

Campos D, Gravato C, Quintaneiro C, Koba O, Randak T, Soares AM, Pestana JL (2016) Are insect repellents toxic to freshwater insects? A case study using caddisflies exposed to DEET. Chemosphere 49:177–182

Collivignarelli MC, Caccamo FM, Bellazzi S, Abbà A, Bertanza G (2023) Assessment of the impact of a new industrial discharge on an urban wastewater treatment plant: proposal for an experimental protocol. Environments 10:108

Ekinci M, Ciftci B, Celik EC, Kose EA, Karakaya MA, Ozdenkaya Y (2020) A randomized, placebo-controlled, double-blind study that evaluates efficacy of intravenous ibuprofen and acetaminophen for postoperative pain treatment following laparoscopic cholecystectomy surgery. J Gastrointest Surg 24(4):780–785

Forghani M, Sadeghi G, Mazyar M (2018) The presence of 17 beta-estradiol in the environment: health effects and increasing environmental concerns. Int J Epidemiol Res 5(4):151–158

Galindo-Miranda JM, Guízar-González C, Becerril-Bravo EJ, Moeller-Chávez G, León-Becerril E, Vallejo-Rodríguez R (2019) Occurrence of emerging contaminants in environmental surface waters and their analytical methodology—a review. Water Supply 19(7):1871–1884

Gosset A, Wiest L, Fildier A, Libert C, Giroud B, Hammada M, Hervé M, Sibeud E, Vulliet E, Polomé P, Perrodin Y (2021) Eco-toxicological risk assessment of contaminants of emerging concern identified by "suspect screening" from urban wastewater treatment plant effluents at a territorial scale. Sci Total Environ 778:146275

Kanama KM, Daso AP, Mpenyana-Monyatsi L, Coetzee MAA (2018) Assessment of pharmaceuticals, personal care products, and hormones in wastewater treatment plants receiving inflows from health facilities in North West province, South Africa. J Toxicol 2018(3751930):1–15

Kock-Schulmeyer M, Ginebreda A, Gonzalez S, Cortina JC, Alda ML, Barcel D (2012) Analysis of the occurrence and risk assessment of polar pesticides in the Llobregat River Basin (NE Spain). Chemosphere 86(1):8–16

Kortesmäki E, Östman JR, Meierjohann A, Brozinski JM, Eklund P, Kronberg L (2020) Occurrence of antibiotics in influent and effluent from 3 major wastewater-treatment plants in Finland. Environ Toxicol Chem 39(9):1774–1789

Kunene PN (2019) Method development and application: Solid phase extraction (SPE), Ultrasonic extraction (UE) and Soxhlet extraction (SE) for the analysis of pesticides in water, soil, sediment and sludge. Masters' Thesis, University of KwaZulu Natal, South Africa (2019). https://researchspace.ukzn.ac.za/handle/10413/17795. Accessed on 03 April 2023)

Lin X, Xua J, Keller AA, Hea L, Gua Y, Zheng W, Sun D, Lu Z, Huang J, Huang X, Li G (2020) Occurrence and risk assessment of emerging contaminants in a water reclamation and ecological reuse project. Sci Total Environ 744:140977

Madikizela LM, Ncube S, Chimuka L (2020) Analysis, occurrence and removal of pharmaceuticals in African water resources: a current status. J Environ Manag 253:109741

Manickum T, John W (2014) Occurrence, fate and environmental risk assessment of endocrine disrupting compounds at the wastewater treatment works in Pietermaritzburg (South Africa). Sci Total Environ 468–469:584–597

Matongo S, Birungi G, Moodley B, Ndungu P (2015) Pharmaceutical residues in water and sediment of Msunduzi River, Kwazulu Natal South Africa. Chemosphere 134:133–140

Morin-Crini N, Lichtfouse E, Fourmentin M, Ribeiro ARL, Noutsopoulos C, Mapelli F, Fenyvesi E, Vieira MGA, Picos-Corrales LA, Moreno-Piraján JC, Giraldo L, Sohajda T, Huq MM, Soltan J, Torri G, Magureanu M, Bradu C, Crini G (2022) Removal of emerging contaminants from wastewater using advanced treatments a review. Environ Chem Lett 20:1333–1375

Mugudamani I, Oke SA, Gumede TP, Senbore S (2023) Herbicides in water sources: communicating potential risks to the population of Mangaung metropolitan municipality South Africa. Toxics 11:538

Nannou C, Kaprara E, Psaltou S, Salapasidou M, Palasantza PA, Diamantopoulos P, Lambropoulou DA, Mitrakas M, Zouboulis A (2022) Monitoring of a broad set of pharmaceuticals in wastewaters by high-resolution mass spectrometry and evaluation of heterogenous catalytic ozonation for their removal in a pre-industrial level unit. Analytica 3:195–212

Necibi MC, Dhiba D, El Hajjaji S (2021) Contaminants of emerging concern in African wastewater effluents: occurrence impact and removal technologies. Sustainability 13:1125

Ofrydopoulou A, Nannou C, Evgenidou E, Christodoulou A, Lambropoulou D (2021) Assessment of a wide array of organic micropollutants of emerging concern in wastewater treatment plants in Greece: occurrence, removals, mass loading and potential risks. Sci Total Environ 802:149860

Peña-Guzmán C, Ulloa-Sánchez S, Mora K, Helena-Bustos R, Lopez-Barrera E, Alvarez J, Rodriguez-Pinzón M (2019) Emerging pollutants in the urban water cycle in Latin America: a review of the current literature. J Environ Manag 237:408–423

Hu J, Zhang Z, Shen WJ, Azhar S (2010) Cellular cholesterol delivery, intracellular processing and utilization for biosynthesis of steroid hormones. Nutr Meta 7(47)

Prior JC (2019) Progesterone is important for transgender women therapy-applying evidence for the benefits of progesterone in ciswoman. J Clin Endocrinol Metab 104(4):1181–1186

Subedi B, Codru N, Dziewulski DM, Wilson LR, Xue J, Yun S, Braun-Howland E, Minihane C, Kannan K (2015) A pilot study on the assessment of trace organic contaminants including pharmaceuticals and personal care products from on-site wastewater treatment systems along Skaneateles Lake in New York State, USA. Water Res 72:28–39

Wang M, Lv J, Deng H, Liu Q, Liang S (2022) Occurrence and removal of triazine herbicides during wastewater treatment processes and their environmental impact on aquatic life. Int J Environ Res Public Health 19:4557

Chapter 4
Effects of Mitigation Scenarios on Water Quality Under Drought-Flood Abrupt Alternation

Wuxia Bi, Dawei Zhang, Baisha Weng, Cheng Zhang, Wenqing Lin, Fan Wang, and Weiqi Wang

Abstract With the increasing of extreme events globally, they had significant adverse impacts on ecosystems according to existing studies. Previous studies pointed out that the drought-flood abrupt alternation (DFAA) events have been increasing in recent years, and they had adverse impacts on eco-environment. To adapt the DFAA, mitigation measures should be applied. However, scant studies concern quantified assessment of these measures. Under these circumstances, this study set drought level degradation and flood level degradation scenarios in history and in the future based on the critical thresholds obtained by our previous field experiments in the

W. Bi (✉) · D. Zhang · B. Weng · C. Zhang · W. Lin · F. Wang · W. Wang
State Key Laboratory of Simulation and Regulation of Water Cycle in River Basin, China Institute of Water Resources and Hydropower Research, Beijing 100038, China
e-mail: biwx@iwhr.com

D. Zhang
e-mail: zhangdw@iwhr.com

B. Weng
e-mail: wengbs@iwhr.com

C. Zhang
e-mail: zhangc@iwhr.com

W. Lin
e-mail: linwq@iwhr.com

F. Wang
e-mail: wangfan@iwhr.com

W. Wang
e-mail: wangwq@iwhr.com

W. Bi · D. Zhang · C. Zhang · W. Lin · F. Wang · W. Wang
Research Center on Flood and Drought Disaster Prevention and Reduction of the Ministry of Water Resources, Beijing 100038, China

B. Weng
Yinshanbeilu Grassland Eco-Hydrology National Observation and Research Station, China Institute of Water Resources and Hydropower Research, Beijing 100038, China

H. Xu (ed.), *The 6th International Symposium on Water Resource and Environmental Management*, Environmental Science and Engineering,
https://doi.org/10.1007/978-3-031-55989-1_4

Northern Anhui Plain. Our findings suggest that the soil P loss increased by 14.5–26.4%, and 0.5–1.3% under drought level degradation scenarios in history and in the future, while decreased by 40.0–41.7%, and −0.5 to 56.2% under flood level degradation scenarios in the summer maize planting areas. This study could provide supports for mitigating the adverse effects of DFAA on water quality, further may guide regulation measures for extreme events.

Keywords Extreme events · Mitigation scenarios · Water pollution indicators · Crop planting areas · Northern Anhui Plain

4.1 Introduction

It can be learned from the report of the United Nations Office for Disaster Risk Reduction (UNDRR) that there were 3,254 floods and 338 droughts occurring from 2000 to 2019, which accounts for about 50% of occurrence of all natural disasters. And the floods and droughts during 2000 to 2019 increased by 1.34 times and 0.29 times than that during 1980 to 1999 (CRED 2020). It can be learned that extreme events have been increasing in recent year, especially the flood and drought events. In recent years, a new type of extreme event called drought-flood abrupt alternation (DFAA), has occurred more and more frequently in recent years. Especially, most of the DFAA events occurred in the southeast of China (Bi et al. 2023). The frequency of DFAA changed from every four years to every 2–3 years in the Huai River Basin (Bi et al. 2020). Therefore, the study on the DFAA events has become one of the research hotspots.

According to the research knowledge map (Fig. 4.1), the key words of previous study on droughts and floods mainly were impact, management, risk, disaster, etc. What are the eco-environmental effects under DFAA? How to evaluate the mitigation measures quantitively? These questions are the key scientific questions to be answered. Certain studies revealed the effects of DFAA events on eco-environment. Xing et al. reveals that the soil nitrogen and phosphorus (P) loss increased first then decreased finally kept stable in both surface runoff and in topsoil of 30 cm (Xing et al. 2015). Hu et al. pointed out that the geochemical province of estuarine samples showed little variation during DFAA (Hu et al. 2014). Tian et al. analyzed the effects of DFAA on the size of wetlands based on the satellite data, the results reveal that there had similar trends between standardized precipitation index and the size of the wetlands (Tian et al. 2016). Ji et al. proposed method for determining the hazard threshold in the Huaihe River Basin (Ji et al. 2017). Niu et al. investigated the impacts of DFAA on agricultural production, human life and ecological aspects in the Huailin Town of Chaohu, which indicates that the DFAA had the most serious impacts on agricultural production, while basically had no effects on other industries (Niu et al. 2013). Bi et al. found that soil P loss increased more than 5 times under DFFA than that in normal conditions, and increased with time (Bi et al. 2022). While seldom

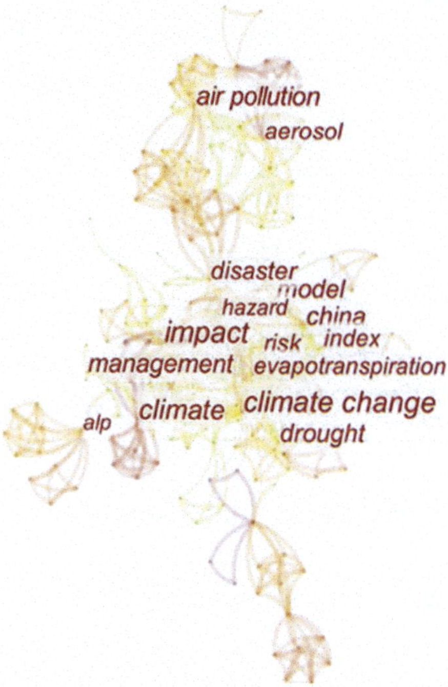

Fig. 4.1 Knowledge map of key words

studies have focused on quantitively estimating the effects of mitigation measures on DFAA.

Under these circumstances, this study aims to quantitively evaluate the mitigation effects on DFAA with setting drought level degradation and flood level degradation scenarios. As one of the important grain bases in China, DFAA events occurred about 85% in summer in the Northern Anhui Plain (Bi et al. 2022). And P is one of the main water pollution indicators there, thus we focused on the summer maize planting areas in the Northern Anhui Plain. The research contains three steps (Fig. 4.2): (i) determining the DFAA events in the study area from 1964 to 2017 (in history) and from 2020 to 2050 (in the future); (ii) calculating the critical thresholds of soil P loss under DFAA; and (iii) analyzing the water pollution indicator under DFAA in the study area with different mitigation scenarios (drought level degradation, flood level degradation). The results could guide targeted regulation facing to DFAA.

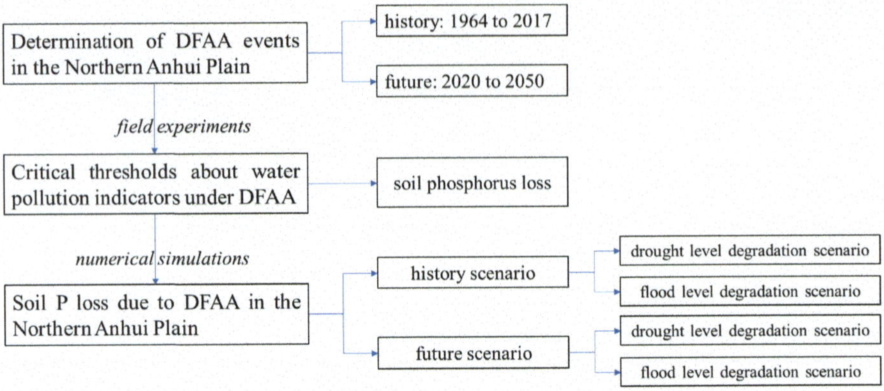

Fig. 4.2 Research outlines

4.2 Materials and Methods

4.2.1 Determination of DFAA Events

A daily method to determine DFAA events was proposed, which considers the indicators in multiple aspects. The detailed classification of DFAA level can be found in our previous studies (Bi et al. 2020, 2022). The drought level is classified by the *Classification of Meteorological Drought* (GB/T 20481-2006) and the *Standard of Classification for Drought Severity* (SL424-2008). To determine the flood level, we calculate the amount of precipitation within 5 days after drought. The DFAA events are classified into three levels (Bi et al. 2022).

4.2.2 Critical Thresholds Obtained from Field Experiments

Our previous paper analyzed the DFAA characteristics in the study area from 1964 to 2017 and from 2020 to 2050 (Bi et al. 2022). To explore the effect of DFAA on water environment, we conducted field experiments with different DFAA scenarios in the summer maize ecosystems. We obtained the soil P loss thresholds in the field experiments, which further help evaluate the key water pollution indicators under DFAA in study area. The experimental design can be found in Bi et al. (2022).

4.2.3 *Numerical Simulation Scenarios*

This study set four types of scenarios, i.e., history scenario, future scenario, drought level degradation scenarios, flood level degradation scenarios.

(1) History scenario
The regional soil P loss from 1964 to 2017 were calculated based on the DFAA frequency and the soil distribution in the study area. The frequency distribution of DFAA events were analyzed with the daily data in the meteorological stations.

(2) Future scenario
It is reported that the best simulation combination was the intermediate stable path RCP4.5 and the climate scenario HADGEM2-ES (Bi et al. 2022). The data resolution is $0.5° \times 0.5°$. The time period was from 2020 to 2050.

(3) Drought level degradation scenarios
To quantitatively assess the effects of mitigation measures on drought level degradation, drought level degradation scenarios were set based on both history and future scenario, i.e., the drought level in DFAA was changed from moderate drought to light drought, and from severe drought to moderate drought, while the flood level remained unchanged.

(4) Flood level degradation scenarios
Similar to the set of drought level degradation scenarios, flood level degradation scenarios were set based on both history and future scenario, i.e., the flood level in DFAA was changed from moderate flood to light flood, and from severe flood to moderate flood, while the drought level remained unchanged.

4.3 Results and Discussion

4.3.1 *Effects of Mitigation Measures with Drought Level Degradation in DFAA on Soil P Loss*

Our field experiments reveal that the soil P loss in different treatments (Bi et al. 2022). The annual mean soil P loss was 7.98–10.34%, and 2.07–14.16% under DFAA in the history scenario (from 1964 to 2017) and in the future scenario (from 2020 to 2050), respectively (Bi et al. 2022). For the drought level degradation scenarios, the average annual soil P loss was 9.14–13.07% during 1964 to 2017 (Fig. 4.3a), and was 2.08–14.34% during 2020–2050 (Fig. 4.3b). Thus, compared with the history and the future scenarios, the drought level degradation scenarios increased the soil P loss by 14.5–26.4%, and 0.5–1.3% from 1964 to 2017 and from 2020 to 2050. The highest soil P loss did not shift in both history scenario and future scenario.

For the history scenarios, the soil P loss increased in both lowest value and highest value, which may be concluded as that when moderate drought-moderate flood events changed to light drought-moderate flood events, the soil P loss increased from 9.9

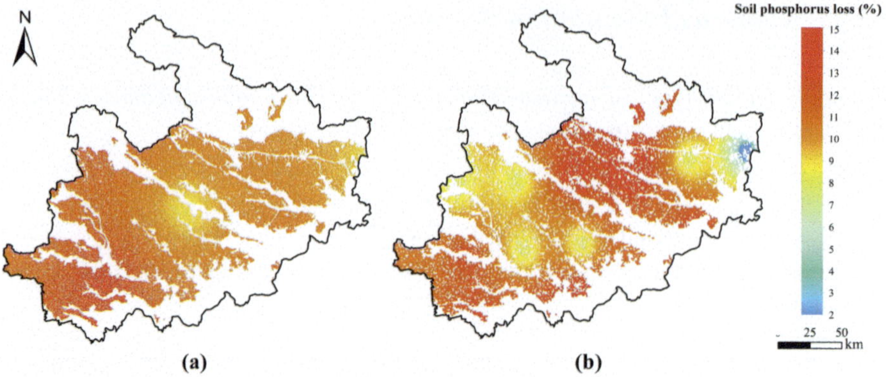

Fig. 4.3 The annual mean soil P loss under drought level degradation scenarios **a** from 1964 to 2017; **b** from 2020 to 2050

to 16.5%, these events occurred almost in whole summer maize planting areas. For the future scenarios, the soil P loss increased slightly in in both lowest and highest values, which indicates that the moderate drought-moderate flood events occurred less compared with the history.

4.3.2 Effects of Mitigation Measures with Flood Level Degradation in DFAA on Soil P Loss

For the flood level degradation scenarios, during 1964 to 2017, the annual soil P loss was 4.65–6.20% (Fig. 4.4a). During 2020–2050, the annual soil P loss was 2.08–6.20% (Fig. 4.4b). Compared with the history and the future scenarios, the flood level degradation scenarios decreased the soil P loss by 40.0–41.7%, and −0.5 to 56.2% from 1964 to 2017 and from 2020 to 2050. The highest soil P loss shifted from northwest to east in the history scenario, while basically no change in the future scenario.

The reasons why flood level degradation scenarios largely decreased the soil P loss were: (i) the DFAA with higher flood level had larger soil P loss; and (ii) lower flood level may reduce the scouring effects on topsoil (Bayad et al. 2022), thus the mitigation measures such as manual storage could reduce the effects of DFAA events on water environment pollution.

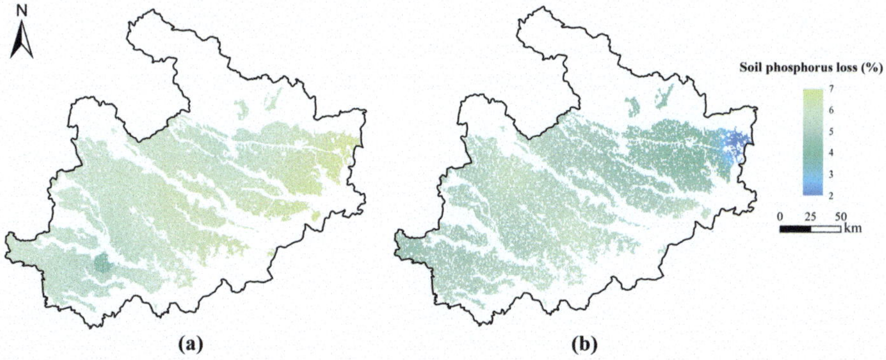

Fig. 4.4 The annual mean soil P loss under flood level degradation scenarios **a** from 1964 to 2017; **b** from 2020 to 2050

4.4 Conclusions

This study set drought level degradation and flood level degradation scenarios based on the history and future scenarios in our previous work. The results shed light on that under the drought level degradation scenario, the soil P loss increased by 14.5–26.4%, and 0.5–1.3% from 1964 to 2017 and 2020 to 2050, respectively. While under the flood level degradation scenario, the soil P loss decreased by 40.0–41.7%, and −0.5 to 56.2% in history and in the future. It can be inferred that the mitigation measures have more significant effects on reducing water pollution under DFAA in the future than that in history. The possible reason could be that the mitigation measures on flood reduced the scouring effects of topsoil in the summer maize farmland systems. As we just consider single change of drought or flood level, further studies could consider more combinations of degradation of drought and flood levels.

Acknowledgements This study was supported by the National Natural Science Foundation of China (No. 52209043), the National Key Research and Development Program of China (No. 2022YFC3080300, 2021YFC2001400), the MWR Major Science and Technology Program (SKS-2022007), and the IWHR Research and Development Support Program (WH0145B022021).

References

Bayad M, Chau HW, Trolove S, Moir J, Condron L, El Gharous M (2022) Surface runoff and losses of phosphorus from hydrophobic pastoral soils. Agric Ecosyst Environ 324:107690

Bi W, Weng B, Yan D, Wang M, Wang H, Wang J, Yan H (2020) Effects of drought-flood abrupt alternation on phosphorus in summer maize farmland systems. Geoderma 363:114147

Bi W, Weng B, Yan D, Wang M, Wang H, Jing L, Yan S (2022) Soil phosphorus loss increases under drought-flood abrupt alternation in summer maize planting area. Agric Water Manag 262:107426

Bi W, Li M, Weng B, Yan D, Dong Z, Feng J, Wang H (2023) Drought-flood abrupt alternation events over China. Sci Total Environ 875:162529

CRED (2020) UN office for disaster reduction. Human cost of disasters: an overview of the last 20 years (2000–2019)

Hu L, Shi X, Wang G, Qiao S, Yang G, Gao J, Bai Y (2014) Study on the geochemical characteristics of the surface sediments from the Yangtze River estuary under the 2011 sharp turn from drought to flood in middle and lower Yangtze. Geochemica 43(1):39–54

Ji Z, Li N, Wu X (2017) Threshold determination and hazard evaluation of the disaster about drought/flood sudden alternation in Huaihe River Basin, China. Theor Appl Climatol

Niu J, He Z, Zhang T, Chen X (2013) Impacts of sudden drought to flood change on manufacture livelihood and ecology. Resour Environ Yangtze Basin 22(1):108–115

Tian R, Cao C, Peng L, Ma G, Bao D, Guo J, Yomwan P (2016) The use of HJ-1A/B satellite data to detect changes in the size of wetlands in response in to a sudden turn from drought to flood in the middle and lower reaches of the Yangtze River system in China. Geomat Nat Hazards Risk 7(1):287–307

Xing D, Zhang Z, Yang J, Chen X, Sheng L (2015) Characteristics of Runoff and Nutrient Loss under the Condition of Rapid Shift from Drought to Waterlogging in Red-soil Sloping Land. Journal of Irrigation and Drainage 34(2):11–15

Chapter 5
Water Quality Assessment and Spatial Analysis of River System in Suburban Area of Hangzhou, China

Hailan Yu, Huabin Shentu, Shaoyi Wang, Linge Xu, Libing Wang, Xiaoyong Zhou, Jianguang Wang, Heng Zheng, Wei Zhao, Jing Chen, Xiaoxu Chen, and Wei Jun

Abstract Water quality degradation caused by excess nutrient inputs is a major issue facing urban wetlands. This study aimed to characterize the spatiotemporal variations in water quality and identify main pollution sources in a degraded urban wetland in Hangzhou, China. Water samples were collected from 14 locations along the Xianlin and Wutong Rivers in from September to March. Concentrations of nitrogen, phosphorus, dissolved oxygen, and other water quality parameters were measured. Single-factor assessment indicated the main pollutants were total nitrogen (TN) and ammonia nitrogen (NH_3-N), with average concentrations of 4.20 mg/L and 2.76 mg/L in the Xianlin River. Nutrient concentrations displayed significant seasonal variations, with higher levels in winter months. Spatial heterogeneity was also evident, with the highest pollution levels observed at convergent and downstream sections influenced by external sources. Based on the results, agricultural runoff and municipal wastewater are likely major contributors to the excessive nitrogen and phosphorus inputs to the wetland. Hydrologic restoration to improve water circulation may help reduce nutrient accumulation. This study provides a detailed characterization of water quality issues and identifies priorities for remediation in the study wetland.

Keywords Water quality · Urban wetlands · Nutrient pollution · Nitrogen · Phosphorus

---

H. Yu · H. Shentu · S. Wang
Huadong Eco-Environmental Engineering Research Institute of Zhejiang Province, Hangzhou 311122, Zhejiang, China

L. Xu · L. Wang · X. Zhou · J. Wang · H. Zheng · W. Zhao · J. Chen · X. Chen · W. Jun (✉)
Department of Ecology and Environment, PowerChina Huadong Engineering Corporation Limited, Hangzhou 311122, Zhejiang, China
e-mail: Wei_j2022@126.com

H. Xu (ed.), *The 6th International Symposium on Water Resource and Environmental Management*, Environmental Science and Engineering,
https://doi.org/10.1007/978-3-031-55989-1_5

5.1 Introduction

Water quality degradation is a major issue facing wetlands and waterways globally. In urbanized watersheds, pollutant inputs from municipal, industrial, and agricultural sources threaten aquatic ecosystems and human health (Morrice et al. 2008; Krall et al. 2014; Mosaffaie et al. 2021). Rapid development in China has exacerbated water pollution, with frequent eutrophication and harmful algal blooms (Li et al. 2021). Effective water quality monitoring and assessment tools are essential to guide management efforts and restoration in impaired urban water bodies (Sun et al. 2022; Ma et al. 2021; Cheng et al. 2021).

This study aims to provide a detailed characterization of water quality in a degraded urban wetland in Hangzhou, China using intensive monitoring and multimetric analysis. The specific objectives are to: (1) Quantify spatiotemporal variations in key water quality parameters, especially nitrogen and phosphorus levels which are indicators of nutrient pollution; (2) Identify hotspots and seasonal patterns to infer major point and nonpoint pollution sources within the watershed. Contrast single-factor and integrated assessment methods to evaluate their effectiveness for comprehensive water quality monitoring. The results will elucidate drivers of water quality impairments and identify priorities for remediation, such as reducing external nutrient loading, improving hydrologic connectivity, and implementing in-situ treatment solutions. Characterizing water pollution dynamics in this urban wetland will provide a scientific basis to guide effective watershed management and restoration programs. The monitoring methodology and analytical framework established in this study will be applicable to degraded urban water bodies globally facing similar anthropogenic pressures.

5.2 Material and Methods

5.2.1 Study Area

The study area is located in the southwestern part of Hangzhou City, Zhejiang Province, China. As shown in Fig. 5.1, the water system characteristics within the study area are as follows: ① Water areas cover over 50% of the area but have experienced significant shrinkage in recent years, resulting in fragmented wetland habitats and increased drainage pressure. The surrounding development has gradually reduced the water storage capacity of polders. ② The internal water system exhibits diverse landscape features such as creeks, harbors, ponds, embankments, and marshes. However, there is limited connectivity and poor internal water circulation among the ponds and channels. ③ The internal river channels are crisscrossed, forming a well-developed water system, but lack systematic organization.

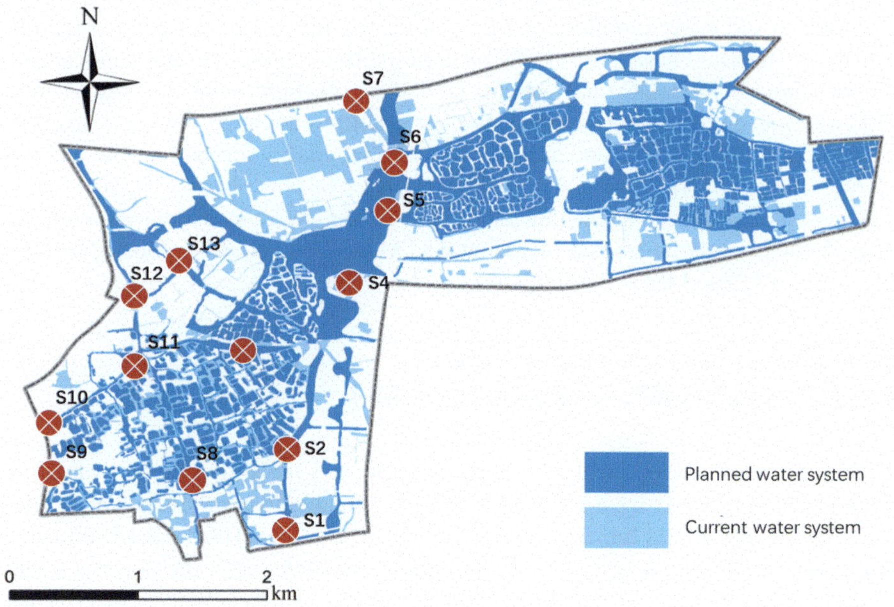

Fig. 5.1 Study area and the sampling point distribution

5.2.2 *Sampling Methods*

Based on the hydrological characteristics, topography, and land use types in the study area, sampling points were established for water quality monitoring in the Wutong River and Xianlin River areas. Along the Xianlin River, sampling points S1 to S7 were designated: S1 represents the upstream inflow section of Xianlin River, controlling the water quality of the upstream inflow; S2 and S4 represent the sections where Xianlin River flows through the study area, before converging with Wutong River; S5 represents the main water area where Xianlin River and Wutong River intersect; S6 represents the section that passes through the Future Science and Technology Cultural Center and its balance block; S7 represents the outflow section of Xianlin River. Sampling points S8 to S14 represent the water quality along the Wutong River: S8 represents the upstream inflow section of Wutong River, controlling the water quality of the upstream inflow; S9, S10, and S11 represent the impact of surrounding residential and construction areas on the water quality of Wutong River; S12 and S13 represent the impact of farmland and dryland areas on the water quality of Wutong River; S14 represents the impact of Hehu Wetland on Wutong River. The sampling was conducted in September 2022, January 2023, and March 2023, with water samples collected from each sampling point for water quality parameter analysis.

River water samples were collected from the middle of the river at a depth of 30 cm below the water surface using water sampling bottles. For each sampling point, a portion of the water sample was analyzed on-site using a portable water quality analyzer to measure basic physicochemical properties, including water temperature, pH, dissolved oxygen (DO) and turbidity (NTU). Another portion of the water sample (500 ml) was collected in polyethylene plastic bottles, acidified with 1–2 drops of concentrated sulfuric acid, and transported to the laboratory for subsequent analysis of total nitrogen (TN), ammonia nitrogen (NH_3-N), total phosphorus (TP), and chemical oxygen demand (COD) within 24 h. The elements were measured according to international standard methods including ASTM D5176 (TN), EPA 350.1 (NH_3-N), EPA 365.1 (TP) and ISO 15705 (COD).

5.2.3 Water Quality Assessment Method

The water quality assessment in this study employed the single-factor evaluation method recommended by the Chinese environmental standard "Technical Guidelines for Environmental Impact Assessment" (HJ 2.1–2011). This method compares the actual monitoring results of each evaluated parameter with the corresponding water quality classification standards to determine the water quality category of each parameter. The worst category among the evaluated parameters is assigned as the overall water quality category for the water body (or section). The single-factor evaluation method is simple, direct, and capable of quickly identifying major pollution factors in the water body. The exceedance of individual pollutants in the study area was observed using the pollution index (Pi), which compares the measured concentrations of selected pollutants with their respective evaluation standards, enabling the calculation of the exceedance ratio and identification of the major pollution factors.

5.3 Result and Discussion

5.3.1 Analysis of Water Quality Characteristics of Xianlin River

As shown in Fig. 5.2, during the sampling period, the average NH_3-N concentration in the inflow section of Xianlin River was 2.76 mg/L, with a minimum concentration of 0.83 mg/L and a maximum concentration of 4.46 mg/L, exceeding the standard by more than 4 times. The average TN concentration was 4.20 mg/L, with a minimum concentration of 1.98 mg/L and a maximum concentration of 6.33 mg/L, exceeding the standard by more than 6 times. The average TP concentration was 0.23 mg/L, with a minimum concentration of 0.12 mg/L and a maximum concentration of 0.35 mg/L. In the outflow section, the average NH_3-N concentration was 3.19 mg/L, with a minimum concentration of 1.27 mg/L and a maximum concentration of 4.60 mg/

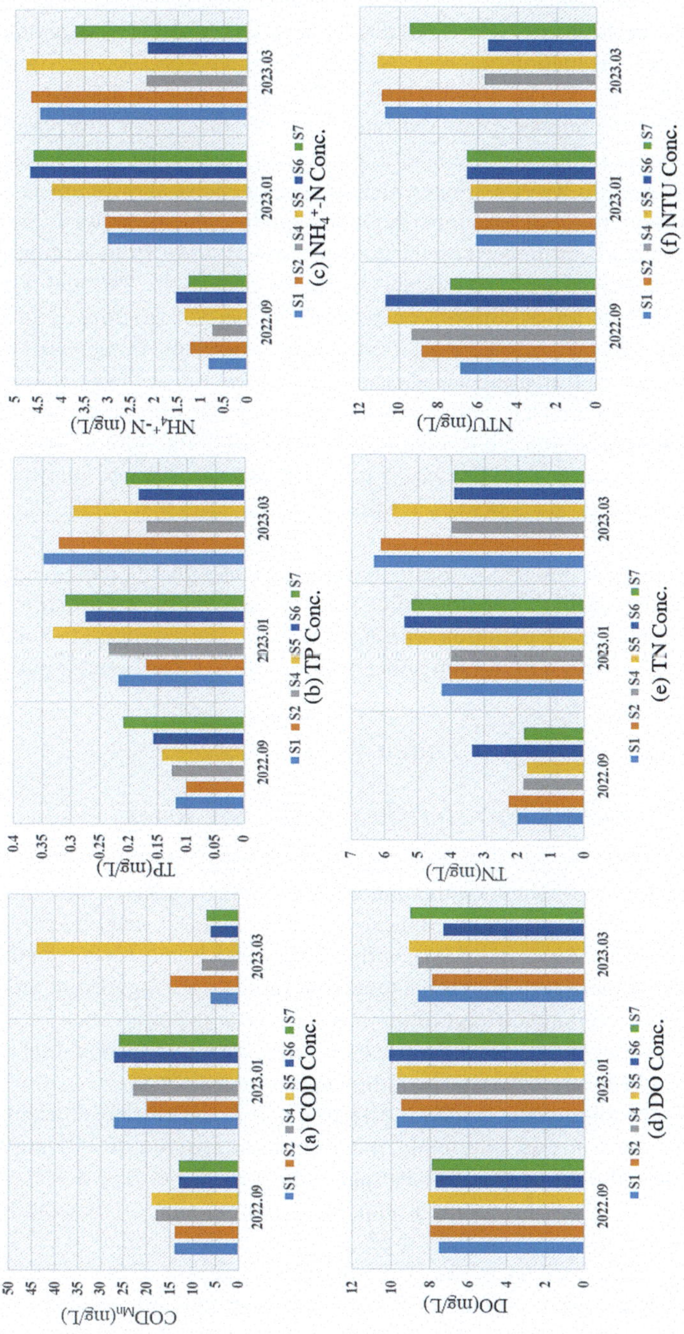

Fig. 5.2 Water quality analysis of Xianlin River

L. The average TN concentration was 3.64 mg/L, with a minimum concentration of 1.78 mg/L and a maximum concentration of 5.21 mg/L.

The seasonal variations of NH_3-N, COD, TP, TN, DO, and NTU were analyzed using single-factor evaluation method. NH_3-N, TP, and TN showed significant seasonal variations ($p < 0.05$), with higher concentrations observed in January and March compared to September. These seasonal variations in water quality parameters indicate a significant influence of temperature on river water quality, especially for NH_3-N and TN. Xianlin River exhibited spatial heterogeneity at different sampling points within study area. For example, the water quality indicators were higher at the S5 section, which is influenced by the convergence of water from Wutong River and the larger catchment area. Significant variations in NH_3-N, TN, and TP were observed, with the maximum differences between S5 and S4 reaching 2.58 mg/L, 1.78 mg/L, and 0.126 mg/L, respectively. Furthermore, due to the influence of human activities in the developed area and its balance block (Morrice et al. 2008), the concentrations of NH_3-N, TN, and TP in this area increased significantly, with NH_3-N and TN concentrations exceeding the standards by 4–5 times in the outflow area. Additionally, the COD concentration increased at the S2 and S4 sections, mainly due to the presence of agricultural residences in Minfeng Village (Li et al. 2021; Saby et al. 2021; Kumar et al. 2021).

In summary, the temporal and spatial differences in water quality parameters indicate that river water quality is mainly influenced by the quality of upstream inflow water and non-point source pollution within the region. The average TP concentration was 0.24 mg/L, with a minimum concentration of 0.21 mg/L and a maximum concentration of 0.31 mg/L.

5.3.2 *Water Quality Characteristics of Wutong River*

The temporal variations of water quality parameters at Wutong River were analyzed using single-factor analysis method. NH_3-N, COD, TP, TN, DO, and NTU exhibited significant seasonal variations ($p < 0.05$). Particularly, NH3-N and TN concentrations showed a pattern of January > March > September, indicating a significant influence of temperature. The temporal differences in COD and TP concentrations were mainly influenced by the water quality of the upstream inflow section and surrounding human activities (Luo and Moiwo 2022).

According to the monitoring results shown in Fig. 5.3, the average concentration of NH3-N at this section was 3.96 mg/L, with a minimum value of 2.06 mg/L and a maximum value of 5.16 mg/L. The average concentration of TN was 5.08 mg/L, with a minimum value of 2.68 mg/L and a maximum value of 7.68 mg/L. The average concentration of TP was 0.48 mg/L, with a minimum value of 0.37 mg/L and a maximum value of 0.64 mg/L. A comparative analysis of the water quality parameter concentrations at S11 and S8 indicated that although residential areas achieved comprehensive coverage of sewage networks, the lack of buffer zones for drainage from green areas and surface flushing drainage facilities along the river

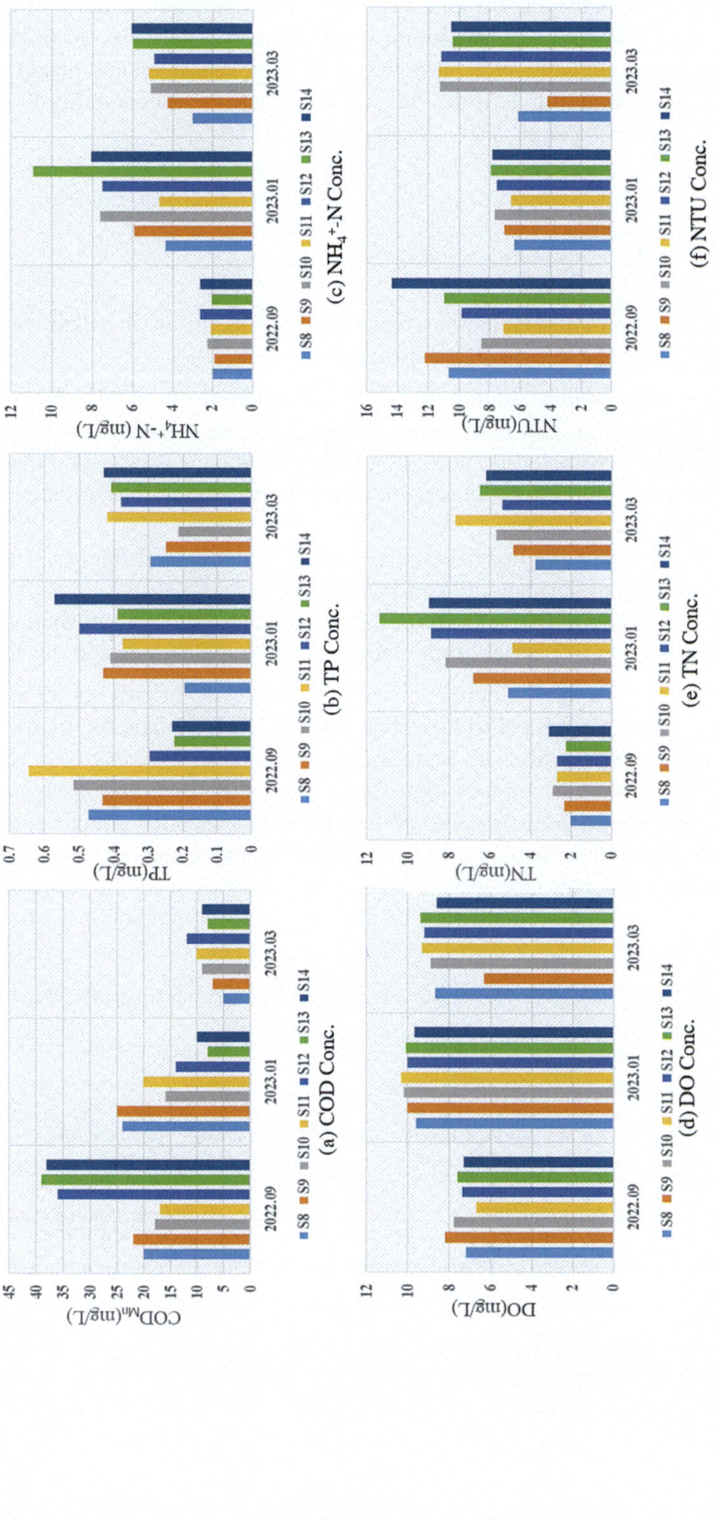

Fig. 5.3 Water quality analysis of Wutong River

poses significant risks to the river's environmental safety (Xu et al. 2021; Zhou et al. 2021). A comparison between S13 and S11 showed a significant increase in NH3-N and TN concentrations, with maximum concentration differences of 6.24 mg/L and 6.52 mg/L, respectively, indicating that the input of nitrogen and phosphorus pollutants from non-point sources is the main contributor to the water quality changes in Wutong River (Marella et al. 2022; Gurung et al. 2013).

5.4 Conclusion

This study provided an integrated assessment of water quality in an urban wetland in Hangzhou, China. The key result are as follows:

(1) Pollutant levels displayed significant seasonal variations, with NH_3-N and TN concentrations up to 2.58 mg/L and 1.78 mg/L higher in winter compared to summer. TP was averaged 0.23 mg/L. Spatial heterogeneity was also evident, with the highest pollution at convergent and downstream sections influenced by external sources.
(2) Spatial heterogeneity pointed to external sources as major contributors, with the highest pollution at downstream confluences.
(3) The severe nitrogen and phosphorus pollution indicates excessive nutrient inputs from likely sources like agricultural runoff and municipal wastewater discharge.

These results suggest that reducing agricultural runoff and improving municipal wastewater treatment are critical to lowering nutrient loads. Enhancing hydrologic connectivity may also help alleviate accumulation issues.

Further research should conduct high-frequency monitoring to capture short-term fluctuations and pinpoint specific pollution events. Detailed source appointment combining chemical speciation, isotopic tracing, and modeling will quantify contributions from different land uses. Cost–benefit analyses of in-situ remediation strategies like constructed wetlands are needed identify optimal solutions for this urban watershed.

References

Cheng L, Tan X, Yao D, Xu W, Wu H, Chen Y (2021) A fishery water quality monitoring and prediction evaluation system for floating UAV based on time series. Sensors (Basel, Switzerland) 21(13)

Gurung DP, Githinji LJM, Ankumah RO (2013) Assessing the nitrogen and phosphorus loading in the Alabama (USA) River Basin using PLOAD model. Air, Soil Water Res 6:S10548

Krall MA, Gundlapalli AV, Samore MH (2014) Chapter 13 - Big Data and Population-Based Decision Support. In: Greenes RA (ed) Clinical decision support, 2nd edn. Academic Press, Oxford, pp 363–381

Kumar, N., Kumar, A., Marwein, B., Verma, D., Jayabalan, I., Kumar, A., Duraisamy, R.: Agricultural activities causing water pollution and its mitigation- a review (2021)

Li X, Zhang W, Wu J, Li H, Zhao T, Zhao C, Shi R, Li Z, Wang C, Li C (2021) Loss of nitrogen and phosphorus from farmland runoff and the interception effect of an ecological drainage ditch in the North China Plain—A field study in a modern agricultural park. Ecol Eng 169:106310

Luo K, Moiwo JP (2022) Sensitivity of the land surface hydrological cycle to human activities in China. Gondwana Res

Ma Y, Song K, Wen Z, Liu G, Shang Y, Lyu L, Du J, Yang Q, Li S, Tao H, Hou J (2021) Remote sensing of turbidity for lakes in Northeast China using sentinel-2 images with machine learning algorithms. IEEE J Sel Top Appl Earth Obs Remote Sens 14

Marella TK, Saxena A, Tiwari A, Datta A, Dixit S (2022) Treating agricultural non-point source pollutants using periphyton biofilms and biomass volarization. J Environ Manage 301:113869

Morrice JA, Danz NP, Regal RR, Kelly JR, Niemi GJ, Reavie ED, Hollenhorst T, Axler RP, Trebitz AS, Cotter AM, Peterson GS (2008) Human influences on water quality in great lakes coastal wetlands. Environ Manage 41(3):347–357

Mosaffaie J, Salehpour Jam A, Tabatabaei MR, Kousari MR (2021) Trend assessment of the watershed health based on DPSIR framework. Land Use Policy 100:104911

Saby L, Nelson JD, Band LE, Goodall JL (2021) Nonpoint Source Water Quality Trading outcomes: Landscape-scale patterns and integration with watershed management priorities. J Environ Manage 294:112914

Sun X, Zhang Y, Shi K, Zhang Y, Li N, Wang W, Huang X, Qin B (2022) Monitoring water quality using proximal remote sensing technology. Sci Total Environ 803

Xu H, Cai C, Du H, Guo Y (2021) Responses of water quality to land use in riparian buffers: a case study of Huangpu River China. Geojournal 86(4):1657–1669

Zhou L, Shen G, Li C, Chen T, Li S, Brown R (2021) Impacts of land covers on stormwater runoff and urban development: a land use and parcel based regression approach. Land Use Policy 103:105280

Chapter 6
Role of Virtual Water Within Generalized Trade Framework

Liu Yong

Abstract The concept of virtual water trade says that a water-scarce economy can import virtual water embodied in commodities from other economies to save domestic physical water for other purposes on the basis of comparative advantage, for example, in Urumqi. In practice, however, it is not always the case, and virtual water can also flow from water-scarce regions to water-rich ones, for example, a huge volume of local water flows out of Xinjiang and into more water-rich economies in virtual senses. This paper aims to investigate these paradoxical cases and finds that virtual water never plays a monodrama but joins in a chorus together with physical trade, virtual energy trade, virtual labor trade, and virtual land trade, and others. These different forms of trade together contribute greatly to an economy, and a term of generalized trade is applied to draw a multiple-perspective portrait of traditional physical trade.

Keywords Virtual water · Commodity exchange · Generalized trade · Urumqi · Xinjiang

6.1 Introduction

Water is an irreplaceable resource, and we consume a great volume of fresh water for different purposes. Some regions have disputes over ownership of water resources, and it is necessary to investigate water consumption and its role in our society. In practice, we not only consume freshwater directly, but also use far more water embodied in various products, indirectly. Water directly consumed is named as physical water, and water embodied in various products is referred to virtual water. They together encompass the total amount of water consumption and are referred as water footprint (Hoekstra and Mekonnen 2012) or generalized water (Liu et al. 2017).

L. Yong (✉)
Qilu University of Technology (Shandong Academy of Sciences), Information Research Institute of Shandong Academy of Sciences, Jinan 250014, China
e-mail: 32516022@qq.com

H. Xu (ed.), *The 6th International Symposium on Water Resource and Environmental Management*, Environmental Science and Engineering,
https://doi.org/10.1007/978-3-031-55989-1_6

The water footprint or generalized water measures our total water consumption.

Fresh water is unevenly distributed on the surface of our planet; however, the concept of virtual water has obvious strategic implications for water-constrained regions on the basis of comparative advantage, buying water-intensive products from other regions and saving local fresh water for other purposes (Hoekstra and Mekonnen 2012). Beijing and Shanghai heavily rely on both physical and virtual water resources from other provinces, and Henan, Hebei and Heilongjiang are virtual water exporters (Cai et al. 2017). Urumqi and Macao buy a huger volume of virtual water from other provinces than that of local fresh water to increase their generalized water supplies (Liu et al. 2017; Chen and Li 2015). The European Union, as a whole, has a net import of virtual water from other continents (Antonelli et al. 2017). In practice, virtual water trade makes a great contribution to sustainable development for both a water-shortage and water-abundant region.

However, virtual water doesn't always work on the basis of comparative advantage. Many Chinese northern water-scarce provinces are water exporters, while southern water-abundant provinces are water importers. For example, Xinjiang suffers from acute water scarcity, but it has a large volume of net virtual outflow of local fresh water embodied in agricultural products to other more water-rich provinces of China (Zhao et al. 2019; Tang and Liu 2023). Liaoning suffers from serious water shortages, but it is a net water exporter (Dong et al. 2012). These paradoxical cases result from many factors (Liu et al. 2019), and virtual water trade is certainly not the only one tool in addressing water scarcity issues (Maxwell 2010). Energy, land, labor, and other factors also have an important influence on trade. Virtual energy trade has greatly improved our ability and efficiency to obtain and use energies (Li et al. 2020). Land resources could be transferred virtually when they are embedded in commodities (Jan et al. 2013). Pollutants can be emitted virtually in region B where products made in region A (Chen et al. 2020). These factors can also influence virtual water trade, and it is necessary to extend the traditional one-dimensional framework of virtual water trade to a multiple-dimensional generalized trade. The rest of this paper is organized as follow: Sect. 6.2 introduces generalized trade in detail, Sect. 6.3 lists two case studies, Sect. 6.4 makes a further discussion, and Sect. 6.5 gives a scientific conclusion.

6.2 Generalized Trade

In a product, there are some amounts of virtual water, virtual energies, virtual land, and virtual labors, and virtual water is a part of the bigger story of generalized trade (Liu 2020). According to physical or virtual senses, generalized trade is divided into physical trade and virtual trade.

Physical trade refers to commodity exchange in barter or currency form. Barter is an exchange system where goods are directly exchanged for others. For example, about twenty years ago, Kazakh herdsmen often exchanged mutton, sheepskin and wool for salt, tea and flour from Han farmers in some mountainous areas far from markets in Xinjiang in northwestern of China. However, there must be a coincidence of wants before two traders could exchange in a barter system, and it became more and more inefficient in practice. Barter trade was doomed to be replaced by a smarter trade form with an exchange medium which could avoid the inconveniences of a pure barter system. The value of the exchange medium is constant and inherent, and the values of various commodities can be assessed against each other with the exchange medium. In this smarter market, one person could sell his products for some exchange medium at his disposal and buy what he wants with his exchange medium. In the modern era, our human society has been deeply monetized, and currency plays an important role as an exchange medium. All kinds of goods, services, assets, intellectual patents and other elements that may bring benefits can be evaluated by currency and traded physically with money. For example, many Arab countries trade oil for dollars that are used to buy commodities from other regions. Physical trade in barter form focuses on the physical exchange of commodity as a whole, and the exchange medium, currency, is a prevailing and smart measurement of physical trade. Physical trade is determined by market laws (Liu 2020).

When talking about trade, we traditionally think about all elements of a final commodity produced in a region and then physically transported to a consumer in a different region. However, various resources embodied in these commodities are also exchanged in virtual senses. For example, there are 1 ton of goods in weight, and they are worth 1 million CNY in currency. They totally consume 2 tons of water, 3 tons of standard coal energy, 1 thousand of labors, and 10 hectares of land in the production process, and these various resources are also transported from one region to another in virtual senses. From another perspective, these goods can be measured by 2 tons of virtual water, 3 tons of standard coal virtual energy, 1 thousand of virtual labors, and 10 hectares of virtual land, respectively. These different expressions offer multiple perspectives of the same goods and gets a more accurate insight into traditional trade. The virtual transfer of these various resources embodied in physical commodities are called as virtual trade (Liu 2020).

Physical and virtual trade together form the extended framework of generalized trade. Generalized trade looks as if more like a conceptual hodgepodge of trade rather than a creative concept; yet the conceptual hodgepodge results from the multivariate production process. In economics, production y is a multivariate function with water, labor, energy, land and other factors as its variables (as Eq. 1 shows).

$$y = f(water,\ labor,\ energy,\ land,\ others) \tag{6.1}$$

Here, y is physical commodities, water, labor, energy, land and other factors are embodied in y. They are transferred at the same time. These different measurements are equal to each other, and they together depict a multifaceted portrait of physical commodity exchange.

6.3 Case Studies

Liu made a long-term investigation into both Xinjiang's and Urumqi's acute water scarcity with an extended noncompetitive input–output model. This model has many attributes well-suited to calculate total water consumption by an industrial sector. Xinjiang and Urumqi have been suffering from acute water scarcity since the 1980s. Urumqi had a net volume of about 3.08 billion m^3 of virtual water from other economies, and it was about 3.14 times the volume of local freshwater available (Liu 2017). Virtual water trade helps to increase its water supply and contributes greatly to its water scarcity alleviation (Liu et al. 2017), population increase (Liu 2018), and economic growth (Liu 2019). Xinjiang had a net volume of about 13.1 billion m^3 of local fresh water that was embodied in local products and transported to other economies, and this volume accounted for 23.7% of its total water supplies (Tang and Liu 2023). In practice, agriculture accounts for as much as about 90% of total human water consumption, and virtual water is mainly embodied in agricultural products. Urumqi buys virtual water embodied in external agricultural products for daily necessities, and Xinjiang sells virtual water embodied in local agricultural products to transform local freshwater resources into economic values. Both in Urumqi and Xinjiang, virtual water trade makes a great contribution to their continual economic growth, although it works in different ways.

In Xinjiang, virtual water trade doesn't work on the basis of comparative advantage, but it is not the case in other forms. Under the framework of generalized trade, water is only one of the variables, and it is interesting to turning from the virtual-water dimension to other dimensions to draw a multiple-perspective portrait. Liu continued his work by estimating Xinjiang 's and Urumqi 's generalized trade in other forms (Liu 2020; Tang and Liu 2023). Urumqi had a net import trade in currency, virtual-water, virtual-labor, virtual-emission, and virtual-wasted-water forms, and they were 17.4 billion CNY, 3.1 billion m^3,1.34 millions of people, 0.54 million tons and 611.6 million tons, respectively. However, it had a net export trade in virtual-energy forms, and it is 9.50 million tons of standard coal. Xinjiang has a net import trade of 698.3 billion CNY in currency, -13.1 billion m^3 in virtual-water form, 55.2 million tons of standard coal in virtual-energy form, 5.21 million of people in virtual-labor form, 352.4 billion CNY in value-added form, 0.41 million tons in virtual-emission form, and 9460 tons in virtual-wasted-water form. As a whole, Urumqi 's generalized trade works well on the basis of comparative advantage; yet Xinjiang 's generalized trade is not the case.

Liu used an elasticity coefficient to make a compare among generalized trade in different forms (Liu 2020; Tang and Liu 2023). In Urumqi, the virtual-water coefficient was 48.9, and the virtual-energy coefficient, the smallest one, was −6.16. It indicates that virtual water trade and virtual energy trade play the most important role in its economy when connected with other economies. In Xinjiang, the virtual-labor coefficient, the largest one, was 2.03, and the virtual-water coefficient, the smallest one, was −0.59. It indicates that virtual water trade and virtual labor trade play the most important role in its economy when connected with other economies.

Both in Xinjiang and Urumqi, virtual water trade plays the most important role, but it works in different ways. In Urumqi, generalized trade occurs mainly because of its populous necessities, for example, physical goods offer their utilities, virtual water increases water supplies, labors create prosperity, virtual energy trade transforms local natural resources into economic benefits. In Xinjiang, generalized trade occurs mainly because its local industrial system is imperfect, for example, many necessary goods are offered by other economies, and many virtual labors and virtual energies are accompanied with these external goods at the same time. In addition, agricultural products are well known, and local freshwater embodied in agricultural products, flows out of Xinjiang, transforming its well-known agricultural products into economic profits.

6.4 Discussions

Xinjiang and Urumqi are water-scarce economies, and virtual water trade plays an important role. However, virtual water trade works in different ways, and this paradoxical phenomenon is worth further discussing.

First, virtual water trade has an important role in an economy's development. Water is unevenly distributed on the global scale, but it is redistributed in virtual senses from a region to another one. A huge number of people live in physically water-scarce areas, and there is not enough water. Trade helps to transport commodities from a region to another one, as well as accompanied with water embodied in these physical commodities. In Urumqi, external virtual water embodied in agricultural products from rural areas brings much water to this arid cities and increases total water supplies. This function of virtual water trade can be found in many other urban areas, for example, Beijing and Shanghai need more agricultural products in which virtual water is embodied and transported. In Xinjiang, local freshwater flows to other economies in virtual senses, embodied in agricultural products. This function of virtual water works in many Chinese northern provinces which are grain production areas. These northern water-scarce provinces sell many agricultural products for economic benefits, and those southern water-rich provinces develop industrial and service industries with more economic benefits. Virtual water trade works well; yet it doesn't always run on the basis of comparative advantage.

Second, virtual water trade doesn't play a monodrama, but joins in a chorus of generalized trade. In an economy, commodity trade helps to buy daily necessary goods to meet its demands and sells what its local products to increase its economic scale, and that is the way physical trade works. Virtual trade is embodied in these commodities, and various resources are redistributed for different purposes. In Urumqi, virtual water trade helps to gain much virtual water to save its limited physical water supplies, virtual energy trade helps to transmit its abundant physical energy resources into energy-intensive products to increase its economic benefits, virtual labor trade helps to bring many labors to produce daily necessities to maintain its continual development, and virtual emission trade can decrease much local emissions. In Xinjiang, virtual labor trade increases its human capitals in different industries to promote its sustainable development, virtual water trade helps to increase its economic benefits by selling its agricultural products, and virtual energy trade helps to bring many necessary products that cannot be produced in local enterprises. Different forms of generalized trade work in different ways, of which some are less important, but they are never absent.

Third, the extended framework of generalized trade offers more scientific suggestions to develop virtual water trade. In Urumqi, virtual water trade makes a great contribution to increase its water supply, and it is essential to improve water infrastructure construction to increase physical water supplies and create convenient conditions to increase virtual water supplies. In Xinjiang, water is not only extremely important for its ecology but also have important economic significance, and it is important to develop large-scale and water-saving agriculture to increase the utilization rate of its limited freshwater. In addition, it is an essential choice for Xinjiang to increase supplies of local industrial products by improving its local industrial system rather than heavily relying on external products.

Forth, generalized trade helps to promote sustainable development. Water, energy, labor, and land are consumed in virtual senses during the production process, and then transported to where they are needed, but these virtual trades are often ignored. Generalized trade happens in any economy that makes the most use of its superior resources to maximize its benefits. Under the framework of generalized trade, our human society not only weaves a physical commodity network, but also knits a multi-dimensional virtual network, binding the global economy together as a viscous organism. Resources are unevenly distributed over time and space at different levels, and generalized trade helps to redistribute resources at different geographical levels. For example, China has exported its surplus labor force in virtual senses through the OEM trade, and provided more and cheaper products for other traders that are in surplus of capital, technology, energy or others. Brazil exports large area of virtual land hidden in a large number of soybeans to China. Policy makers not only pay a lot of attention on physical trade, but also on virtual trade under the framework of generalized trade to gain a more sustainable development.

6.5 Conclusions

The term of virtual water refers to the sum of the water use along with each step of the production chain, and it has gained weight in the scientific and political debates. A popular viewpoint holds that virtual water flows from water-scarce regions to water-rich regions, but some studies show that virtual water flows from water-rich regions to water-scarce regions. After a long-term of investigation, it is found that virtual water trade doesn’t play a monodrama but joins in a chorus of generalized trade. In economics, production is a multivariate function with water, labor, energy, land and other factors as its variables. Generalized trade includes physical trade, virtual energy trade, virtual labor trade, and virtual land trade, and others, and they contribute greatly to an economy's sustainable development. Both in Urumqi and Xinjiang, virtual water trade makes a great contribution to their economic growth, although it works in different ways. In Urumqi, generalized trade occurs mainly because of its populous necessities. In Xinjiang, generalized trade occurs mainly because its local industrial system is imperfect. Our human society not only weaves a physical commodity network, but also knits a multi-dimensional virtual network. The role of virtual water trade can be much clearer under the generalized trade framework, and the concept of generalized trade offers an amplified framework to investigate trade itself as well as our globalization.

6.6 Declaration

This study was approved and supported by Information Research Institute of Qilu University of Technology (Shandong Academy of Sciences). Liu, the only one author, declared no potential conflicts of interest with respect to the research, and approved the final manuscript and the submission to the journal for publication. This paper was supported by no funding project support. The datasets generated during the current study are publicly available.

Acknowledgements This research was supported by the author, and all errors are our responsibility.

References

Antonelli M, Tamea S, Yang H (2017) Intra-EU agricultural trade, virtual water flows and policy implications. Sci Total Environ 587:439–448

Cai, Beiming, Wang, Chencheng, Zhang, Bing (2017) Worse than imagined: Unidentified virtual water flows in China. J Environ Manag

Chen GQ, Li Jiashuo (2015) Virtual water assessment for Macao, China: Highlighting the role of external trade. J Clean Prod. 93

Chen Shaoqing, Chen Bin, Feng Kuishuang, Liu Zhu, Fromer Neil, Tan Xianchun, Alsaedi Ahmed, Hayat Tasawar, Weisz Helga, Schellnhuber Hans, Klaus Hubacek (2020) Physical and virtual carbon metabolism of global cities. Nat Commun 11. 182. China. Ecol. Econ. 61 (1), 159e170

Dong Huijuan, Geng Yong, Sarkis Joseph, Fujita Tsuyoshi, Okadera Tomohiro, Xue Bing (2012) Regional water footprint evaluation in China: a case of liaoning. Sci Total Environ. 442C. 215–224

Hoekstra Arjen, Mekonnen Mesfin (2012) The water footprint of humanity. In: Proceedings of the National Academy of Sciences of the United States of America. 109. 3232–7

Jan Weinzettel, Hertwich Edgar, Peters Glen, Steen-Olsen Kjartan, Galli Alessandro (2013) Affluence drives the global displacement of land use. Glob Environ Chang. 23. 433–438

Li Yilin, Chen Bin, Chen Guoqian, Meng Jing, Tasawar Hayat (2020) An embodied energy perspective of urban economy: A three-scale analysis for Beijing 2002–2012 with headquarter effect. Sci Total Environ. 732. 139097

Liu Y (2018) Why an urban population continues to grow under intensifying water scarcity: an answer from generalized water resources. Urban Water J 15(10):961–965

Liu Y, Yang DG, Lü GH (2017) Water scarcity and generalized water resources management in Urumqi: contributions from a noncompetitive water input-output model. J Water Resour Plan Manag. 143, 8

Liu Yong (2019) External virtual water contributes greatly to continual development in Urumqi. Hydrol. 7(4), pp. 56–61

Liu Yong (2020) Sustainable development in Urban Areas: contributions from generalized trade. Sustain Cities Soc. 61, p 102312

Liu Xi, Du Huibin, Zhang Zengkai, Crittenden John, Lahr Michael, Moreno-Cruz Juan, Guan Dabo, Mi Zhifu, Zuo Jian (2019) Can virtual water trade save water resources?. Water Res. 163

Liu Y (2017) Study on generalized water resources: contributions from A Non-competitive water Input-Output Model in Urumqi. Ph.D. Thesis. Univ Chin Acad Sci/Xinjiang Univ

Maxwell S (2010) The concept of virtual water: understanding our real water use. Am Water Work Association J 102(12):88–90

Tang CH, Liu Y (2023) Role of virtual water trade under the generalized trade framework in xinjiang's economic sustainable development. Water Econ Policy

Zhao Dandan, Klaus Hubacek, Feng Kuishuang, Sun Laixiang, Liu Junguo (2019) Explaining virtual water trade: A spatial-temporal analysis of the comparative advantage of land, labor and water in China. Water Res. 153

Chapter 7
Algae-Bacterial Aerobic Granular Sludge for Surfactant-Laden Saline Industrial Wastewater Treatment: A Sustainable Approach

Hani Ezz, Mahmoud Nasr, Manabu Fujii, and Mona G. Ibrahim

Abstract In recent years, interest in algae-bacteria aerobic granular sludge (ABGS) for industrial wastewater treatment has grown. This study compares ABGS and bacterial granular sludge (BGS) in treating surfactant-laden wastewater with salinity from 1 to 3%. Two lab-scale reactors were employed, one with ABGS and the other with BGS. Higher salinity positively influenced Sodium Dodecyl Sulfate (SDS) removal, with ABGS outperforming BGS at 93.5% versus 88.4% at 3% salinity. Organic matter removal was moderately impacted, with ABGS achieving 96.4% and BGS 91.15% at 3% salinity. ABGS consistently surpassed BGS in ammoniacal-nitrogen ($NH_4^+{-}N$), total inorganic nitrogen (TIN), and phosphate-phosphorus (PO_4^{-3}-P) removal, reaching max removals of 97.45%, 77.78%, and 76.94%, respectively, compared to 94.41%, 72.98%, and 71.45% for BGS. ABGS aligns with Sustainable Development Goals (SDGs), contributing to water pollution mitigation, biodiversity conservation, climate action, and more. This study emphasizes ABGS's potential in treating surfactant-contaminated wastewater under varying salinity conditions,

H. Ezz (✉) · M. Nasr · M. G. Ibrahim
Environmental Engineering Department, Egypt-Japan University of Sciences and Technology (E-JUST), New Borg Al Arab City, Alexandria 21934, Egypt
e-mail: hani.ezz@ejust.edu.eg

H. Ezz
Environmental Engineering Department, Faculty of Engineering, Zagazig University, Zagazig 44519, Egypt

M. Nasr
Sanitary Engineering Department, Faculty of Engineering, Alexandria University, Alexandria 21544, Egypt

M. Fujii
Department of Civil and Environmental Engineering, Tokyo Institute of Technology, Tokyo, Japan

M. G. Ibrahim
Environmental Health Department, High Institute of Public Health, Alexandria University, Alexandria, Egypt

H. Xu (ed.), *The 6th International Symposium on Water Resource and Environmental Management*, Environmental Science and Engineering,
https://doi.org/10.1007/978-3-031-55989-1_7

particularly excelling in nutrient and surfactant removal compared to BGS, aligning with multiple SDGs and promoting sustainable wastewater treatment practices.

Keywords Algal–bacterial aerobic granular sludge · Surfactant-laden industrial wastewater · Sustainable development goals (SDGs)

7.1 Introduction

Surfactants play a pivotal role in a wide range of industries, from detergents and textiles to paints and pharmaceuticals (Badmus et al. 2021; Nasr et al. 2021). However, when untreated wastewater containing surfactants is released into the environment, it can have dire consequences for both human health and aquatic ecosystems (Dadebo et al. 2023). These substances have a tendency to generate foam and intricate compounds upon entering water bodies, thereby triggering eutrophication in aquatic environments (Bindes and Franco 2015). Moreover, as surfactants become more hydrophobic, their toxicity escalates. Additionally, it's important to highlight that industrial wastewater often carries high salinity levels (Bella et al. 2015). This saline wastewater is laden with salts and other harmful chemicals, posing risks such as the depletion of aquatic species, harm to ecosystems, and soil pollution (Niu et al. 2023). Furthermore, high salinity can detrimentally affect the efficiency of wastewater treatment processes (Li et al. 2023). Prior to being released into water systems, saline wastewater contaminated with surfactants must be effectively treated in order to shield the environment from the harmful impacts of these pollutants.

Various techniques have been employed to combat surfactant contamination in wastewater, encompassing methods such as adsorption (Bindes and Franco 2015), activated sludge (González et al. 2007), and coagulation (Dadebo et al. 2023) Nevertheless, these treatments often entail substantial operational and maintenance costs, especially in the context of adsorption and coagulation (Ezz et al. 2022). Moreover, the unique foam structure generated during bioremediation can impede the settling and dewatering processes of activated sludge, a commonplace component in conventional wastewater treatment plants (Guan et al. 2017). Furthermore, these technologies exhibit limited efficiency in nutrient (N and P) removal, necessitating additional treatment steps (Fallahi et al. 2021). As such, there is a pressing need for the development of a more cost-effective and efficient treatment approach.

Thick granules are produced using the BGS method by using self-immobilized cells that are joined by an exo-polymeric matrix (Nancharaiah et al. 2018). By significantly cutting down on settling time and doing away with the need for large tanks, this method outperforms conventional techniques and minimises the amount of space needed for typical systems while still efficiently treating an equivalent volume of wastewater (Sarma et al. 2017). Furthermore, the composition of communities within these granules varies concerning their depth and oxygen availability. The concurrent processes of aerobic nitrification and anoxic denitrification augment the overall efficiency of nutrient removal (Chen et al. 2019). The presence of a diverse and highly active microbial community enhances resilience to exposure from toxicity, salinity, and abrupt influxes of pollutants. The exceptional feature of reducing space, in conjunction with its high effectiveness in eliminating contaminants including chemical oxygen demand (COD), nutrients, and hazardous materials from different kinds of wastewater, results in a noteworthy enhancement of water quality (Li et al. 2014).

On the other hand, ABGS offers an attractive and economically efficient option that not only tackles wastewater treatment but also substantially improves the retrieval of nutrients (Zhang et al. 2021). In the mutualistic relationship of the algal–bacterial symbiosis system, bacteria, and algae establish mutual benefits by exchanging substances and transmitting signals (Liu et al. 2022a). The algae in these algal–bacterial consortia are capable of absorbing inorganic nutrients like phosphate and nitrogen. Additionally, they generate oxygen through photosynthesis to support aerobic metabolism, consequently reducing the expenses associated with aeration. As a result, it significantly lessens the reliance of wastewater treatment facilities on conventional mechanical aeration, which results in significant cost savings and reduced environmental effects (Liu et al. 2023). This comprehensive method maximizes the utilization of carbon sources, leading to increased biomass productivity and the facilitation of improved nutrient reclamation. Studies conducted earlier have also exhibited outstanding pollutant removal from different kinds of wastewater using ABGS. Remarkable efficiency has been demonstrated in leachate wastewater treatment, as evidenced by Ilmasari et al., with removal rates of 88% $\pm$ 14% for COD, 56% $\pm$ 16% for NH_4^+–N, 57% $\pm$ 14% for total phosphorus (Ilmasari et al. 2023). Through the utilisation of bacteria and algae's complementary properties, this innovative method opens the door to a sustainable and efficient future in wastewater treatment.

Recently, the United Nations (UN) introduced 17 sustainable development goals (SDGs) accompanied by 169 specific targets aimed at fostering both socio-economic growth and safeguarding the environment (United Nations 2019). ABGS, owing to its capacity to shield the environment from a spectrum of organic and inorganic contaminants, aligns with SDGs that are pertinent to the well-being of human and aquatic ecosystems. Additionally, the utilization of granules with their rich lipids, carbohydrates, and protein content boasts a multitude of practical applications spanning bioenergy, animal feed, fertilization, and medicinal preparation. This multifaceted utility intersects with the SDGs that are economically oriented. Furthermore, due to its eco-friendly nature and aesthetically pleasing attributes, the implementation of ABGS in industrial wastewater treatment has the potential to sustain public acceptance, thereby contributing to the attainment of SDGs that hold social relevance.

Although ABGS in wastewater treatment has been widely studied, there is still a requirement for a thorough evaluation of how well it aligns with the three aspects of sustainability. Therefore, this study has two primary objectives. Firstly, it aims to conduct a comparative analysis of the performance of ABGS and BGS in the removal of SDS, COD, NH_4^+–N, TIN, and PO_4^{-3}-P, in industrial wastewater characterized by varying salinity levels (0 to 3%). Secondly, it seeks to establish a direct connection between the utilization of ABGS and the three sustainability pillars.

7.2 Materials and Methods

7.2.1 *Synthetic Surfactant-Laden Wastewater and Seed Granular Sludge*

The artificial solution was painstakingly created to replicate the features of actual industrial wastewater that has been tainted with surfactants (Dadebo et al. 2023). The primary component of this solution was the anionic surfactant Sodium Dodecyl Sulfate (SDS) ($NaC_{12}H_{25}SO_4$), with a consistent concentration of 20 mg/L throughout the experiments. To introduce varying levels of salinity, different amounts of NaCl (w/v) were carefully added to the influent. This resulted in salinity levels of 1% during Stage II and 3% in Stage III, respectively. The synthetic solution also included other essential components, including approximately 300 mg/L of COD (as NaAc), around 38 mg/L of NH_4^+−N (as NH_4Cl), and roughly 10 mg/L of PO_4^{-3}-P (as KH_2PO_4). Furthermore, it contained 30 mg/L of $CaCl_2 \cdot 2H_2O$, 25 mg/L of $FeSO_4.7H_2O$, 25 mg/L of $MgSO_4 \cdot 7H_2O$, and 1 mL/L of a trace element solution (Meng et al. 2019). To ensure optimal conditions, we carefully adjusted the pH values within the range of 7.2 to 7.5.

Mature BGS and ABGS were grown in a laboratory environment that had been consistently operational for over 90 days. The seed ABGS and BGS primarily fell within the size range of 1.0 to 5.0 mm. In the two reactors (R_1 for BGS and R_2 for ABGS), the initial biomass concentrations were approximately 4000 ± 0.03 mg/L of mixed liquor volatile suspended solids (MLVSS), with a sludge volume index (SVI_{30}) of 36.4 ± 0.3 ml/g for ABGS and 40.25 ± 0.3 ml/g for BGS.

7.2.2 *Reactors Set-Up and Operation*

Two identical sequencing batch reactors made of clear acrylic plastic were used in the experiment. With an inner diameter of 8.40 cm, a height of 20 cm, and a working volume of 1000 mL, both reactors had the same dimensions (see Fig. 7.1). The initial reactor, denoted as R_1, functioned as the control and received exclusive inoculation with BGS). Conversely, the second reactor, named R_2, received inoculation with ABGS. Throughout the experiment, reactor R_2 was exposed to white LED tubes emitting light at an intensity of 4000 lx. In contrast, reactor R_1 was covered with a plastic bag to prevent light penetration. Through a porous diffuser at the bottom of the reactors, air was introduced into the reactors at a rate of 4000 mL/min to achieve aeration. The reactor ran for four hours in total, including two minutes of feeding time, sixty minutes of non-aeration time, two hours of aeration time, and ten minutes of settling time (Meng et al. 2019). Furthermore, a discharge point was added at the reactor's middle to allow for a 50% exchange ratio and an 8-h hydraulic retention period. There were three steps in the entire experiment. In Stage I, which ran from days 1 to 30, there was no additional salinity (0%). Stage II, spanning days 30 to

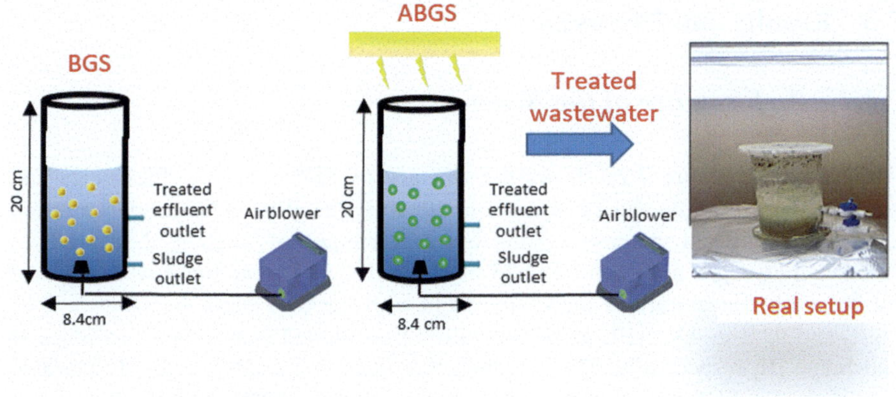

Fig. 7.1 Schematic diagram for the reactors

50, witnessed a increase in salinity to 1%. Stage III, from days 50 to 70, witnessed increase of salinity to 3%.

7.2.3 Analytical Methods

Water samples were taken at the end of each operating cycle, every three days. After passing through a 0.22 m membrane filter, these samples were subjected to analysis. The analysis of the wastewater samples included the determination of parameters such as COD, NH_4^+–N, TIN, and PO_4^{-3}-P in accordance with standard methods (USEPA 2012). The pH of the bulk liquid was measured using a pH meter (Jenway 3510 pH meter, USA). The methylene blue technique was used to determine the SDS levels according to Yasin et al. (2021).

7.2.4 Statistical Analysis

Microsoft Excel 2019 was used for the statistical analysis, and one-way analysis of variance (ANOVA) was used. A significance threshold of $p < 0.05$ was used. The results are shown as the mean ± standard deviation, and the analyses were carried out in triplicate.

7.3 Results and Discussion

7.3.1 Performance of the Reactors

SDS removal. Figure 7.2a shows the changes of SDS removal efficiencies and effluent concentrations during operation. Notably, the effluent SDS concentration exhibited significant increases, rising from an average of 74.22 ± 2.56 mg/L to 79.56 ± 1.86 mg/L and 84.62 ± 2.25 mg/L for R_1. The same trend was observed for R_2, with concentrations increasing from 82.84 ± 3.41 mg/L to 87.37 ± 2.92 mg/L and 92.2 ± 4.9 mg/L as the operation transitioned from Stage I to Stage II and Stage III, respectively. This suggests that increased salinity created favorable conditions for SDS removal. It's important to note that SDS removal in granular sludge occurs in two steps: first, adsorption on the granule's surface, followed by biodegradation (Wang et al. 2023a). Therefore, the enhanced removal efficiency can be attributed to the higher biosorption of SDS in the presence of elevated salinity. The diverse functional groups in the extracellular polymer substances (EPS) of the granules provide an ideal surface for anionic surfactant sorption. Additionally, higher salinity can compress the electrical double layer, resulting in increased adsorption of anionic surfactants with rising salt concentrations (Belhaj et al. 2020). Furthermore, the presence of algae may contribute to increased functional group diversity in ABGS granules, facilitating rapid surfactant adsorption on the granule surface, followed by biodegradation (Wang et al. 2023b).

COD removal. Throughout the trials, both reactors continuously produced acceptable COD removal efficiencies, taking into account the unique properties of the wastewater they were treating (Fig. 7.2b). Initially, in the first two weeks, COD removal exhibited fluctuations and instability, which can be attributed to the adaptation process required for surfactant-laden wastewater. Subsequently, during stage I, both the R_1 and R_2 reactors showed improvements in COD removal, reaching 96.30 ± 0.97% and 94.46 ± 1.1%, respectively, by the end of this stage. However, in stages II and III, a slight decrease in COD removal was observed in both reactors as salinity levels increased, with average removal rates of 93.76 ± 1.6% and 91.46 ± 2.1% during stage II and 91.92 ± 1.1% and 90.21 ± 1.55% for R_1 and R_2, respectively. The slight COD superiority exhibited by ABGS over BGS can be attributed to the co-culture system. In this setup, heterotrophic bacteria, which are the primary consumers of organic carbon, break down complex carbon compounds into simpler molecules (Wang et al. 2023a). These simpler molecules serve as nutrients for the growth of microalgae. Simultaneously, microalgae, through photosynthesis, supply oxygen, which fosters the growth of bacteria (Ilmasari et al. 2023). This symbiotic and mutually beneficial relationship enhances the efficiency of organic matter removal. These results suggest that both BGS and ABGS are efficient in removing organic matter from surfactant-laden wastewater under different salinity conditions, aligning with observations by Meng et al. (2020) regarding the high adaptability of BGS and ABGS for organic matter removal across varying salinity levels.

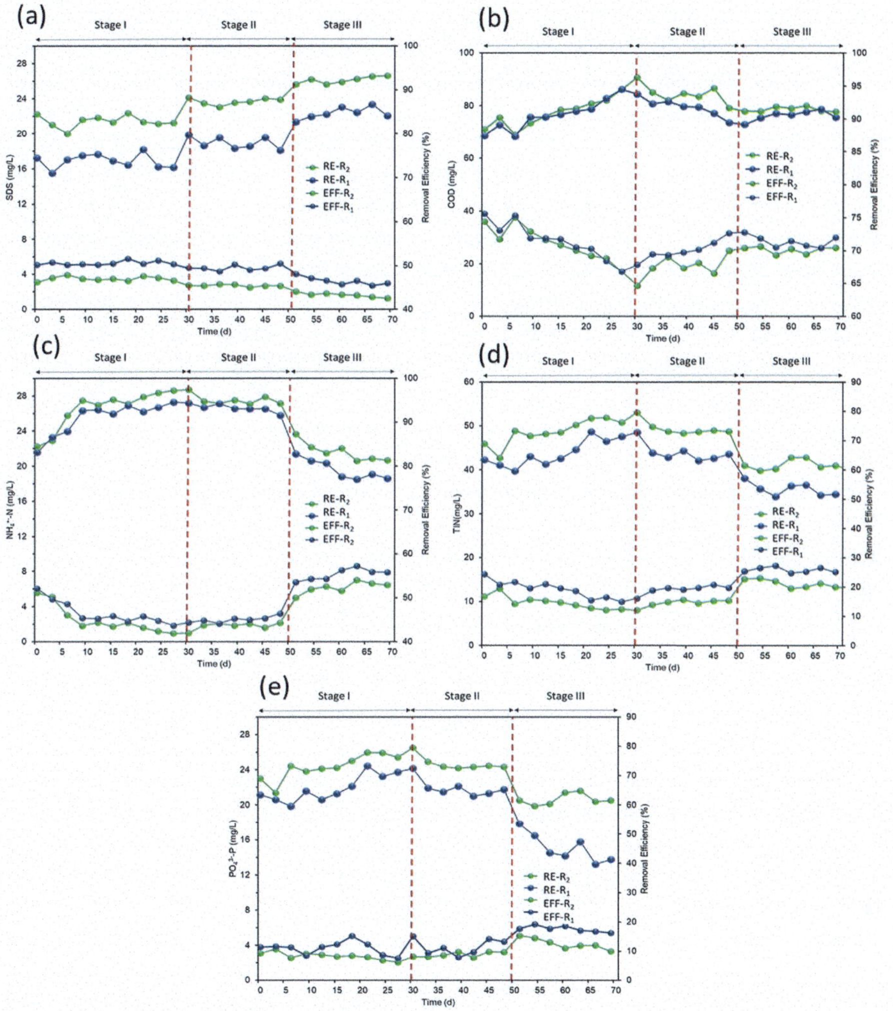

Fig. 7.2 Variations of SDS (**a**), COD(a), $NH_4^+ - N$ (**b**), TIN (**c**) and PO4-3-P (**d**) in removal efficiencies and concentrations

Nitrogen removal. Figure 7.2c illustrates the changes of NH_4^+−N concentration and removal efficiency in both reactors. In the initial stage, there was a rapid decline in ammoniacal nitrogen concentration in the effluent within the first five days of the experiment, consistently maintaining levels below 5 mg/L throughout this phase. Notably, both reactors exhibited high removal rates, reaching a maximum of 94.4 ± 1.75% for R_1 and an impressive 97.45 ± 1.1% for R_2. This swift enhancement in removal efficiency during phase I indicates the rapid establishment of effective nitrification in response to surfactant-laden wastewater treatment, aligning with a previous study on surfactant-contaminated wastewater by granular sludge (Wang et al. 2023a). As the experiment progressed into the second stage, both reactors sustained commendable NH_4^+-N removal rates, averaging 93.23 ± 2.2% for R_1 and 95.15 ± 1.72% for R_2. However, in the third stage, the removal of NH_4^+-N experienced a sharp decline, ultimately reaching 77.2 ± 2.15% and 81.7 ± 1.65% in R_1 and R_2, respectively. This observation strongly suggests that elevated salinity levels hindered the activity of ammonia oxidizing-bacteria, a phenomenon corroborated by Dong et al. (2021). Furthermore, it's worth highlighting that ABGS outperformed BGS in terms of NH_4^+-N removal efficiency during operation, possibly due to the presence of algae that contributes to nitrogen assimilation. This aligns with findings from Liu et al. (2022b), suggesting that algae can directly assimilate ammoniacal nitrogen from wastewater, complementing the ammonia oxidation process carried out by ammonia-oxidizing bacteria, which converts it into nitrite NO_2^- and nitrate NO_3^-.

Figure 7.2d shows the changes of TIN removal efficiency in both reactors throughout the experiment. Initially, during Stage I, a rapid increase was observed in TIN removal efficiency. Specifically, in R_1, it increased from 63.28 ± 1.15% to 72.79 ± 1.61%, while in R_2, it reached 79.57 ± 1.35%. The boost in TIN removal efficiency observed in both reactors during Phase I corresponded with changes in NH_4^+-N removal efficiency in this phase. Moving on to Stage II, TIN removal efficiency showed a remarkable stability of around 66% in R_1 and an even higher 74% in R_2. However, as we transitioned into Stage III, there was a decline in TIN removal efficiency, and the average removal was 53.46 ± 1.36% in R_1 and 61.82 ± 1.74% in R_2. This decline in TIN removal efficiency could be attributed to the inhibition of nitrification, which is more pronounced in high salinity and surfactant-laden conditions (Wang et al. 2023a). This decrease was also concurrent with an increase in NH_4^+-N concentration in the effluents. These results suggest the presence of algae played a pivotal role in elevating TIN removal efficiency for ABGS compared with BGS by promoting nitrification, denitrification, and enhancing nitrogen assimilation (Meng et al. 2019).

Phosphorus removal. The variations in the concentration and removal rates of PO_4^{-3}-P in the effluent are displayed in Fig. 7.2e. A pattern becomes apparent: R_2 regularly removed more PO_4^{-3}-P than R_1. Because the granules were adjusting to their new environment, both reactors first saw a minor drop in the removal of phosphorus. Subsequently, PO_4^{-3}-P removal rates increased, reaching 72.35 ± 2.9% for R_1 and 79.54 ± 2.1% for R_2. In stage II, where salinity was increased to 1%, there was a minor impact on total phosphorus PO_4^{-3}-P removal. During this phase, average removal was 66.32 ± 3.1% for R_1 and 74.19 ± 1.92% for R_2. However, in Stage III, with salinity elevated to 30 g/L, PO_4^{-3}-P removal rates showed a significant decline. On average, R_1 achieved 45.36 ± 1.82% removal, while R_2 achieved 61.9 ± 2.42%. This decline in PO_4^{-3}-P removal may be attributed to the inhibition of the growth or activity of polyphosphate accumulating organisms (PAOs) under high salinity conditions. Additionally, the high removal of surfactants during this stage could have adversely affected the activity of PAOs. It's worth noting that algae growth in R_1 resulted in superior P uptake and removal efficiency. This observation is in line with prior research which has suggested that the growth of algae plays a significant role in enhancing TP removal (Fan et al. 2021).

7.3.2 Sustainable Development Goals (SDGs) Linked to Industrial Wastewater with ABGS

Environmental-related SDGs. The application of ABGS into industrial wastewater treatment contaminated with surfactants seamlessly aligns with several critical Environmental-related SDGs and their corresponding targets. Firstly, it serves as a significant contributor to SDG 6, which centers on Clean Water and Sanitation. ABGS systems play a pivotal role in directly addressing Target 6.3, where they enhance water quality, reduce pollution, and effectively curtail the release of contaminants into water bodies. Moreover, their remarkable adaptability in treating wastewater under varying salinity conditions significantly reduces groundwater contamination, thereby actively contributing to the mitigation of groundwater pollution. This aligns well with the findings of a previous by Saqr et al. (2021), emphasizing the adoption of appropriate water treatment methods to prevent the deterioration of groundwater quality. Additionally, ABGS systems further support Target 6.4 by optimizing water usage efficiency and ensuring a sustainable supply of freshwater, effectively addressing concerns related to water scarcity. Moving to SDG 13, dedicated to Climate Action, ABGS systems play a crucial role. They actively assist in the reduction of greenhouse gases, primarily through algae's remarkable capacity to capture and stabilize carbon dioxide (CO_2), perfectly aligning with Target 13.2 focused on climate change policy. Notably, research by Zhang et al. (2021) highlights that the adoption of ABGS for domestic wastewater treatment plants could lead to a substantial reduction in greenhouse gas emissions, estimated at an impressive 77%. Furthermore, these systems contribute to Target 13.3 by enhancing education and awareness regarding climate change mitigation and adaptation through the sustainable wastewater treatment practices they employ. In the context of SDG 14, which pertains to Life Below Water, the application of ABGS for industrial wastewater treatment makes significant strides toward achieving Target 14.1. This involves the significant reduction of marine pollution and active participation in the preservation of marine environments. Furthermore, research indicates that granules with their high capacity for nutrient removal and elements recovery can be effectively harnessed to produce biochar with high quality. These biochar products have the potential to be applied to farmlands, leading to improved soil quality and enhanced agricultural productivity, thereby contributing to SDG 2 (Zero hunger) Target 2.1 (End hunger) and Target 2.3 (Double agricultural productivity (Ezz et al. 2021).

In the realm of Economic-related Sustainable Development Goals (SDGs), the integration of ABGS technology not only addresses environmental considerations but also contributes significantly to the progress of SDG 8, specifically Decent Work and Economic Growth. By generating employment opportunities across the spectrum, encompassing the design, construction, operation, and maintenance of these advanced wastewater treatment systems, it closely aligns with Target 8.3. This target underscores the significance of fostering sustained, inclusive, and sustainable economic growth by ensuring full and productive employment along with decent work for all.

Moreover, ABGS technology assumes a pivotal role in fulfilling Target 8.4, which centers on resource efficiency in consumption and production. Through the optimization of resource utilization, these systems actively promote sustainable economic growth and advocate for responsible resource management—crucial elements of SDG 8. Transitioning to SDG 9, which addresses Industry, Innovation, and Infrastructure, the application of ABGS systems serves as a catalyst for innovation within the wastewater treatment sector. This alignment is particularly evident in Target 9.2, which underscores the promotion of inclusive and sustainable industrialization and the fostering of innovation. Furthermore, in keeping with Target 9.4, by updating wastewater treatment facilities and infrastructure, these systems support sustainable industrialization. Thus, the overall objectives of SDG 9 are supported by increased resource use efficiency and the integration of clean technologies.

The utilization of ABGS in the treatment of industrial wastewater significantly contributes to various Social-related SDGs and their corresponding targets. Primarily, ABGS systems play a crucial role in advancing SDG 3, which focuses on Good Health and Well-being. Through efficient wastewater treatment and the reduction of hazardous chemicals and pollution, these systems actively contribute to achieving Reduce the number of diseases and fatalities associated with contaminated water and soil in order to meet target 3.9. This not only enhances public health but also improves overall well-being. Furthermore, ABGS technology supports SDG 6, addressing Clean Water and Sanitation objectives. These systems immediately contribute to ensuring that critical resources are accessible to all, which is in line with SDG 10, Reduced Inequalities (Target 10.2), by supporting universal access to safe drinking water (Target 6.1) and facilitating appropriate sanitation and hygiene facilities (Target 6.2). This strategy is centred on meeting the needs of marginalised groups and encouraging universal access to water and sanitation services, regardless of socioeconomic standing. Adopting ABGS systems also supports SDG 11, which is focused on Sustainable Cities and Communities. Through the mitigation of the environmental effects of wastewater management, which includes waste creation (Target 11.6), these systems actively support the growth of environmentally friendly and sustainable urban areas. This improves the general well-being of the people living in cities and towns in addition to improving their quality of life.

7.4 Conclusions

This study underscores the remarkable effectiveness of ABGS systems in treating saline industrial wastewater contaminated with surfactants. As salinity levels increased to 3%, SDS removal improved significantly, reaching a maximum removal of 93.11 ± 1.21% for ABGS and 86.62 ± 1.56% for BGS. Both ABGS and BGS exhibited outstanding performance in reducing COD, with a slight impact observed with the increase in salinity. ABGS, in particular, demonstrated superior removal efficiency for NH_4^+−N, TIN, and PO_4^{-3}-P compared with BGS due to the pivotal role of algae in nutrient assimilation. However, the elevated salinity levels led to a decline in the removal of nitrogen and phosphorus, primarily due to the inhibition of ammonia-oxidizing bacteria and phosphate-accumulating organisms. The implications of this study extend across the three pillars of sustainability. Environmentally, the findings contribute to water pollution reduction and the protection of ecosystem biodiversity. Economically, they open avenues for wastewater treatment technologies and biomass-based industries. Socially, they foster greater public awareness of ABGS technology.

Acknowledgements The principal author expresses sincere gratitude to the Ministry of Higher Education (MoHE) of Egypt for providing kind financial support in the form of a Ph.D. scholarship. Additionally, we would like to express our gratitude to the Japan International Cooperation Agency (JICA) for providing us with the facilities and equipment that allowed us to successfully complete this research project.

References

Badmus SO, Amusa HK, Oyehan TA, Saleh TA (2021) Environmental risks and toxicity of surfactants: overview of analysis, assessment, and remediation techniques. Environ Sci Pollut Res 28:62085–62104

Belhaj AF, Elraies KA, Mahmood SM, Zulkifli NN, Akbari S, Hussien OS (2020) The effect of surfactant concentration, salinity, temperature, and pH on surfactant adsorption for chemical enhanced oil recovery: a review. J Pet Explor Prod Technol 10:125–137 (2020)

Bindes MMM, Franco MR (2015) Surfactant removal from aqueous solutions onto activated carbon using UV spectroscopy. Desalin Water Treat 56:2890–2895

Chen C, Ming J, Yoza BA, Liang J, Li QX, Guo H, Liu Z, Dang J, Wang Q (2019) Characterization of aerobic granular sludge used for the treatment of petroleum wastewater. Biores Technol 271:353–359

Dadebo D, Ibrahim MG, Fujii M, Nasr M (2023) Sequential treatment of surfactant-laden wastewater using low-cost rice husk ash coagulant and activated carbon: modeling, optimization, characterization, and techno-economic analysis. Bioresour Technol Rep 22:101464

Di Bella G, Di Prima N, Di Trapani D, Freni G, Giustra MG, Torregrossa M, Viviani G (2015) Performance of membrane bioreactor (MBR) systems for the treatment of shipboard slops: assessment of hydrocarbon biodegradation and biomass activity under salinity variation. J Hazard Mater 300:765–778

Dong X, Zhao Z, Yang X, Lei Z, Shimizu K, Zhang Z, Lee DJ (2021) Response and recovery of mature algal-bacterial aerobic granular sludge to sudden salinity disturbance in influent wastewater: granule characteristics and nutrients removal/accumulation. Biores Technol 321:124492

Ezz H, Ibrahim MG, Fujii M, Nasr M (2021) Dual biogas and biochar production from rice straw biomass: a techno-economic and sustainable development approach. Biomass Convers Biorefinery. https://doi.org/10.1007/s13399-021-01879-y

Ezz H, Ibrahim MG, Fujii M, Nasr M (2022) Enhanced removal of methylene blue dye by sustainable biochar derived from rice straw digestate. Key Eng Mater 932:119–129

Fallahi A, Rezvani F, Asgharnejad H, Nazloo EK, Hajinajaf N, Higgins B (2021) Interactions of microalgae-bacteria consortia for nutrient removal from wastewater: a review. Chemosphere 272:129878

Fan S, Zhu L, Ji B (2021) Deciphering the effect of light intensity on microalgal-bacterial granular sludge process for non-aerated municipal wastewater treatment. Algal Res 58:102437

González S, Petrovic M, Barceló D (2007) Removal of a broad range of surfactants from municipal wastewater-comparison between membrane bioreactor and conventional activated sludge treatment. Chemosphere 67:335–343

Guan R, Yuan X, Wu Z, Wang H, Jiang L, Li Y, Zeng G (2017) Functionality of surfactants in waste-activated sludge treatment: a review. Sci Total Environ 609:1433–1442

Ilmasari D, Yuniarto A, Khen C, Purba LDA, Lei Z, Yuzir A (2023) Microalgal-bacterial aerobic granular sludge for old leachate treatment: development, performance, and lipid production. J Clean Prod 417:138053

Li Y, Zou J, Zhang L, Sun J (2014) Aerobic granular sludge for simultaneous accumulation of mineral phosphorus and removal of nitrogen via nitrite in wastewater. Biores Technol 154:178–184

Li Y, Jiang J, Chen Y, Qie W, Zhu W, Xu N, Zhao J (2023) Effects of salinity on the performance, microbial community, and functional genes among 4-chlorophenol wastewater treatment. Biores Technol 384:129282

Liu Z, Ning F, Hou Y, Zhang D, Yang R, Wang J, Zhang A, Chen Y, Liu Y (2022b) Deciphering the effect of algae sources on the formation of algal-bacterial granular sludge: endogenous versus exogenous algae. J Clean Prod 363:132468

Liu Z, Wang J, Zhang S, Hou Y, Wang J, Gao M, Chen X, Zhang A, Liu Y, Li Z (2023) Formation characteristics of algal-bacteria granular sludge under low-light environment: from sludge characteristics, extracellular polymeric substances to microbial community. Biores Technol 376:128851

Liu Z, Liu J, Zhao Y, Zhang S, Gao M, Wang J, Zhang A, Zhang T, Liu Y (2022) Understanding the effects of algae growth on algal-bacterial granular sludge formation: from sludge characteristics, extracellular polymeric substances, and microbial community. J Clean Prod 410:137327

Meng F, Liu D, Huang W, Lei Z, Zhang Z (2019) Effect of salinity on granulation, performance and lipid accumulation of algal-bacterial granular sludge. Bioresour Technol Rep 7:100228

Meng F, Huang W, Liu D, Zhao Y, Huang W, Lei Z, Zhang Z (2020) Application of aerobic granules-continuous flow reactor for saline wastewater treatment: granular stability, lipid production and symbiotic relationship between bacteria and algae. Biores Technol 295:122291

Nancharaiah YV, Kiran Kumar Reddy G (2018) Aerobic granular sludge technology: mechanisms of granulation and biotechnological applications. Bioresour Technol 247:1128–1143

Nasr M, Attia M, Ezz H, Ibrahim MG (2021) Recent applications of downflow hanging sponge technology for decentralized wastewater treatment. Elsevier

Niu X, Han X, Jin Y, Yue J, Zhu J, Xie W, Yu J (2023) Aerobic granular sludge treating hypersaline wastewater: impact of pH on granulation and long-term operation at different organic loading rates. J Environ Manage 330:117164

Saqr AM, Ibrahim MG, Fujii M, Nasr M (2021) Sustainable development goals (SDGs) associated with groundwater over-exploitation vulnerability: geographic information system-based multi-criteria decision analysis. Nat Resour Res 30:4255–4276

Sarma SJ, Tay JH, Chu A (2017) Finding knowledge gaps in aerobic granulation technology. Trends Biotechnol 35:66–78

USEPA (2012) Standard methods for the examination of water and wastewater 22th Edition. Am Public Heal Assoc Am Water Work Assoc Water Environ Fed

Wang H, Liu T, Ding Y, Wang Z, Zhang Z, Lei Z, Shimizu K, Yuan T (2023a) Enhanced performance of algal-bacterial aerobic granular sludge in comparison to bacterial aerobic granular sludge for treating surfactant-containing wastewater. Bioresour Technol Rep 22:101462

Wang Z, Wang H, Nie Q, Ding Y, Lei Z, Zhang Z, Shimizu Z, Yuan T (2023b) Pb(II) bioremediation using fresh algal-bacterial aerobic granular sludge and its underlying mechanisms highlighting the role of extracellular polymeric substances. J Hazard Mater 444:130452

Yasin M, Tauseef M, Zafar Z, Rahman M, Islam E, Iqbal S, Afzal M (2021) Plant-microbe synergism in floating treatment wetlands for the enhanced removal of sodium dodecyl sulphate from water. Sustainability 13:1–11

Zhang M, Ji B, Liu Y (2021) Microalgal-bacterial granular sludge process: a game changer of future municipal wastewater treatment? Sci Total Environ 752:141957

Chapter 8
Flows with Return of Water Turbined by a Hydroelectric Power Plant in a Transfer Receiving Basin Using WEAP

Ortiz De Orue Moreno Kimberly Madelen and **Sedano Jesus Roosevelt Edwin**

Abstract The purpose of this study is to assess the water return flows generated by a hydroelectric power plant that receives water from a transfer system. To achieve this, various tasks were undertaken, including the collection of hydrometeorological data, calculations related to water supply and demand, energy production, and the integration of water flows originating from the transfer. We utilized the WEAP software for this purpose. In an effort to comprehend the current state of the receiving basin for the transfer, an analysis was conducted under different scenarios, including the present state, the driest and wettest years, as well as the scenario of Aricota lagoon with and without water contributions from the Kovire tunnel. This study also underscores the critical importance of managing water resources in transboundary basins for the preservation of essential ecosystems in the region. Furthermore, an evaluation of the water returns flows from the Aricota I and II hydroelectric power plants to the Curibaya River was conducted. As a validation measure for the model, calibration was carried out for the Callazas and Salado rivers, yielding Nash indices of 0.69 and 0.75, respectively. These results affirm the robustness of the model and its capacity to provide reliable data in hydrological simulations.

Keywords Water turbined · Hydroelectric plant · Receiving basin · Transfer · Calibration · Water · Aricota lagoon · WEAP

O. D. O. Moreno Kimberly Madelen (✉) · S. J. Roosevelt Edwin
Peruvian University of Applied Sciences, Lima, Peru
e-mail: U201420180@upc.edu.pe

S. J. Roosevelt Edwin
e-mail: U201518464@upc.edu.pe

H. Xu (ed.), *The 6th International Symposium on Water Resource and Environmental Management*, Environmental Science and Engineering,
https://doi.org/10.1007/978-3-031-55989-1_8

8.1 Introduction

Currently, freshwater resources are facing increasing pressure on a global scale, even in areas that are typically unaccustomed to water scarcity (Wang et al. 2018). Since the availability of freshwater is crucial for ensuring a sustainable future in a rapidly changing climate, it is important to estimate the degree of future water security (Jia et al. 2017). The sustainability of water resources needs to be studied in rural areas. Within our context, mountainous regions make a significant contribution to stabilizing and sustaining ecological and agricultural systems in arid and semi-arid zones (Bao et al. 2018).

In this study, it will explore the impact of water extraction for hydroelectric power generation and the influence of the flow rates from the Kovire tunnel on the water balance. This analysis focuses on a specific case: Aricota Lagoon, located in the Tacna region. This region is characterized as one of the driest in Peru, as it is situated at the headwater of the Atacama Desert.

Since 1977, under Resolution No. 00306-77-AG-DGA issued by the Ministry of Agriculture, Sub-Regional Agricultural Directorate of Tacna, Electro Perú, later renamed as the Southern Electric Generation Company (EGESUR), has been granted a water extraction license for a flow rate of 4.6 m^3/s for energy purposes for the Aricota I and II hydroelectric power plants, as shown in the diagram, Fig. 8.1. (Ministerio de Agricultura Dirección Sub-Regional Agraria Tacna Administración Técnica Distrito de Riego Locumba—Sama 1996).

From 1977 until 1996, the same extraction flow rate was maintained, resulting in overexploitation of the lagoon, causing its volume to decrease from 700 MMC to 27.02 MMC. Consequently, under Administrative Resolution No. 166-96-DISRAGT-ATDRL/S, it was decided to authorize EGESUR to reduce the pumping

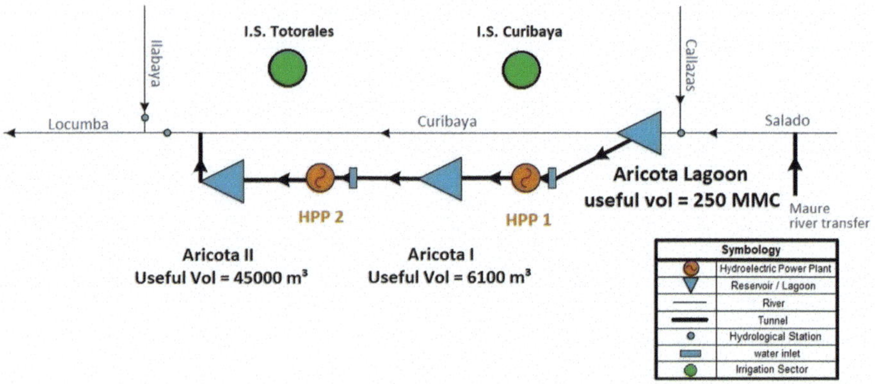

Fig. 8.1 Illustrates the hydraulic scheme for water extraction employed in both hydroelectric power generation and irrigation

flow rate from 1.5 m^3/s to 1.1 m^3/s to mitigate the impact caused by this mismanagement (Ministerio de Agricultura Dirección Sub-Regional Agraria Tacna Administración Técnica Distrito de Riego Locumba—Sama 1996). Due to this overexploitation, in 1996, the Maure River transfer operation began, originating in Peru and entering Bolivian territory (a transboundary basin) through hydraulic works such as the KOVIRE and ICHICOLLO tunnels. On average, these tunnels transfer a volume of 5.3 MMC per year, achieving a balance between exploitation and water inflow to Aricota lagoon (Ministerio de Agricultura Dirección Sub-Regional Agraria Tacna Administración Técnica Distrito de Riego Locumba—Sama 1996). By Administrative Resolution No. 173-96-DISRAGT-ATDRL/S dated October 9, 1996, EGESUR was granted a water use license for up to 1.5 m^3/s from the waters of Aricota lagoon for energy production at the Aricota I and II power plants (Ministerio de Agricultura Dirección Sub-Regional Agraria Tacna Administración Técnica Distrito de Riego Locumba—Sama 1996).

8.2 Methodology

In order to conduct the analysis of return flows from turbined water, historical precipitation data for the watershed was collected, and its location was modeled using ArcGIS software. The upstream water supply for Laguna Aricota was identified, and information regarding water demand for irrigation in the Curibaya sector was gathered. Finally, the power production at the Aricota 1 Hydroelectric Power Plant (HPP1), the transfer flow rate, and lastly, model calibration was performed.

8.2.1 Softwares

Water Evaluation and Planning System (WEAP). It is a software developed by the Stockholm Environment Institute for the comprehensive assessment of water systems. This program provides a user-friendly graphical interface to model hydrological aspects, water supply, and quality, allowing users to analyze scenarios, manage water demand, and strategically plan water resource management. Its benefits include facilitating informed decision-making, adaptability to various contexts, resource optimization, and collaboration among stakeholders in sustainable water planning.

The proposed approach offers several advantages. Firstly, it facilitates the automatic determination of the balance. Secondly, it allows for the conversion of precipitation into runoff. Additionally, the WEAP software enables the consideration of return flow from irrigation sectors and the delineation of watersheds. Furthermore, it allows for simulations within specific time intervals. However, there are some drawbacks to consider. The program is not freely available as licenses are required. The level of analysis is constrained to monthly averages, which may be limiting, given the preference for a daily resolution. Lastly, the software does not support the transit of flows from one point to another.

Geographic Information System (ARCGIS). It provides tools for mapping, spatial analysis, and data management. Its versatility and constant evolution contribute to its adoption in various disciplines related to geographic information.

8.2.2 A Collection of Precipitation Information and Watershed Location

Precipitation data, temperature, wind, etc., were obtained from the hydrometeorological stations in the Locumba watershed, as well as from the flow measurement station of major rivers, which will be processed in WEAP. The study watershed is located in South America, Peru, in the Tacna region, specifically in the Jorge Basadre province, as shown in Fig. 8.2.

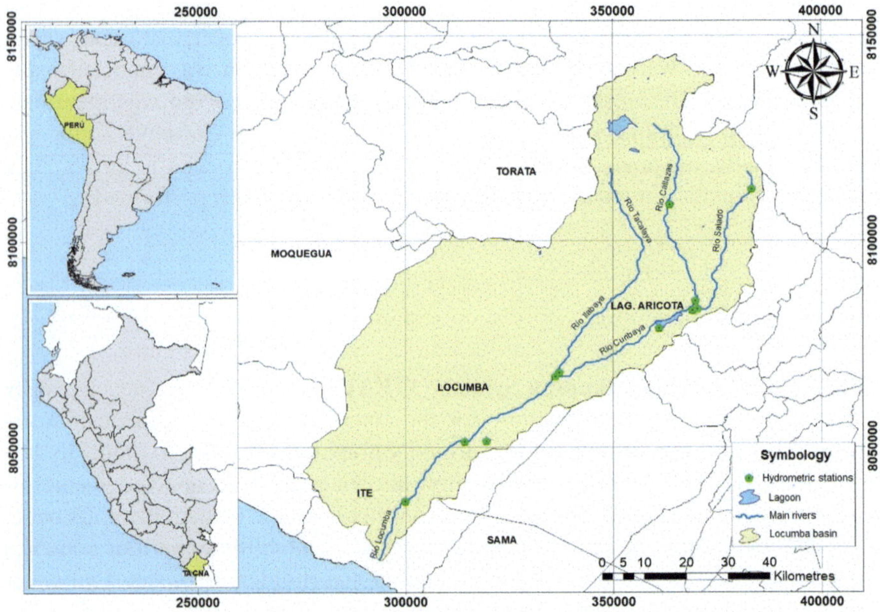

Fig. 8.2 Location of the Aricota lagoon and main rivers in the Locumba watershed

Table 8.1 Contribution flows to the Aricota lagoon (m^3/s)

	Jan	Feb	Mar	Apr	May	Jun	Jul	Aug	Sep	Oct	Nov	Dec
Salado & Kovire	1.63	3.22	2.81	1.88	1.32	1.15	1.00	0.86	0.72	0.58	0.53	0.64
Callazas	1.99	2.62	2.71	0.94	0.74	0.73	0.66	0.65	0.64	0.59	0.61	0.74
Total	3.62	5.84	5.52	2.82	2.06	1.88	1.66	1.51	1.36	1.17	1.14	1.38

Table 8.2 Statistical data of the series of discharges entering the Aricota lagoon

Average m^3/s	2.070
Standard deviation	0.822
Max m^3/s	17.677
Min m^3/s	0.960
Asymmetry coefficient	1.624

8.2.3 Calculation of Water Supply

The water supply from the Callazas and Salado rivers has been determined by converting precipitation into runoff using a hydrological model developed with WEAP software. Additionally, the water supply from the Kovire transfer was directly recorded and processed for flow.

Table 8.1 shows the water contribution from the tributaries Callazas and Salado to Laguna Aricota, with peak values occurring during the rainy period from January to March, reaching a total of approximately 5.84 m^3/s.

The following provides details on some statistics derived from the records of the Callazas and Salado rivers that flow into the Aricota reservoir (Table 8.2).

8.2.4 Calculation of Water Demand

Consumptive Demand. Irrigation sectors (I.S.), such as Curibaya and Totorales, present a consumptive water demand for irrigation of approximately 270 hectares, where various types of agricultural crops are cultivated, including alfalfa, chili peppers, and corn (Table 8.3).

Table 8.3 Monthly consumptive demand (m^3/s)

I.S	Jan	Feb	Mar	Apr	May	Jun	Jul	Aug	Sep	Oct	Nov	Dec
Curibaya	0.12	0.12	0.11	0.09	0.07	0.07	0.07	0.08	0.09	0.11	0.12	0.12
Totorales	0.09	0.09	0.09	0.07	0.05	0.06	0.05	0.06	0.07	0.08	0.09	0.09
Total	0.22	0.22	0.20	0.17	0.13	0.13	0.12	0.13	0.16	0.19	0.21	0.21

Table 8.4 Flow Extraction from Aricota lagoon and its derivations (m^3/s)

	Jan	Feb	Mar	Apr	May	Jun	Jul	Aug	Sep	Oct	Nov	Dec
HPP 1	1.31	1.28	1.29	1.32	1.33	1.37	1.36	1.35	1.34	1.31	1.30	1.31
I.S. Curibaya	0.10	0.11	0.10	0.09	0.07	0.07	0.06	0.07	0.08	0.09	0.09	0.09
Total	1.42	1.39	1.39	1.41	1.40	1.43	1.42	1.42	1.41	1.40	1.39	1.36

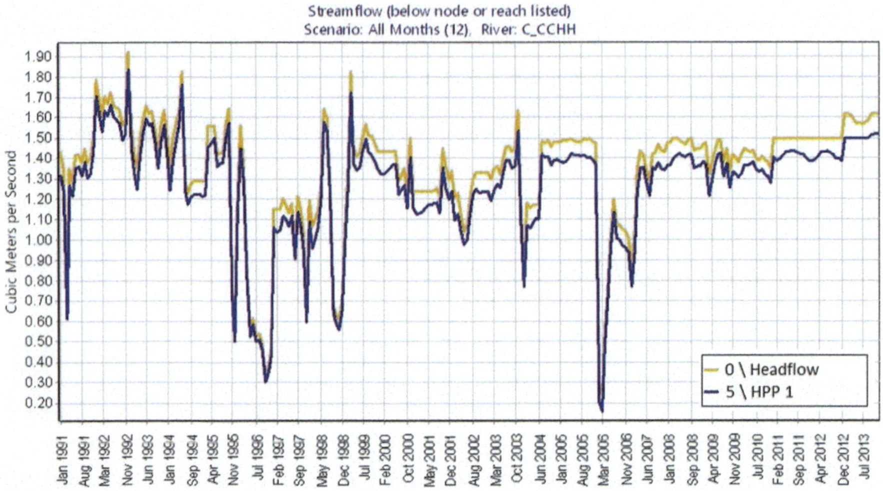

Fig 8.3. Flow extracted from the Laguna (yellow) and flow entering HPP1 (blue)

This table displays the water demand for agricultural use in the mentioned sectors, ranging from 0.12 to 0.22 m^3/s.

No-consumptive Demand. In the section recorded from the lagoon to the hydroelectric power plant, there is a diversion of flow to benefit a small cultivation area. Therefore, the flow entering HPP1 is reduced in small quantities, as shown in Fig. 8.7.

In the following Figure, generated from the data in Table 8.4, it can be seen how over time the extraction of water has been carried out for both hydroelectric power generation and irrigation purposes (Fig. 8.3).

8.2.5 Analysis of Power Production

The water extraction from Aricota lagoon is 1.5 m^3/s on average per day. However, it's worth noting that within these 24 h of HPP1 operation, there is a peak water extraction period (14:00–21:00) of 4.6 m^3/s and in the rest of the day an approximately extraction of 1.1 m^3/s. This increase in water extraction is reflected in the power generation

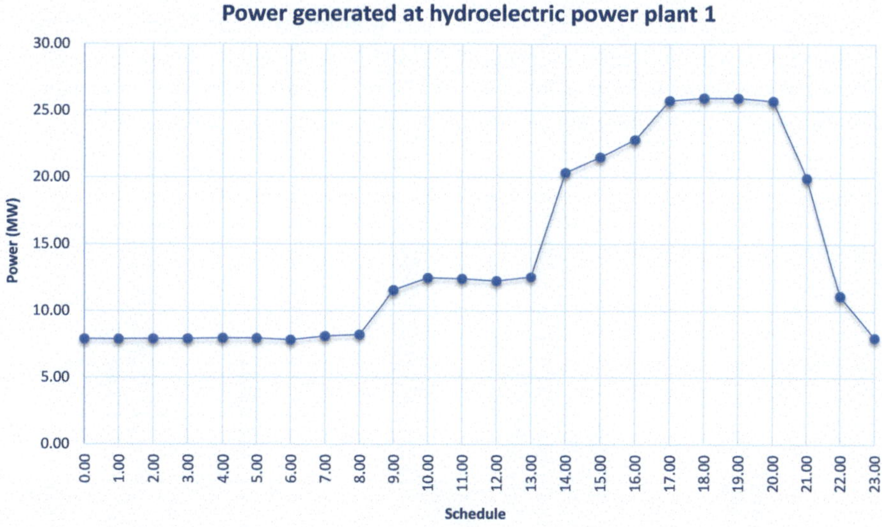

Fig. 8.4 Variability in energy production throughout the day by HPP1 in the year 2020; the peak production hours are from 5:00 PM to 8:00 PM, with a maximum power output of 25 MW

Table 8.5 Monthly average power generated by the two turbines from HPP I in MW

	Jan	Feb	Mar	Apr	May	Jun	Jul	Aug	Sep	Oct	Nov	Dec
Turbine 1 (P1)	7.12	7.31	7.89	7.11	7.15	7.16	7.14	7.02	6.58	7.14	7.10	6.79
Turbine 2 (P2)	7.12	7.31	7.38	7.11	7.15	7.15	7.13	7.02	6.57	7.13	7.09	6.79

produced at HPP1 (Comité de Operación Económica del Sistema Interconectado Nacional (COES SINAC) 2020), as shown in Fig. 8.4 (Table 8.5).

8.2.6 Analysis of Contributions from the Kovire Transfer

During the initial years of water extraction from Aricota lagoon, there was a decrease in its volume that nearly led to its depletion. Hence, the Peruvian government, through the Special Project Tacna, constructed the Kovire tunnel to transfer water from the Maure River to replenish Aricota lagoon. The mentioned transfer project began its construction in 1990 and was completed in 1996 (Instituto Nacional and de Recursos Naturales (INRENA) 1995), contributing an average volume of 16.6 MMC annually to Aricota lagoon and an average flow rate of 0.58 m^3/s.

In Figs. 8.5 and 8.6, a summary of the water transfer from the Maure River basin to enhance the Aricota Lagoon is presented. March stands out as the month with the highest contribution, yielding a flow rate of 1.5 m^3/s. In dry years, the water transfer decreases to values on the order of 4 million cubic meters per year (4 MMC).

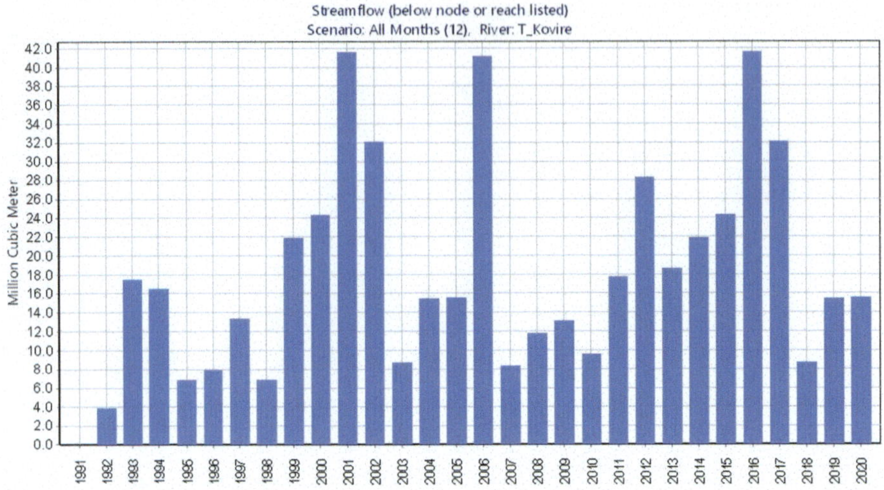

Fig. 8.5 Water volumes provided by the Kovire transfer in MMC/year

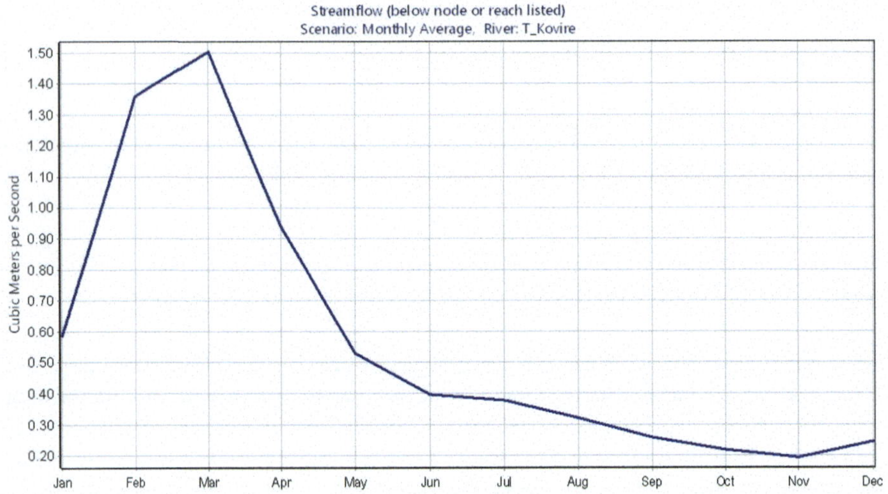

Fig. 8.6 Average flow rates contributed to the Salado River

8.2.7 *Model Calibration*

To verify the project's consolidation and its validity, the model was calibrated for the Callazas and Salado rivers, which have been calibrated with historical records of the aforementioned rivers for the period 1991–2013 using the Nash–Sutcliffe Efficiency (NSE) in the WEAP software.

Nash–Sutcliffe Efficiency (NSE). This index is particularly useful in hydrology to assess the ability of a model to replicate patterns and variations in observed data, aiding modelers in refining and improving their models (Nash and Sutcliffe 1970).

$$NSE = 1 - \frac{\sum_{t=1}^{n}(O_t - S_t)^2}{\sum_{t=1}^{n}(O_t - \overline{O})^2} \tag{8.1}$$

O_t = Observed data in time t.

S_t = Data simulated by the model at time t.

$\overline{O}$ = Is the mean of the observed data.

n = Is the total number of time points.

Coefficient of Determination (R2). The coefficient of determination is a crucial measure for assessing the quality of fit of a model to observed data, providing information about the relationship between independent variables and the dependent variable (Asghar et al. 2019).

$$R^2 = 1 - \frac{\sum_{i=1}^{n}(y_i - \hat{y}_i)^2}{\sum_{I=1}^{n}(y_i - \overline{y})^2} \tag{8.2}$$

y_1 = Are the observed values of the dependent variable

$\hat{y}_i$ = Are the values predicted by the model.

$\overline{y}$ = Is the mean of the observed values.

n = total number of applicants.

8.3 Results

8.3.1 *Scenario Design Using WEAP Software*

After processing the data in the WEAP software, various scenarios were designed to assess the return flow of the turbinated waters (Fig. 8.7).

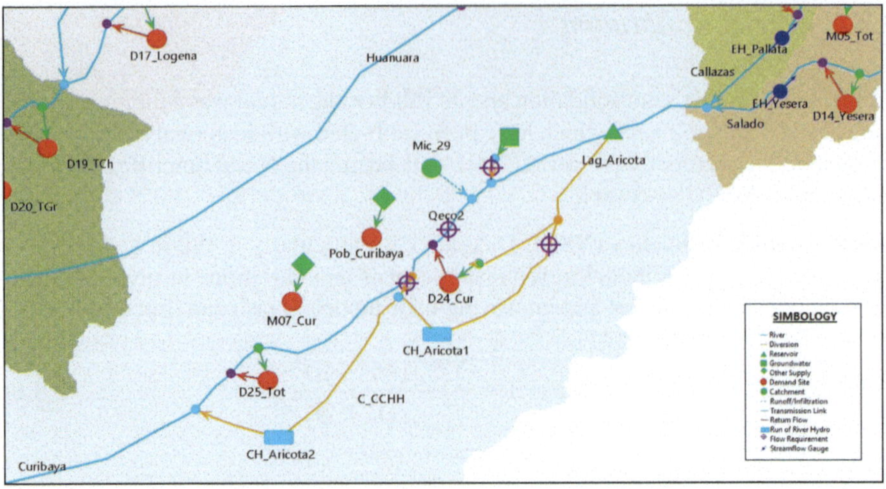

Fig. 8.7 Scheme modeled in WEAP software

Table 8.6 Average flow rate—Curibaya river mouth (m^3/s)

	Jan	Feb	Mar	Apr	May	Jun	Jul	Aug	Sep	Oct	Nov	Dec
Curibaya	1.87	1.94	1.88	1.93	1.94	1.98	1.96	1.94	1.93	1.89	1.86	1.81

Analysis of the current scenario. All the information described in the methodology was processed, and the following results were obtained using WEAP software for the current scenario at the mouth of the Curibaya River (Table 8.6).

As seen in the graph and the previous table, due to the return of turbined water, the average flow rate at the Curibaya River outlet point is 1.90 m^3/s.

Scenario design for the driest and wettest years. Aricota lagoon is a natural multi-year regulation reservoir with an average volume of around 250 MMC. HPP1 does not take into account the presence of dry and wet years, so it extracts the same flow rate of 1.5 m^3/s.

In Table 8.7 and Fig. 8.8, the scenario design for the driest year in the Callazas and Salado rivers is presented, with the lowest contribution occurring in the month of December, registering a flow rate of 0.67 m^3/s.

Table 8.7 Inflow to Aricota lagoon—dry year scenario 1991 (m^3/s)

River	Jan	Feb	Mar	Apr	May	Jun	Jul	Aug	Sep	Oct	Nov	Dec
Callazas	1.45	1.16	1.21	0.65	0.51	0.52	0.48	0.46	0.46	0.43	0.45	0.45
Salado	0.79	0.66	0.58	0.47	0.40	0.39	0.34	0.32	0.30	0.26	0.25	0.23
Total	2.23	1.82	1.79	1.12	0.91	0.91	0.82	0.78	0.76	0.69	0.70	0.67

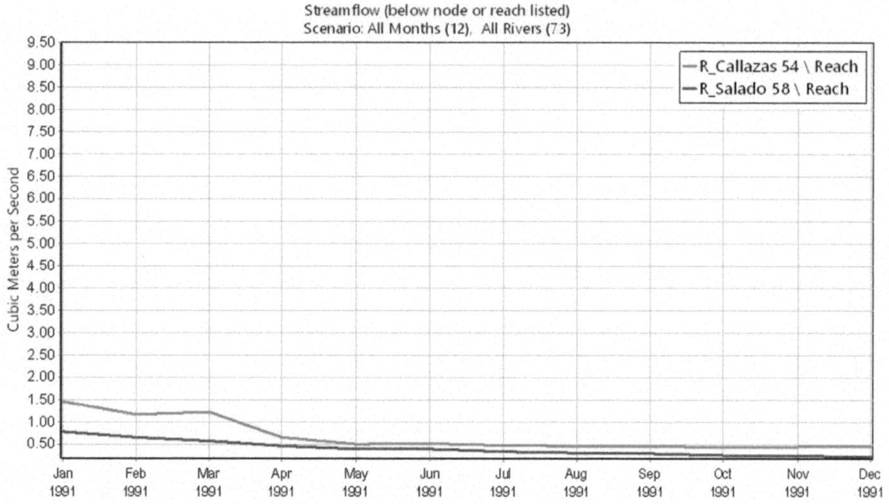

Fig. 8.8 Inflows to lagoon Aricota in the year 1991

Table 8.8 and Fig. 8.9 depict the scenario design for the wettest year in the Callazas and Salado rivers, with the highest contribution occurring in the month of February, reaching a flow rate of 18.14 m^3/s.

Volume level of Aricota lagoon without Kovire transfer contributions. Without the contribution from the Kovire transfer, in the period 1991–2020 the lagoon would maintain an average volume of 100 MMC. Consequently, HPP1 would have to reduce its extraction flow to avoid affecting the availability of water resources downstream of the lagoon. Figure 8.10.

Volume level of Aricota lagoon with Kovire transfer contributions. With flow contributions from the Kovire transfer, in Fig. 8.11 illustrates the evolution of the water storage volume in the Aricota Lagoon from 1991 to 2020, showing an average volume of 250 million cubic meters (MMC). Currently, there is a balance between water extraction from the Aricota Lagoon and contributions from its tributaries.

Table 8.8 Inflow to Aricota lagoon—wet year scenario 2001 (m^3/s)

River	Jan	Feb	Mar	Apr	May	Jun	Jul	Aug	Sep	Oct	Nov	Dec
Callazas	2.70	8.94	6.33	1.64	1.14	1.08	0.95	0.89	0.86	0.78	0.76	0.72
Salado	2.87	9.20	7.20	4.28	2.97	2.66	1.81	1.62	1.29	1.00	0.94	1.01
Total	5.57	18.14	13.53	5.91	4.11	3.73	2.76	2.51	2.14	1.78	1.70	1.73

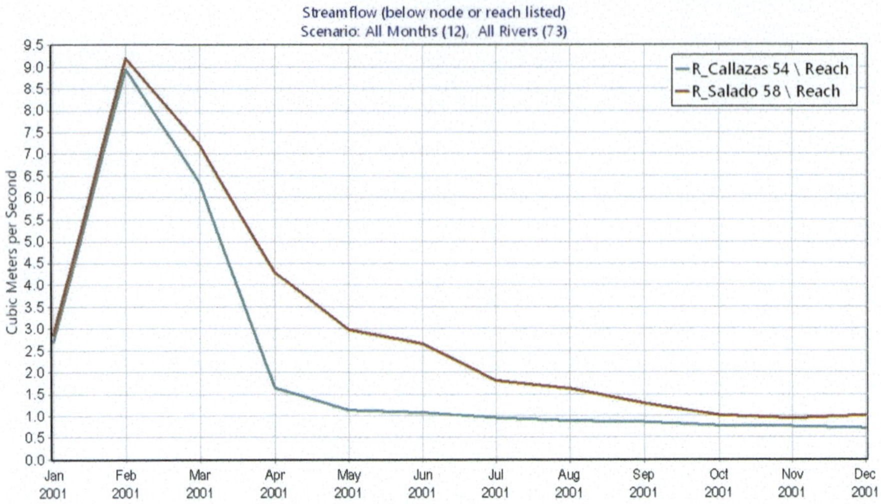

Fig. 8.9 Inflows to lagoon Aricota in the year 2001

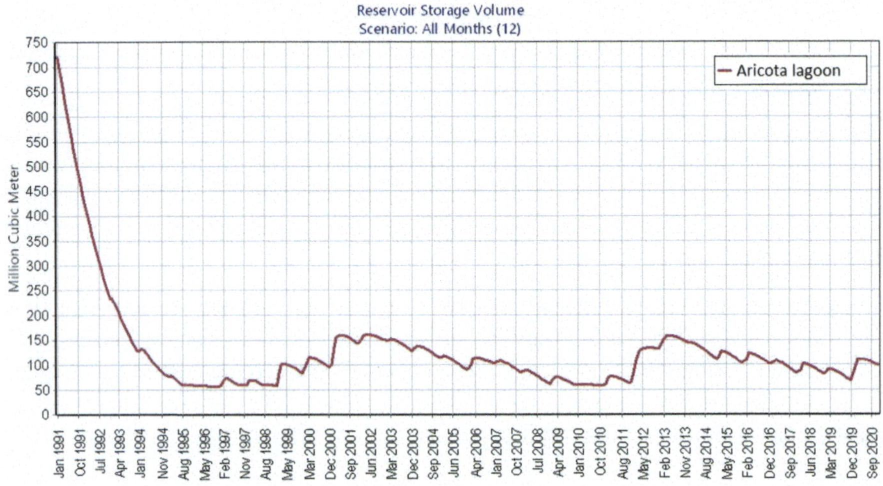

Fig. 8.10 Annual volume levels of Aricota lagoon in MMC

8.3.2 Utilization of Water from Aricota Lagoon for Hydroelectric Power Generation and Irrigation of the Locumba Valley

According to the power production of HPP1, the flow rates processed by each turbine were determined, and the approximate extraction flow rate for the year 2020 was

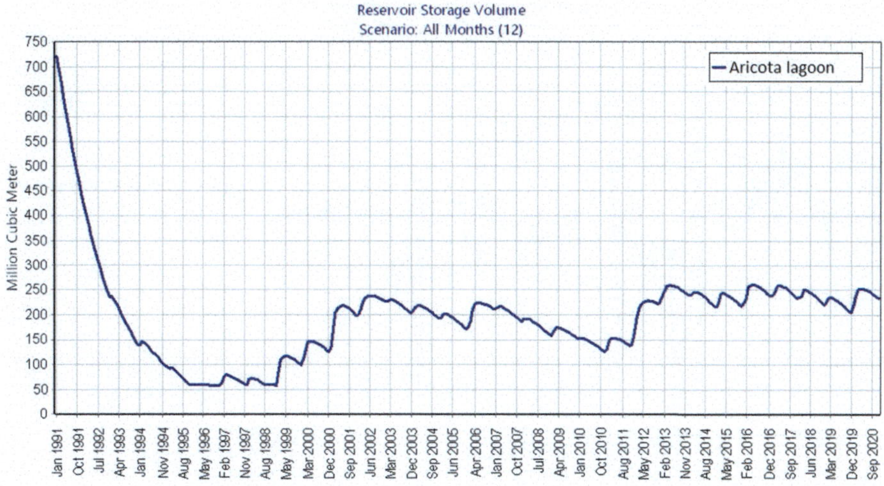

Fig. 8.11 Annual volume levels of Aricota lagoon in MMC

Table 8.9 Approximate flows of turbinated water based on generated power (m^3/s)

	Jan	Feb	Mar	Apr	May	Jun	Jul	Aug	Sep	Oct	Nov	Dec
Turbine 1 (P1)	1.13	1.16	1.17	1.12	1.13	1.13	1.13	1.11	1.04	1.13	1.12	1.07
Turbine 2 (P2)	1.12	1.15	1.17	1.12	1.13	1.13	1.13	1.11	1.04	1.13	1.12	1.07
Total flow	2.25	2.31	2.33	2.25	2.26	2.26	2.26	2.22	2.08	2.25	2.24	2.15

found, such as is depicted in Table 8.9, wherein the value experiences an increase in the month of March, reaching a magnitude of 2.33 m^3/s.

8.3.3 *Flow Return After HPP1 and HPP2 to the Curibaya River*

The extraction of water from Aricota Lagoon for hydroelectric purposes averages about 1.40 m3/s, enabling power generation in HPP1 and diversion to a small cultivated area. In addition to this flow, contributions from Laguna filtrations with an average flow rate of 0.63 m3/s are included. This results in a combined flow rate of 1.95 m3/s, which is then turbinated at the Aricota 2 Hydroelectric Power Plant (HPP2). Further details can be found in Fig. 8.12. Table 8.10 shows the average monthly flows entering the HPP2.

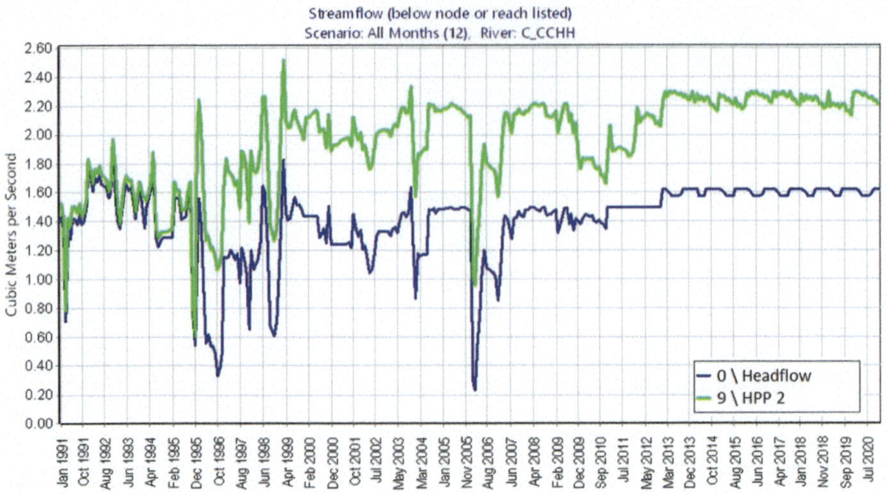

Fig. 8.12 Annual extraction flow from the Aricota lagoon to HPP 1 (blue). Outflow from HPP 2 to Curibaya river (green)

Table 8.10 Flow returns from the HPP 2 to the Curibaya river (m^3/s)

	Jan	Feb	Mar	Apr	May	Jun	Jul	Aug	Sep	Oct	Nov	Dec
Flow ext	1.42	1.39	1.39	1.41	1.40	1.43	1.42	1.42	1.41	1.40	1.39	1.36
Leaks	0.63	0.68	0.65	0.65	0.63	0.64	0.63	0.62	0.63	0.61	0.61	0.60
HPP 2	1.95	1.97	1.94	1.97	1.97	2.00	1.99	1.97	1.97	1.93	1.91	1.87

8.3.4 Validation

Callazas River. In order to validate the data for the Callazas River, the model was calibrated using the flow measurement data from the Pallata station. Table 8.11 and Fig. 8.13 present the validation of recorded and simulated discharges using the WEAP software, yielding a NSE value of 0.69. This value indicates a high degree of similarity between the observed and simulated data.

Salado River. In order to validate the data for the Salado River, the model was calibrated using the flow measurement data from the Yesera station. Table 8.12 and Fig. 8.14 depict the validation of recorded and simulated discharges using the WEAP

Table 8.11 Modeled and observed flows of the Callazas River (m^3/s)

	Jan	Feb	Mar	Apr	May	Jun	Jul	Aug	Sep	Oct	Nov	Dec
Modeled	1.98	2.77	2.43	0.95	0.75	0.73	0.67	0.65	0.65	0.60	0.61	0.76
Observed	1.68	2.84	2.45	0.93	0.76	0.74	0.78	0.74	0.61	0.57	0.54	0.70

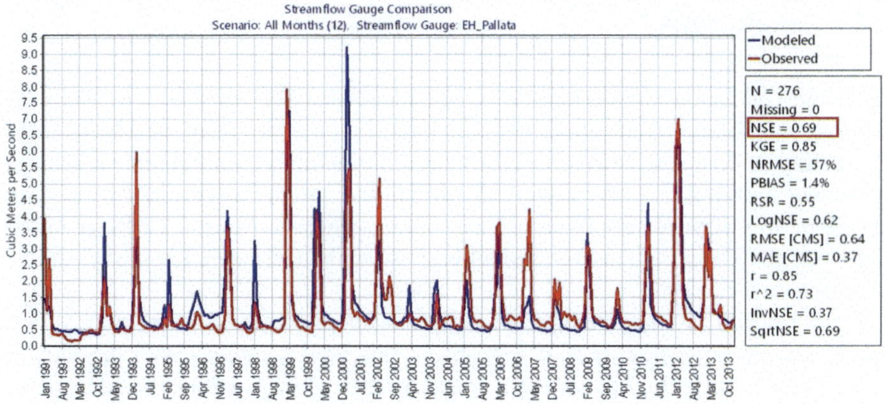

Fig. 8.13 Comparison of Callazas River flow rates

Table 8.12 Modeled and observed flows of the Salado River (m^3/s)

	Jan	Feb	Mar	Apr	May	Jun	Jul	Aug	Sep	Oct	Nov	Dec
Modeled	1.62	3.22	2.56	1.77	1.27	1.12	0.98	0.86	0.73	0.59	0.55	0.65
Observed	1.60	3.26	2.89	1.25	1.00	0.91	0.86	0.83	0.71	0.62	0.58	0.80

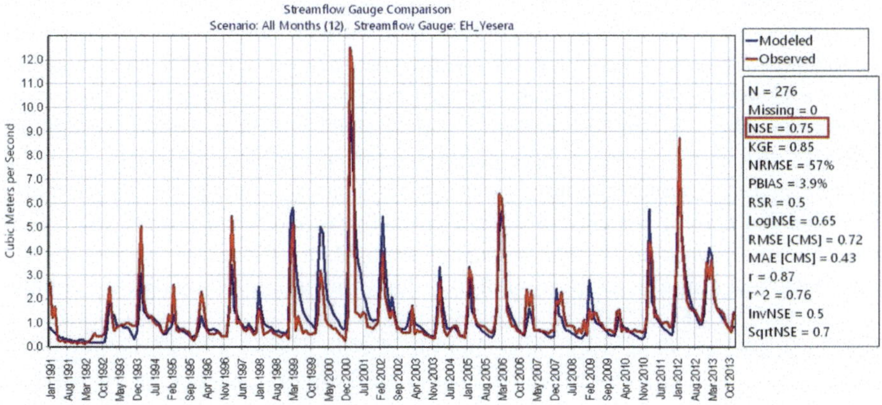

Fig. 8.14 Comparison of Salado River flow rates

software, yielding a NSE value of 0.75. This value indicates a high level of similarity between the observed and simulated data.

8.4 Conclusions

From the conducted simulations, it is concluded that the contributions from the Callasas and Salado rivers, along with the interbasin transfer from the Maure River originating in Peru and flowing into Bolivia, are crucial for maintaining an average volume of 250 million cubic meters (MMC) in the Aricota reservoir. This ensures the extraction of water for hydroelectric power generation in the HPP1 & HPP2 and meets the water demand for agricultural use in the Curibaya, Locumba, and Ite irrigation sectors.

Curibaya River, as the primary source of water supply for the Locumba–Ite Valley, provides an average flow rate of 1.91 m^3/s. This flow combines with the Ilabaya River's flow to form the Locumba River and meet the downstream demand from the lagoon Figure 8.1.

The variability of dry and wet years is clearly reflected in the water contributions to Aricota lagoon through the Callazas and Salado rivers. In the Locumba River basin, where the hydroelectric power plants are located, there is a cyclical occurrence of extreme events, including storms and droughts. However, Aricota lagoon, as a multi-year regulation reservoir, plays a crucial role in allowing water extraction for hydroelectric power generation without being significantly affected by these mentioned climatic events. This underscores its importance as a reliable source of water resources for energy production in the region.

The Kovire transfer is a project that captures water from the Maure River, which originates in Peru and crosses the border into Bolivia, belonging to an essential transboundary basin. It provides an average flow rate of 12 MMC/year and has played a fundamental role in preserving Aricota lagoon. Historical records indicate that in 1991, the lagoon was in a critical state, with a minimum volume of 21 MMC. However, thanks to the contributions from the Maure River through the Kovire transfer, Aricota lagoon has managed to achieve a balance between water extraction and natural contributions from the Callazas and Salado rivers. In a hypothetical scenario without the water transfer from Maure River, the Aricota lagoon would have lost a significant portion of its volume and would currently only contain around 100 MMC (Fig. 8.10). This is because water extraction from the lagoon by the hydroelectric power plant would have far exceeded the contributions from its natural sources. This study underscores the crucial importance of transboundary watershed management or the conservation of vital ecosystems in the region.

Hydroelectric power generation at the HPP1 and HPP2, with an average flow of 1.5 (m^3/s) granted by the National Water Authority, plays a fundamental role in providing water for the irrigation sectors of Locumba and Ite located downstream of Aricota lagoon. In the absence of these hydroelectric plants, the water supply would be limited solely to the lagoon's filtrations and the contributions from the Ilabaya River, which would be insufficient to meet the growing water demand of these agricultural sectors.

To carry out the hydrological simulations presented, it was essential to develop a hydrological model using the WEAP software. This model was meticulously calibrated using data from the hydrometric stations at Yesera, which monitor the Salado River, and Pallata, which is responsible for the Callazas River. The comparison between historical records and model-generated data yielded highly satisfactory results, with NSE of 0.69 and 0.75, respectively. These results support the validity of the model and its ability to provide reliable outcomes in the hydrological simulations conducted.

The proposed method facilitates the integration of water supply and demand, yielding a hydrological balance expressed in terms of "unmet demand in MMC" and "percentage coverage of demand." Additionally, it enables the analysis of scenarios through simulation.

A balanced utilization of Aricota Lagoon waters has been achieved for both hydroelectric power generation and agricultural demands, leveraging contributions from the Callazas and Salado rivers, as well as the interbasin transfer from the Maure River originating in Peru and flowing into Bolivia.

References

Asghar A, Iqbal J, Amin A, Ribbe L (2019) Integrated hydrological modeling for assessment of water demand and supply under socio-economic and IPCC climate change scenarios using WEAP in Central Indus Basin. J Water Supply Res Technol 68:136–148

BaoYh, He XB, Wen AB, Gao P, Tang Q, Yan DC, Long Y (2018) Dynamic changes of soil erosion in a typical disturbance zone of China's three gorges reservoir. Catena 169:128–39

Comité de Operación Económica del Sistema Interconectado Nacional (COES SINAC) (2020). Datos de energía generada en la central hidroeléctrica Aricota I. Recuperado de. https://www.coes.org.pe/Portal/search?k=ARICOTA

Comparación de diferentes algoritmos ANN (FFBP, GRNN, RBF) y regresión lineal múltiple para la predicción diaria del caudal en el río Kocasu, Turquía (2022) Boletín medioambiental de Fresenius 31(5):4699–4708

Instituto Nacional de Recursos Naturales (INRENA) (1995) Evaluación de Impacto Ambiental del Proyecto Derivación Kovire. Recuperado de, Perú. https://repositorio.ana.gob.pe/handle/20.500.12543/3851

Impactos potenciales sobre la hidrología y la producción hidroeléctrica bajo el calentamiento climático de Sierra Nevada (2011). Revista de Agua y Cambio Climático 2(1):29–43

Jia XX, Shao MA, Zhu YJ, Luo Y (2017) Soil moisture decline due to afforestation across the Loess Plateau, China. J Hydrol 546:113–122

Ministerio de Agricultura Dirección Sub-Regional Agraria Tacna Administración Técnica Distrito de Riego Locumba—Sama (1996). Resolución Administrativa N. 173-96-DISRAGT-ATDRL/S. Ministerio de Agricultura Dirección Sub-Regional Agraria Tacna, Perú-Tacna

Nash JE, Sutcliffe JV (1970) River flow forecasting through conceptual models part I—a discussion of principles. Hydrol J 10:282–290

Riverware: una herramienta generalizada para el modelado de sistemas de yacimientos complejos (2001) Revista JAWRA de la Asociación Estadounidense de Recursos Hídricos 37(4):913–929

Wang S, Fu BJ, Chen HB, Liu Y (2018) Regional development boundary of China's Loess Plateau: water limit and land shortage. Land Use Policy 74:130–136

Chapter 9
Application of Image Processing Technology in the Identification of Sediment Dosing and Flocculation Process

He Bai, Jiachen Zeng, Yuchi Hao, Bing Guo, Xiaowei Yan, and Zheng Lu

Abstract Aiming at the composition and nature of river and lake mud in space and time have greater complexity and uncertainty, there are dosing treatment of intelligent difficulties and other issues, this paper uses image processing technology to identify the mud flocculent, image edge detection and image segmentation technology as the image processing method, respectively, to extract the flocculent group contour and the size of the area and other parameter information, to explore the advantages and applicability of different image processing methods in the process of river and lake mud dosing and identification of flocculent groups. To investigate the advantages and applicability of different image processing methods in the identification process of flocculating groups in river and lake mud dosing, to provide a basis for the realization of intelligent processing of bottom mud dosing, and to improve the efficiency and effect of dewatering treatment, the results show that the edge detection methods based on Canny, Sobel and Laplacian are ineffective in the process of flocculating images of bottom mud dosing in river and lake mud, and the maximal Precision, Recall and F1 are 0.38, 0.57, 0.4 and 0.4, respectively., 0.57, and 0.4, and the threshold-based image segmentation method has better applicability, with its Precision, Recall, and F1 reaching 0.88, 1.00, and 0.94, respectively.

Keywords Flocculation · Image processing · Edge detection · Image segmentation · Recognition

H. Bai · J. Zeng (✉) · Y. Hao · B. Guo · X. Yan · Z. Lu
CCCC National Engineering Research Center of Dredging Technology and Equipment Co., Ltd., Shanghai 200082, China
e-mail: zengjiachen@cccc-drc.com

H. Xu (ed.), *The 6th International Symposium on Water Resource and Environmental Management*, Environmental Science and Engineering,
https://doi.org/10.1007/978-3-031-55989-1_9

9.1 Introduction

The flocculation and sedimentation process of river and lake sediment is the key to its reduction, stabilization, harmlessness and resourcefulness Fang (2022), which especially involves the precise addition of drug dosage and timely regulation of the process, efficient and accurate identification of the size of flocculating groups under different dosage is the key to realize the intelligence of the dosing process Gao (2018).

With the development of artificial intelligence technology, significant breakthroughs have been made in the research of deep learning, neural networks, and computer vision Huang (2014), which provides technical support for advancing the intelligent treatment process of river and lake remediation. However, unlike other industries, methods based on statistical mathematical models such as machine learning are not applicable to realize the intelligent treatment process of river and lake substrate dewatering Jayaprakash (2016), which is mainly attributed to the fact that the composition and nature of river and lake substrate have a large complexity and uncertainty in space and time, and there is a scarcity of monitoring means, making it difficult to realize real-time Jia (2010). On the contrary, computer vision technology can monitor the bottom mud treatment process in real time Kasmi (2017). By deploying vision sensors and image processing algorithms, the operational status of the substrate dewatering equipment and the treatment effect of the substrate can be monitored in real time, and the operating parameters can be adjusted with the feedback of the monitoring results, so as to realize the intelligent treatment of substrate dosing and improve the efficiency and effect of the dewatering treatment Li (2014).

Innovative and applicable image processing algorithms are the key to realize the intelligent and high precision of subsoil dosing. Image processing techniques include gray scale conversion, filter, edge detection, image segmentation, feature extraction, etc. (Li 2019; Li and Zhang 2019; Lü and Luo 2019), which can improve the image quality, enhance the image details, mark the image features, visualize the image processing results, etc. Marini (2018), which helps to analyze and evaluate the effect of bottom mud treatment and enhance the interpretability of intelligent treatment. Among them, edge detection and image segmentation can play an important role in identifying bottom mud flocculating groups Nobuyuki (1979). Edge detection can accurately locate the edges of mud flocculent groups and provide shape and contour information Song (2022). Image segmentation separates the flocs from the background to form a separate region, which allows measurement and analysis of the flocs, including the calculation of their area, volume, and shape characteristics Wahab Fazal (2022). However, such image processing techniques are mostly applied in the process of automatic driving and target tracking Xu (2015), and there is a lack of research in the dosing flocculating groups of river and lake sediments, and the effect of its practical application is still unknown.

This study is based on a substrate disposal project, selecting the alum flower flocculation images taken in the indoor test as the main body of the study, using image processing technology to process the alum flower flocculation images, comparing the differences and applicability of different image processing algorithms in the identification of river and lake substrate dosing flocculation, and analyzing the more feasible river and lake substrate dosing flocculation image recognition technology, which will provide a basis for the intelligent process of substrate dosing.

9.2 Algorithm

In order to obtain more accurate flocculent size, area and other parameter information in the image, and to provide methods for the intelligent process of dosing of subsoil, this program is based on the characteristics of the engineering characteristics of the complexity and variability of the mud, the characteristics of the characteristics are difficult to distinguish, and so on, and chose to use the clearer edge of the image information characteristics as the target of distinguishing to achieve the process of identifying the alum flocculent, and the method includes the edge detection algorithm and the image segmentation algorithm.

9.2.1 *Edge Detection Algorithm*

Image Edge Detection Yan (2022) aims to extract the edge information of an object from the localized intensity change significant parts of the image based on the measure and localization of the gray scale change, the edge is usually represented in the image as a sudden change in luminance or color, and the gradient represents the rate of change in the image, so calculating the gradient of an image can help us to find the location of the edge, the use of the first-order and the second-order derivative to detect discontinuities in luminance values is a commonly used method of calculating the gradient, in which first-order derivatives help to detect the rate of change of luminance, and the second-order derivatives can provide more information on edges.

For an image $f\prime(x, y)$, its computation process at pixel point (x, y) can be expressed as follow:

First order directional derivatives in x, y and θ directions.

$$F_x = \frac{f\prime(x, y)}{x\prime} \tag{9.1}$$

$$F_y = \frac{f\prime(x, y)}{y\prime} \tag{9.2}$$

$$F_\theta = \frac{f\prime(x, y)}{x\prime} cos\theta + \frac{f\prime(x, y)}{y\prime} \sin\theta \tag{9.3}$$

Definition of gradient vector.

Table 9.1 Characterization of the three types of operators

Name	Advantages	Disadvantages	Applications
Canny	High accuracy, multi-threshold	Computational, complexity	Extracting image contours and edge features
Sobel	Adjustable sensitivity	Lack of directional sensitivity	Extracting texture and shape features
Laplacian	Detecting edges and textures	Sensitivity to noise	Extracting texture information and features

$$G_{(x,y)} = [F_x F_y]^T \tag{9.4}$$

Magnitude Definition.

$$|G_{(x,y)} = \sqrt{F_x^2 + F_y^2}| \tag{9.5}$$

Direction Definition.

$$\theta = arctan(\frac{F_x}{F_y}) \tag{9.6}$$

In practical calculations, it is common to express the gradient magnitude values in terms of approximations.

$$|G_{(x,y)}| = |f(x+1) - f(x)| + |f(y+1) - f(y)| \tag{9.7}$$

For different detection methods of images can be realized by using different operators, the operators used in this scheme include Canny, Sobel and Laplacian operators, the properties of which are shown in Table 9.1.

9.2.2 *Image Segmentation Algorithm*

Image segmentation Yan (2010) is the process of extracting the target from the background in an image and aims to divide the image into different regions or objects for better understanding and processing of the image. In this global thresholding method based on global thresholding is used to binarize the image by calculating to get a suitable threshold value, which is defined as shown in the following equation:

$$R_{ij} = \begin{cases} 0 I_{ij} \leq T \\ 1 I_{ij} \geq T \end{cases} \tag{9.8}$$

I_{ij} denotes the magnitude of the eigenvalue corresponding to the pixel point in the image;

T is the threshold value obtained by calculating;

R_{ij} denotes the output binary image.

Different from the edge detection algorithm, the image segmentation algorithm focuses more on segmenting the image into different regions or objects, the pixels in each region have similar characteristics, the segmented image is more recognizable visually, which facilitates the subsequent process of calculating the size, area, etc., was the edge detection algorithm focuses mainly on extracting pixels from the boundaries of the objects in the image, and the search of the image boundaries is more delicate, the image segmentation technique used in this paper is based on the threshold segmentation method of Otsu.

9.3 Experimental

9.3.1 *Data Sources*

The river and lake substrate used in this test was taken from a substrate disposal project in Xiamen, the water content of 11.97% of the river and lake substrate was placed in a glass tank, the top of the tank was placed in a high-definition waterproof camera, the camera structure is shown in Fig. 9.1. Add the appropriate amount of dewatering chemicals to make the substrate produce alum flocculent condensate and slowly stir the liquid, so that the alum flower uniform floating flow, when passing under the camera for rapid shooting to collect alum flower images. The flocculation image used in this image processing process is shown in Fig. 9.2, the flocculation group particle size is different, the image size is 285 mm × 285 mm, in which the number of larger flocs that can be clearly seen is 8, and the number of smaller flocs is a number of smaller flocs, and the image information obtained from the processing should be completely recognized as much as possible to identify the larger flocs, while ignoring the smaller flocs.

9.3.2 *Preprocessing*

A certain degree of pre-processing operations on the flocculated image before processing helps to reduce the effect of noise, enhance the contrast of the picture, make the edge and target more obvious, etc. In this experiment, the Gaussian filter noise reduction and histogram equalization are carried out on the basis of image grayscaling to enhance the contrast, and the resulting processed image is shown in Fig. 9.3, where (a) represents the image after grayscaling, (b) represents the image after noise reduction, and (c) represents the image after contrast enhancement.

Fig. 9.1 Underwater high-definition camera unit

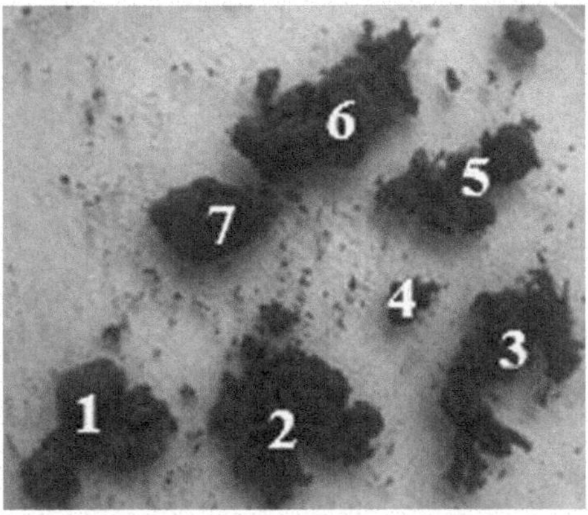

Fig. 9.2 Bottom mud flocculation original picture

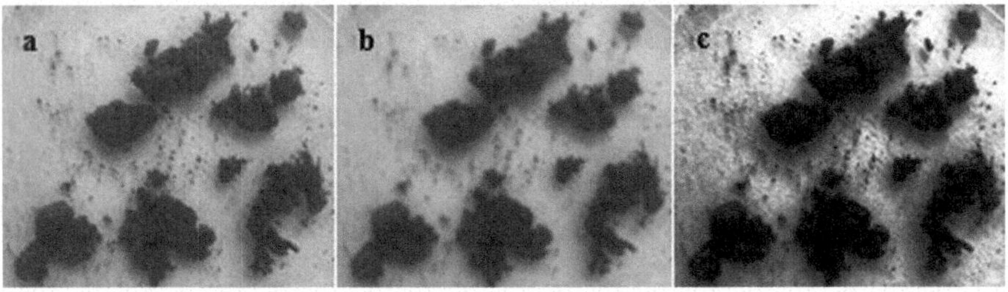

Fig. 9.3 Image preprocessing

9.3.3 Contour Extraction

On the basis of image preprocessing, Canny, Sobel, Laplacian operator and Threshold image segmentation methods are utilized for different preprocessing means for contour extraction, and the resulting processed contour information is shown in Fig. 9.4.

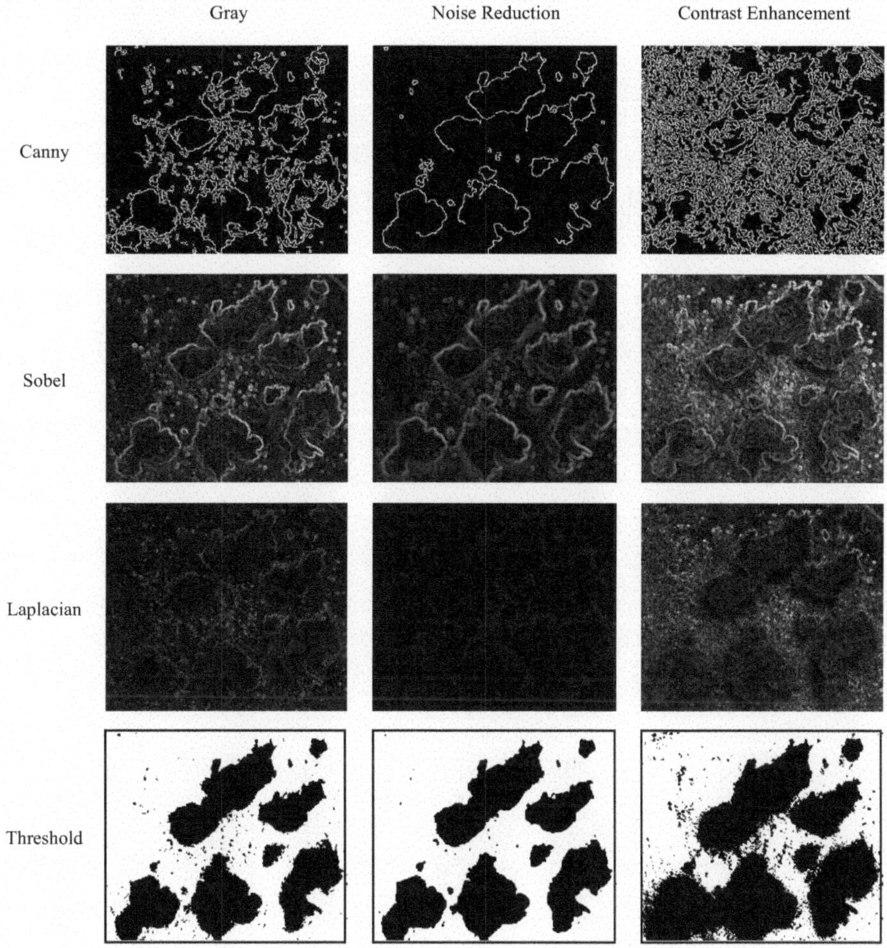

Fig. 9.4 Contour information extracted by different image processing means

9.3.4 Target Identification

The contour feature extraction results of the edge curve drawing and visualization process to show the applicability of different algorithms in the river and lake sediment flocculation image recognition, and its contour is located in the histogram to calculate the area, this calculation parameters as decision-making parameters for the recognition process selection. For different contour extraction methods, the image results of target recognition are shown in Fig. 9.5. Its image recognition information is shown in Table 9.1.

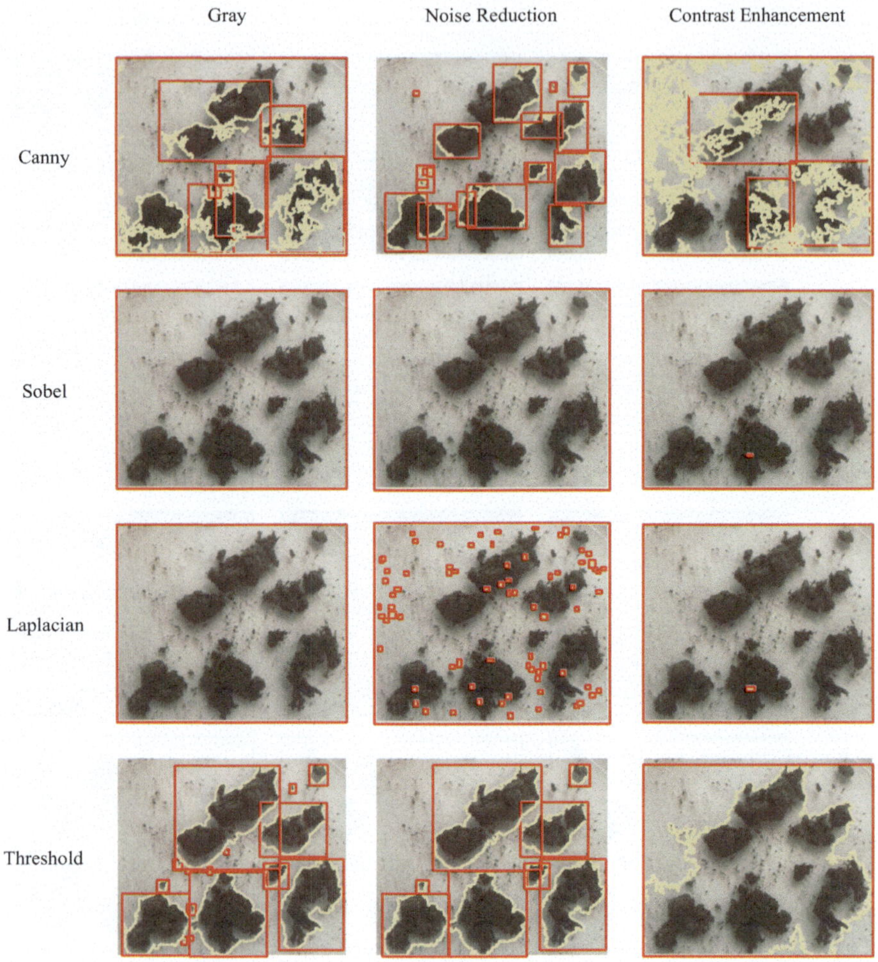

Fig. 9.5 Flocculation image target recognition results

9.3.5 Performance Evaluation

The recognition results of the substrate flocs of different image processing methods were evaluated by Precision, Recall, and F1 values, which were expressed by the formula: ns:

$$P = E/(E + Ne) \tag{9.9}$$

$$R = E/O \tag{9.10}$$

$$F1 = 2 \times ((P \times R)/(P + R)) \tag{9.11}$$

P denotes Precision;
R denotes Recall;
E denotes the number of flocs identified;
Ne denotes the number of unrecognized flocculants;
O denotes the number of initial flocculants;

9.4 Results

9.4.1 Contour Extraction Effect Analysis

As can be seen in Fig. 9.4, after different preprocessing means are processed by different edge detection algorithms, the effect of the extracted contours of the bottom mud flocculent varies greatly, and the contour extraction effect of Sobel and Laplacian is obviously poorer than that of Canny and Threshold, which is mainly attributed to the fact that the outputs of Sobel and Laplacian are all grayscale images. This is mainly attributed to the fact that both Sobel and Laplacian outputs are grayscale images, and their image features are extracted by sliding the convolution kernel used by the operator through the image and calculating the derivative value of each pixel, from which the edge intensity change of the pixel is calculated and reacted on the grayscale image. Whereas the effect of Canny and Threshold depends on the change in threshold value, Canny algorithm marks the edge pixels as white and non-edge regions as black and the output image behaves as a black and white color image. Threshold method divides the image into black and white colors to segment the flocculent from the background. These differences make them different in the output image results in terms of brightness, sharpness and edge connection. From the results they show, it can be learned that Canny and Threshold algorithms have better applicability in contour extraction of river and lake bottom mud.

For the image processing results of Canny and Threshold algorithms, different means of image preprocessing have a greater impact on the algorithm contour extraction. From the figure, it is not difficult to find that when the image is grayed out on the basis of noise reduction processing, the resulting contour map has better results with less impurities, which facilitates the analysis of the subsequent decision-making system. When only grayed out, the image will contain the contour of the original fine particles within, is not conducive to the subsequent identification process of the main flocculants, when in the grayed out process on the basis of histogram equalization processing, found that the picture contour information suddenly increased, impurity noise factors become more. From this, it can be concluded that for the processing of river and lake sediment flocculation images, only a certain parameter of noise reduction on the basis of grayscaling can achieve a better effect of contour extraction.

9.4.2 *Target Recognition Effectiveness Analysis*

Figure 9.5 shows the visualization results of different contour feature extraction algorithms on the river and lake sediment flocculation image, from the visualization results, it can be clearly seen that, for the Sobel and Laplacian algorithms, due to the extraction of contour features does not have a large difference, so its recognition effect is the worst. For the Canny algorithm, the Gray and Contrast Enhancement preprocessing have large noise, and the contour curves in the recognition process are more heterogeneous, and the more accurate recognition results cannot be obtained; the Noise Reduction preprocessing optimizes the extraction of the contour features to a certain extent, but it also performs poorly in the subsequent recognition process, and it cannot identify the entire contour of the flocculent in full. The flocculent contours were not recognized completely.

On the contrary, the Threshold image segmentation process performs better under Gray and Noise Reduction preprocessing results, and it can be clearly seen that all the flocs are recognized, but Gray also recognizes the smaller particles of flocs, which is poor in applicability, while the recognition process under Contrast Enhancement preprocessing In the Contrast Enhancement pre-processing process, the recognition effect was poor because the contour features could not be well extracted. Therefore, it can be seen that the Noise Reduction pre-processing with Threshold image segmentation can be used for the recognition of river and lake sediment flocs to achieve excellent floc recognition results.

Table 9.2 records the data information of each algorithm in the recognition process under different preprocessing situations, in which Decision Parameters indicates that in the recognition process, when its Area is smaller than the value of the parameter, the information obtained from the recognition will be discarded, and the inconsistency of the Decision Parameters corresponding to different algorithms is due to the differences in the extraction of contour features, so that different parameter values can make the recognition results as good as possible. The reason for the inconsistency of the Decision Parameters of different algorithms is the difference in the extraction of contour features. It should be noted that the Identification Number data listed in the table only represents the number of correctly identified positions in the image to a certain extent, it does not represent the availability of the identification results, and the applicability of the corresponding algorithms needs to be further analyzed with the results of image identification.

As can be seen from the data in the table, the number of flocculated clusters in the initial sediment image is considered to be 7. In the recognition process under different algorithms and preprocessing conditions, the Threshold algorithm achieves the maximum recognition accuracy under the Noise Reduction preprocessing condition, and the number of identified clusters is 8, while the number of unrecognized clusters is 0, which achieves the optimal effect, and is consistent with the results of its visualization.

Table 9.2 Different algorithms for recognizing information about the number of flocs

Algorithm	Decision parameters	Identification number						
		Original	Gray		Noise reduction		Contrast enhancement	
			Extraction	Not extracted	Extraction	Not extracted	Extraction	Not extracted
Canny	100	7	8	1	19	0	4	2
Sobel	5		1	7	1	7	2	7
Laplacian	10		1	7	Error	7	2	7
Threshold	2000		15	0	8	0	1	7

9.4.3 Algorithm Accuracy

Evaluation of the calculation results of different algorithms can be more detailed to quantify the performance effect of different algorithms in the process, can verify the applicability of algorithms, compare and select the best algorithms, and facilitate the tuning of algorithm parameters and the subsequent improvement process. In this experiment, different image processing techniques were evaluated by Precision, Recall and F1 indicators to assess the results of river and lake sediment flocculation, and the results are shown in Table 9.3.

As can be seen in the table, the two types of image processing methods, Sobel and Laplacian, have a value of 0 regardless of whether they are Precision or Recall evaluation methods, and they are also 0 after F1 comprehensive evaluation, which is consistent with the visualization of the recognition results. From Fig. 9.6, it can be found that for the Canny method, its overall accuracy is lower than 50%, which does not reach the goal of practical application. On the contrary, for the Threshold method, the Precision, Recall and F1 after Noise Reduction preprocessing in the field of river and lake sediment flocculation image recognition reached 0.88, 1.00 and 0.94, which showed a better recognition effect. It shows that whether it is to evaluate the accuracy of the positive samples, or the situation of leakage detection, or to comprehensively consider the double factors of accuracy and leakage detection rate, the Threshold method can show better application performance, in which the evaluation indexes under the premise of Noise Reduction preprocessing have the highest values and the best performance effect.

Figure 9.7 plots the distribution of area calculation results of different methods in the process of flocculation image recognition to compare the applicability of different methods in the process of flocculation image recognition with the distribution of their area calculation results compared to the real area. The red dashed line represents the upper and lower limits of the true area value, and it is easy to find that the results of Threshold-Noise Reduction are basically all within the range of this interval, i.e., it shows a better application performance for accurately identifying the real flocs. On the contrary, the other methods show a large deviation from the real values, which is positively correlated with the visualization results, i.e., the identification ability is lacking. In addition, Fig. 9.7 plots the change curves of different evaluation indexes,

Table 9.3 Evaluation results of different image recognition methods

	Gray			Noise reduction			Contrast enhancement		
	P	R	F1	P	R	F1	P	R	F1
Canny	0.38	0.43	0.40	0.21	0.57	0.31	0.25	0.14	0.18
Sobel	0.00	0.00	0.00	0.00	0.00	0.00	0.00	0.00	0.00
Laplacian	0.00	0.00	0.00	0.00	0.00	0.00	0.00	0.00	0.00
Threshold	0.47	1.00	0.64	0.88	1.00	0.94	0.00	0.00	0.00

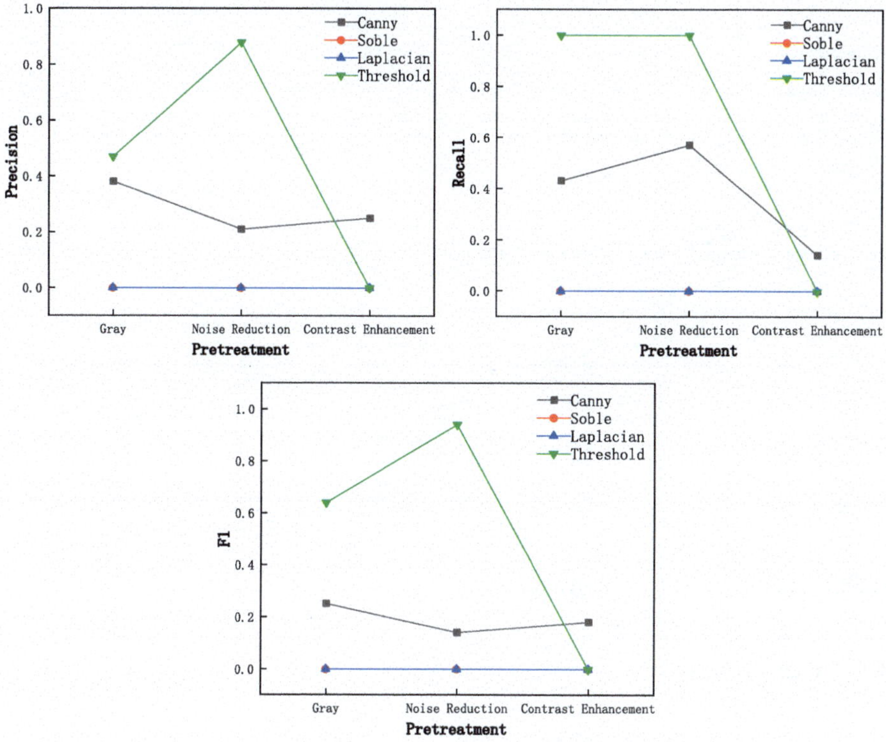

Fig. 9.6 Comparison curve chart of different assessment indicators

so that the advantages and disadvantages of different methods can be compared more clearly and intuitively.

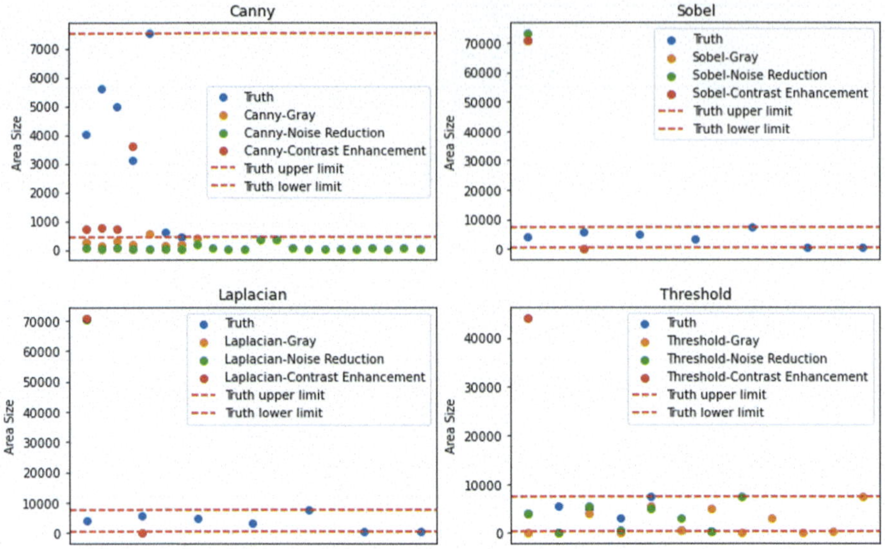

Fig. 9.7 Distribution of area calculation results

9.5 Conclusion

Because of the complexity and variability of river and lake mud, real-time detection difficulties and other characteristics, resulting in dosing treatment of intelligent difficulties, treatment costs are high. In this regard, this paper proposes a technology that can be applied to the intelligent dosing of river and lake mud. Based on the characteristics of the mud management process, this paper utilizes image processing technology to identify the mud flocs, studies the applicability of different edge detection algorithms and image segmentation technology in the field of mud floc image recognition in rivers and lakes, and evaluates and analyzes the recognition results through different evaluation indexes. It is found that the traditional edge detection algorithms, such as Canny, Sobel and Laplacian, are not applicable to the recognition process of river and lake mud flocculent images, and the scores obtained by the three types of evaluation indexes are low, especially the evaluation index value of the recognition results of the Sobel and Laplacian methods is 0. This is mainly attributed to the fact that the flocculent image has more noise and lower recognition, and the traditional algorithms are also affected by the noise in the recognition process. algorithms are also affected by the above factors in the recognition process. For this kind of situation, this paper uses the analysis of chromaticity difference and other factors to take the image processing method based on threshold segmentation to carry out image recognition, and the results are found to have excellent river and lake bottom mud recognition effect, which can recognize the real flocculent and its size contour better, and its three assessment indexes of Precision, Recall, and F1 have reached 0.88, 1.00, and 0.94 respectively, which have the ability to meet the requirements of river and lake bottom mud dosing. It has the ability to meet the intelligent identification link of river and lake bottom mud dosing flocculation, and to a certain extent, it provides certain technical support for the intelligent development of river and lake bottom mud disposal.

References

Fang XS, Zhai ZG, Zang J (2022) An intelligent dosing algorithm model for wastewater treatment plant. J Phys: Conf Ser 2224:012027

Gao MH, Liu YM, Wang G (2018) Comparison and analysis of image edge detection algorithms based on MATLAB. Beijing Surv Mapp 7(1):48–51

Huang JT (2014) Comparative study of several edge detection operators. MechatronS Technol 3(5):155–157

Jayaprakash S, Veni S (2016) Open CV based disease identification of mango leaves. Int J Eng Technol 7(8):1990–1998

Jia L, Jiao SH (2010) A review of research on image edge detection techniques. Sci Technol Wind, 226–234

Kasmi A, Abriak NE, Benzerzour M, Azrar H (2017) Effect of dewatering by the addition of flocculation aid on treated river sediments for valorization in road construction. Waste Biomass Valorization 8(3):585–597

Li W (2014) Research on image edge detection algorithm in MATLAB. Beijing Surv Mapp 3:3–7

Li CY, Guo CL, Ren WQ (2019) An underwater image enhancement benchmark dataset and beyond. IEEE Trans Image Process 29:4376–4389

Li GH, Zhang LL, Xu B (2019) Image analysis and processing of skin cell injury based on openCV. J Phys: Conf Ser (1237), 032003

Lü T, Luo C, Qi D (2019) Efficient treatment of emulsified oily wastewater by using amphipathic chitosan-based flocculant. React Funct Polym 139:133–141

Marini S, Fanelli E, Sbragaglia V (2018) Tracking fish abundance by underwater image recognition. Sci Rep 8(1), 13748

Nobuyuki O (1979) A threshold selection method from gray level histograms. IEEE Trans Syst Man Cybern 9(1):62–66

Song JY, Jeong H, Jeong J (2022) Performance optimization of object tracking algorithms in OpenCV on GPUs. Appl Sci (15), 7801

Wahab Fazal, Ullah Inam, Shah Anwar (2022) Design and implementation of real-time object detection system based on single-shoot detector and Open CV. Front Psychol 13, 1039645

Xu LJ, Cao WT, Liu CJ (2015) The comparison and analysis of sediment contour extraction based on image processing. J China Inst Water Resour Hydropower Res 13:28–33

Yan S (2022) Design of chemical phosphorus dosing system based on intelligent algorithm. Integr Circuit Appl 39(07):28–31

Yan H, Jiang P, Chen B (2010) Digital image processing techniques based on MATLAB toolbox. Microcomput Inf 11(9):214–216

Chapter 10
Photocatalytic Activity of Metal-Doped MgO Nanoparticles for Dye Removal with Cost Estimation

Khumbolake Faith Ngulube, Amal Abdelhaleem, Manabu Fujii, and Mahmoud Nasr

Abstract Metal-doping is a technique used to modify the surface, crystalline, textural, and/or optical properties of the photocatalytic semiconductors, giving greater photocatalytic degradation performances. In this study, a MgO semiconductor catalyst material was doped using copper (Cu) (3% wt) and zinc (Zn) (3% wt), respectively, via a simple co-precipitation synthesis technique. The synthesized 3% Cu-MgO and 3% Zn-MgO were then tested for the photocatalytic degradation of methylene blue dye (MB) as a model pollutant under neutral pH for 120 min. The catalyst dosage used was 500 ppm and the initial MB concentration was maintained at 11 ppm in both cases. After 120 min, the 3% Zn-MgO revealed a 68.85 $\pm$ 2.31% MB degradation efficiency, whereas the 3% Cu-MgO revealed a 63.91 $\pm$ 2.17% MB removal efficiency. Moreover, the synthesized pure MgO material was also employed for the photocatalytic degradation of MB under the same conditions (as a control), revealing only a 41.12 $\pm$ 1.34% MB removal efficiency. These MB removal efficiencies suggest the importance of metal-doping in improving the photocatalytic degradation process. Moreover, the unit treatment cost of MB solution using the proposed photocatalytic process reached \$7.95/m^3. This study directly addressed UN SDG 6 "Clean Water and Sanitation" and 14 "Life below water" through the conservation of the marine environment by effectively treating wastewater prior to its exposure to the marine environment.

Keywords Cost estimation · Magnesium oxide · Metal-doping · Photocatalysis · UN SDGs

K. F. Ngulube (✉) · A. Abdelhaleem · M. Nasr
Egypt-Japan University of Science and Technology (E-JUST), Alexandria 21934, Egypt
e-mail: khumbolake.ngulube@ejust.edu.eg

M. Fujii
Tokyo Institute of Technology, Meguro-Ku, Tokyo 152-8552, Japan

M. Nasr
Sanitary Engineering Department, Alexandria University, Alexandria 21544, Egypt

H. Xu (ed.), *The 6th International Symposium on Water Resource and Environmental Management*, Environmental Science and Engineering,
https://doi.org/10.1007/978-3-031-55989-1_10

10.1 Introduction

It is the collective responsibility of mankind to preserve the natural environment including the marine ecosystem. The United Nations organization spells this out under the sustainable development goal 14 of "life below water" (Mishra and Padhi 2021). In this regard, many researchers have devoted their work to address the issues of water pollution resulting from the disposal of wastewater into surface water bodies. These studies aim to combat public health threats such as endocrine perturbations and chronic toxicity which are imminent, especially in effluents from pharmaceuticals and personal care products' manufacture (Bello et al. 2019).

The existing conventional techniques that are currently being applied for wastewater treatment have been superseded by emerging superior and more advanced techniques that can treat recalcitrant wastewater pollutants. One such group of advanced treatment techniques that has shown great promise is advanced oxidation process (AOP) (Arora et al. 2022). AOPs have displayed an ability to mineralize pollutants by efficiently oxidizing and thereby degrading recalcitrant pollutants (Bello et al. 2019).

Heterogenous photocatalysis is a type of AOP technique that can degrade toxic and recalcitrant organic pollutants (Khlyustova et al. 2020). However, it is noted that heterogeneous photocatalysts are limited by low quantum efficiency, resulting from the issue of electron/hole (e^-/h^+) recombination against the transfer of charge carriers as well as the poor use of solar energy (Xu et al. 2012). Moreover, the semiconductor photocatalysts may fall short due to reduced efficiency for wide band gaps (Xu et al. 2012). To mitigate these shortfalls, several studies have shown that it is important to modify the physiochemical properties of the photocatalyst materials to yield the desired properties. Metal-doping is one such modification technique that is being explored.

The doping of narrow band-gap materials with metal dopants can greatly improve the photocatalytic activity of a photocatalyst material by altering the morphology and electrical structure of the material (Mohtar et al. 2021). Metal-doping has the potential to effectively manage the band gap challenge and improve the degradation performance of the semiconductor photocatalytic materials (Mohtar et al. 2021). For instance, the combining of nano MgO and Cu conceived optimal band gaps of 4.28 and 3.35 eV as opposed to the original 6.38 eV of nanostructured MgO and 7.8 eV of the bulk MgO (Nor Fadilah et al. 2019). A study by Vasanthi V. doped MgO with Zn and this resulted in a band gap reduction from 4.6 eV for MgO to 3.9 eV for Zn doped MgO (Vasanthi et al. 2017).

It has been highlighted in the literature that MgO nanoparticles have presented attractive physiochemical properties, especially for photocatalysis. These properties of MgO have shown great promise with regard to their application in environmental remediation. For instance, MgO nanoparticles have been doped with 2% Ag to enhance their photocatalytic performance (Alaizeri et al. 2021). The 2% Ag-enhanced MgO revealed an increased pollutant removal efficiency of about 75% in 180 min compared with 52% using the pure MgO material (Alaizeri et al. 2021). Another study reported that the use of 2 – 12% Cu to dope MgO, achieving a dye degradation rate of up to 93% with optimized conditions (Shaheen et al. 2021). Moreover, MgO and Fe/MgO synthesized nanoparticles were compared for photocatalytic activity, depicting that the Fe/MgO had better efficiency for the photocatalytic degradation of acid blue dye (Borhade et al. 2012). Therefore, metal-doping is revealed to have the potential of producing highly efficient photocatalysts with more desirable bandgaps and photocatalytic activity, which have inspired the efforts of the current research.

In this study, MgO was doped with 3% Cu and Zn, respectively, via a simple co-precipitation method. The synthesized 3% Cu-MgO and 3% Zn-MgO photocatalysts were applied for photocatalytic degradation using the methylene blue dye (MB) as the model pollutant. A techno-economic analysis was also performed to determine the treatment cost per m^3 of treated wastewater. Finally, the study highlights the achievable sustainable development goals (SDGs) related to the application of metal-doped MgO for the treatment of textile effluents.

10.2 Materials and Methods

10.2.1 Materials

Magnesium sulfate heptahydrate ($MgSO_4 \cdot 7H2O$; $\geq$ 99% purity) and copper (II) sulphate pentahydrate ($CuSO_4 \cdot 5H_2O$, 98.5% purity) were purchased from Tekkim Kimya Labs, Turkey. Zinc sulfate heptahydrate ($ZnSO_4 \cdot 7H_2O$) was purchased from the El Nasr pharmaceutical chemicals company in Egypt. Sodium hydroxide pellets (NaOH; 95% purity) was purchased from SD-Fine Chemicals Limited (SDFCL), India.

10.2.2 Photocatalyst Material Synthesis

Synthesis of MgO. To synthesize the MgO catalyst material, a 0.5 M solution of the precursor solution was prepared by dissolving an appropriate mass of $MgSO_4 \cdot 7H_2O$ in 100 mL distilled water at room temperature. Once fully dissolved, a prepared 1 M solution of NaOH was added to the precursor solution dropwise under constant stirring conditions until the pH ranged between 10–12. A white precipitate began to form. This adjustment of the solution pH is necessary to facilitate the formation of precipitates. When the solution pH reaches the isoelectric point (pHI) (point where the overall charge on the surface of precipitates is zero), precipitates begin to form and aggregate (Yang et al. 2014). Therefore, in nanoparticle synthesis, the solution pH needs to be adjusted upwards of the pHI (Yang et al. 2014) in order to prevent agglomeration of the formed precipitates as a result of repulsive forces due to precipitates having the same overall surface charge. This is important especially when no surfactants are used in the synthesis process. When the desired pH was achieved, the addition of NaOH was stopped and the precipitate was allowed to stir at room temperature for an additional 30 min, allowing for the complete formation of the precipitate. The formed precipitate was then separated from the liquid by filtration and washed several times with distilled water until the pH was neutral. The separated precipitate was dried overnight at 105 °C, crushed into a fine powder, and finally calcined in a muffle furnace at 350 °C for 3 h to obtain the MgO.

Preparation of 3% Zn-MgO and 3% Cu-MgO. The 3% Zn-MgO was synthesized by first preparing a 0.5 M solution of $MgSO_4{\cdot}7H_2O$. In a separate beaker, the desired amount of $ZnSO_4{\cdot}7H_2O$ to achieve a 3% ratio of Zn:Mg was dissolved in distilled water. Once fully dissolved, the $ZnSO_4{\cdot}7H_2O$ solution was added to the $MgSO_4{\cdot}7H_2O$ solution and kept under continuous stirring to allow for the complete mixing of the prepared solutions. Once fully mixed, a 1 M solution of NaOH was added to the mixture dropwise under constant stirring conditions until the pH was between 10–12. A white precipitate began to form. The formed precipitate was processed in the same way as the MgO to finally obtain the 3% Zn-MgO. The 3% Cu-MgO was then synthesized following a similar procedure, where a solution of $CuSO_4{\cdot}5H_2O$ was made by dissolving the desired mass of $CuSO_4{\cdot}5H_2O$ required to achieve a 3% ratio of Cu:Mg in distilled water. Once fully dissolved, the $CuSO_4{\cdot}5H_2O$ solution was mixed with an already dissolved solution of $MgSO_4{\cdot}7H_2O$ and allowed to mix. When the solutions were fully mixed, again, a 1 M solution of NaOH was added to the mixture under constant stirring until a pH of 10–12 was maintained. A light blue precipitate began to form. The formed precipitate was processed in the same way as the synthesized MgO to yield the 3% Cu-MgO.

Characterization. The concentrations of the MB dye utilized throughout the study were measured using a UV spectrophotometer (Jasco V-630 spectrophotometer, Jasco Inc., Tokyo, Japan) fixed at a wavelength of 664 nm.

Photocatalytic Tests. To assess the photocatalytic performance of the synthesized catalyst materials, a set of photocatalytic degradation runs were carried out using MB as a model pollutant. The synthesized MgO, 3% Zn-MgO, and 3% Cu-MgO were applied for the photocatalytic degradation of MB. In each case, the initial pH was maintained at a neutral pH, the initial MB concentration was fixed at 11 mg/L and the catalyst dosage was 500 mg/L. Moreover, a control experiment was also carried out, in which no catalyst was added to the MB solution. In the first step, the four beakers were placed into the photocatalytic reactor (comprised of two metal halide lamps of a light intensity of 25,000 lx, 400 W, $\lambda = 510$ nm UV–visible light), without light and kept under constant stirring for 1 h to achieve adsorption–desorption equilibrium. Once this equilibrium was achieved, 2 mL samples were extracted from each beaker and their concentrations were measured. The light in the reactor was then switched on and photocatalysis was allowed to proceed for 2 h, extracting 2 mL samples every 30 min. For each sample extracted, the MB removal efficiency was calculated using the equation given by Eq. 10.1:

$$R\% = [(C - C_0)/C_0] \times 100 \tag{10.1}$$

10.3 Results and Discussions

10.3.1 Photocatalytic Tests

Figure 10.1. shows the photocatalytic degradation of MB by MgO, 3% Zn-MgO, and 3% Cu-MgO. After 120 min of photocatalytic activity, the pure MgO displayed an MB removal of about 41.12%. When the MgO was doped with 3% Cu and 3% Zn, the MB removal increased to 63.91% and 68.85%, respectively. The increased photocatalytic performance was indicative of successful metal doping and enhanced performance of MgO. Moreover, the enhanced performance could have been due to the reduced energy band-gap resulting from the metal-doping of the MgO photocatalyst (Nor Fadilah et al. 2019), facilitating an increased generation of e^-/h^+ pairs under UV–visible light irradiation (Eq. 10.2):

$$3\%\ Zn/Cu - MgO + hv \rightarrow e^- + h^+ \tag{10.2}$$

A similar trend was observed for Zn/MgO prepared via sol–gel synthesis technique and applied for photocatalytic degradation of MB (Sierra-Fernandez et al. 2017).

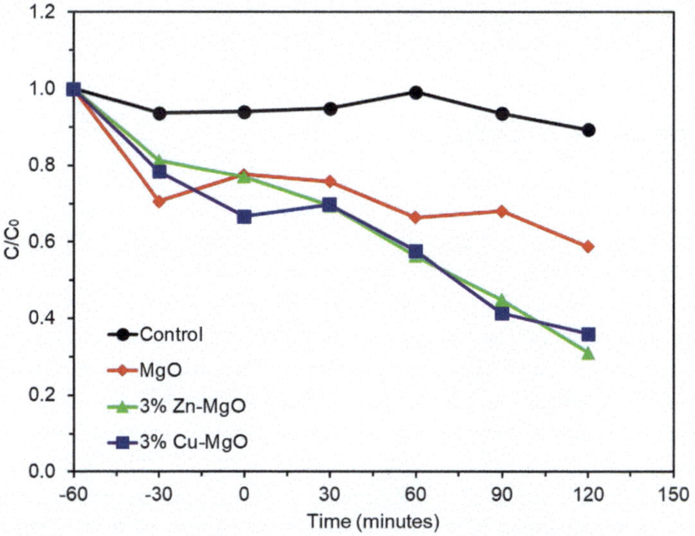

Fig. 10.1 MB degradation curves of MgO, 3% Zn-MgO, and 3% Cu-MgO [catalyst dose = 500 mg/L, MB concentration = 11 mg/L, pH = neutral, and degradation time = 120 min]

10.3.2 Cost Estimation

The results obtained from the initial photocatalytic degradation tests were used to estimate the capital and investment costs associated with the application of metal-doped MgO photocatalyst materials for the treatment of textile effluents (Table 10.1). The cost estimation considered a treatment facility receiving a daily influent of 10 m^3. This inflow was suitable for a small- to medium-scale treatment plant.

The capital costs included all the one-time expenses encountered in the initial phases of setting up the treatment plant. These expenses comprised items, such as machinery and equipment, buildings, and construction fees among others. Moreover, the operational costs represented the expected costs to be incurred to run the treatment facility. The operational costs included the prices related to chemicals, electricity, repair and maintenance fees, workers' salaries, and taxes.

The economic analysis revealed a running cost of treating textile wastewater of about \$5.21/m^3. Moreover, the unit capital cost was also estimated as \$2.74/m^3, giving a total unit cost of plant set-up and wastewater treatment of \$7.95/m^3. This treatment cost was compared with the treatment cost of other recently trending treatment techniques (Table 10.2). The treatment cost by photocatalysis was about \$1/m^3 higher than other treatment techniques. However, the degradation efficiencies by photocatalysis were much higher and the process is also much more environmentally friendly as there is less waste generated.

10.3.3 Related UN SDGs

To achieve world peace and prosperity both now and in the future, the United Nations (UN) member states outlined and adopted 17 interlinked sustainable development goals (SDGs) (Fig. 10.2).

To achieve world peace and prosperity, the SDGs integrate strategies to deal with the improvement of health and education, reduction of inequalities, and increasing economic growth. The SDG agenda also addresses the climate change issues and the preservation of oceans and forests (https://sdgs.un.org/goals).

The application of metal-doped MgO for photocatalytic treatment of textile effluents is directly linked to the achievement of some of the 17 UN SDGs. For instance, goal 2 "*Zero Hunger*" and goal 14 "*Life Below Water*" can be achieved as a result of the ability of metal-doped MgO photocatalysis to effectively treat textile effluents prior to their exposure to marine environments. This results in reduced exposure to the toxicity of dyes by fish, preserving their health and helping the fishing industry. Goal 3 "*Good Health and Well-being*", goal 6 "*Clean Water and Sanitation*" and goal 15 "Life on Land" could also be met by the application of metal-doped photocatalysis. When water bodies are protected from pollution by toxic pollutants, such as dyes, the number of water pollution-related deaths is directly reduced. Furthermore,

Table 10.1 Cost estimates for the application of metal-doped MgO for photocatalytic treatment of textile wastewater (flow = 10 m^3/day)

	Description	Justification	Cost
Capital costs	Photocatalytic Reactor Installation (PRI)	Reactors ($1504.46 × 2 = $3008.93), UV lamps ($1029.84 × 2 = $2059.69), stirrer motors ($528.35 × 2 = $1056.71), pumps ($1799.98 × 2 = $3599.97), and rotary calcination furnace ($2865.65)	$12,590.95
	Piping, valves, and wiring	0.3 PRI	$3777.29
	Site work	0.1 PRI	$1259.10
	Subtotal (1)	1.4 PRI = Q	$17,627.33
	Contractors O&P	0.15 Q	$2644.10
	Subtotal (2)	1.15 Q = R	$20,271.43
	Engineering fees	0.15 R	$3040.72
	Subtotal (3)	1.15 R = S	$23,312.15
	Contingency charges	0.2 S	$4662.43
	Subtotal (4)	1.2 S = T	$27,974.58
	Other costs	0.2 T	$5594.92
	Total capital cost	1.2 T	$33,569.50
	Capital cost per annum	$33,569.50/project lifetime (5 years)	$6713.90/yr
	Interest consideration	$6713.90/yr × $(1 + i)^n$	$8220/yr
	Unit capital cost	$8220/yr ÷ (10 m^3/d × 300 d/yr)	$2.74/$m^3$
Operational costs	Chemicals	15 kg/month × $6.5/kg	$97.50/month
	Electrical consumption	7523.57 kWh/month × $0.13/kWh	$978.06/month
	Water Usage	164 m^3/month × $0.24/m3	$39.36/month
	Workers' Salaries		$120/month
	Land rentals	5% of total capital cost ($822/month)	$41.1/month
	Repair and maintenance	15% of total capital cost ($822/month)	$123.3/month
	Other	20% of total capital cost ($822/month)	$164.4/month
	Total operating cost	Sum of total operating items	$1563.7/month
	Unit total operating cost	$1563.7/month ÷ (10 m^3/d × 30 d/month)	$5.21/$m^3$
	Unit Total cost	Unit capital cost + Unit total operating cost	$7.95/$m^3$

Table 10.2 Cost comparison of photocatalysis and other treatment techniques for textile wastewater treatment

Treatment technique	Treatment cost ($/m^3)	References
Photocatalysis	7.95	This study
Coagulation/flocculation—adsorption	6.37	Mahmoud et al. (2021)
Nanofiltration	6.00	Samhaber and Nguyen (2014)
Anaerobic moving bed biofilm reactor	6.21	Shahzad et al. (2023)

Fig. 10.2 United Nations sustainable development goals (UN SDGs)

the high-quality effluent produced after textile wastewater is subjected to photocatalytic treatment could also be reused in various textile processes, increasing water-use efficiency. Goal 8 "Decent Work and Economic Growth" and goal 9 "Industry, Innovation and Infrastructure" could also be achieved as a result of the application of metal-doped MgO photocatalysis for textile effluent treatment. More jobs will be created, directly reducing unemployment levels. Moreover, the introduction of such innovative technologies could promote economic productivity and continuous technological advancement with regard to wastewater treatment.

10.4 Conclusions

The study successfully developed a smart metal doping-based material used for the photocatalytic degradation of MB dye. These materials were 3% Cu-MgO and 3% Zn-MgO, maintaining more than 60% of MB removal efficiency. The treatment cost was estimated, giving a price of \$7.95 for treating 1 m^3 of dye-laden wastewater. Despite the cost of treatment per m3 being higher than that of conventional biological treatment, the high efficiencies and eco-friendliness related to photocatalytic treatment guarantee more profits as well as shorter payback periods, making their application a more attractive approach. Moreover, the study outputs were also directly linked to multiple SDGs associated with pollution reduction, good human health, and aquatic culture preservation.

References

Alaizeri ZM, Alhadlaq HA, Aldawood S, Akhtar MJ, Amer MS, Ahamed M (2021) Facile synthesis, characterization, photocatalytic activity, and cytotoxicity of Ag-doped MgO nanoparticles. Nanomaterials 11(11)

Arora I, Chawla H, Chandra A, Sagadevan S, Garg S (2022) Advances in the strategies for enhancing the photocatalytic activity of TiO_2: conversion from UV-light active to visible-light active photocatalyst. Inorg Chem Commun 143:109700

Bello MM, Abdul Raman AA, Asghar A (2019) A review on approaches for addressing the limitations of Fenton oxidation for recalcitrant wastewater treatment. Process Saf Environ Prot 126:119–140

Borhade AV, Kanade KG, Tope DR, Patil MD (2012) A comparative study on synthesis, characterization and photocatalytic activities of MgO and Fe/MgO nanoparticles. Res Chem Intermed 38(8):1931–1946

Khlyustova A, Sirotkin N, Kusova T, Kraev A, Titov V, Agafonov A (2020) Doped TiO_2: the effect of doping elements on photocatalytic activity. Mater. Adv. 1(5):1193–1201

Mahmoud AS, Mostafa MK, Peters RW (2021) A prototype of textile wastewater treatment using coagulation and adsorption by Fe/Cu nanoparticles: techno-economic and scaling-up studies. Nanomater Nanotechnol

Mishra US, Padhi B (2021) Sustaining the sustainable development goals. Econ Polit Wkly 56(34):30–31

Mohtar SS, Aziz F, Ismail AF, Sambudi NS, Abdullah H, Rosli AN, Ohtani B (2021) Impact of doping and additive applications on photocatalyst textural properties in removing organic pollutants: a review. Catalysts 11:1160–1169

Nor Fadilah C, Norlida K, Nurhanna B, Kelimah E (2019) Effect of Cu doped in MgO on nanostructures and their band gap energies. Solid State Phenom 290 SSP:323–328

Samhaber WM, Nguyen MT (2014) Applicability and costs of nanofiltration in combination with photocatalysis for the treatment of dye house effluents. Beilstein J Nanotechnol 476–84

Shaheen M et al (2021) Synthesis of Cu-doped MgO and its enhanced photocatalytic activity for the solar-driven degradation of disperse red F_3BS with condition optimization. Zeitschrift Fur Phys Chemie 235(11):1395–1412

Shahzad HMA, Khan SJ, Khan M, et al (2023) Performance and cost-benefit analysis of anaerobic moving bed biofilm reactor for pretreatment of textile wastewater. Korean J Chem Eng 1389–1400

Sierra-Fernandez A, De La Rosa-García SC, Gomez-Villalba LS, Gómez-Cornelio S, Rabanal ME, Fort R, Quintana P (2017) Synthesis, photocatalytic, and antifungal properties of MgO, ZnO and Zn/Mg Oxide nanoparticles for the protection of calcareous stone heritage. ACS Appl Mater Interfaces 9(29):24873–24886, 7

THE 17 GOALS|Sustainable Development. https://sdgs.un.org/goals. Accessed 3 Nov 2023

Vasanthi V, Kottaisamy M, Anitha K, Ramakrishnan V (2017) Near UV excitable yellow light emitting Zn doped MgO for WLED application. Superlattices Microstruct 106:174–183

Xu J, Wang W, Sun S, Wang L (2012) Enhancing visible-light-induced photocatalytic activity by coupling with wide-band-gap semiconductor: a case study on Bi_2WO_6/TiO_2. Appl Catal B Environ 111–112:126–132

Yang X, Shang Y, Li Y, Zhai J, Foster NR, Li Y, Zou D, Pu Y (2014) Synthesis of monodisperse iron oxide nanoparticles without surfactants. J Nanomater 2014:5

Chapter 11
Groundwater Markets in the Era of Environmental Challenges: Pathways to Sustainability

Amar Razzaq and Yewang Zhou

Abstract This paper investigates the growing significance of groundwater markets in the context of agricultural water allocation, especially in regions grappling with water scarcity exacerbated by climate change. It critically analyzes the historical development, governance structures, and the dual challenges of sustainability and social equity inherent in these markets. The research highlights how groundwater markets act as adaptive strategies to climate change, focusing on their potential to maintain water security under evolving environmental conditions. The need for policy changes to incorporate effective governance models that align market mechanisms with regulatory frameworks is stressed. This involves exploring ways policies can tackle equitable water access and sustainable use, ensuring groundwater markets benefit agricultural systems. By consolidating recent studies and worldwide trends, this study offers valuable perspectives on changing groundwater markets, pushing for their evolution into tools that support sustainable, fair, and resilient agricultural methods. The comprehensive analysis presented aims to guide policymakers, stakeholders, and researchers in understanding and enhancing the functionality of groundwater markets within the broader scope of sustainable water resource management.

Keywords Groundwater management · Aquifer management · Water resource policy · Climate adaptation · Equity in water access · Groundwater markets · Integrated governance

A. Razzaq · Y. Zhou (✉)
Business School, Huanggang Normal University, No. 146, Xinggang 2nd Road, City Development Zone, Huanggang 438000, China
e-mail: zhouyewang@hgnu.edu.cn

H. Xu (ed.), *The 6th International Symposium on Water Resource and Environmental Management*, Environmental Science and Engineering,
https://doi.org/10.1007/978-3-031-55989-1_11

11.1 Introduction

Groundwater markets have become increasingly vital for agricultural water allocation, particularly in areas facing water scarcity exacerbated by climate change. Recent trends in groundwater market governance emphasize the need for adaptive capacity, integrating horizontal and vertical resource management strategies, and utilizing adaptive management frameworks (Holman & Trawick, 2011). These markets allow for the trading of groundwater rights, redistributing water from less productive to more valuable uses. However, concerns over unregulated extraction and issues of equity and efficiency challenge the sustainability of these markets.

This paper argues for the necessary reform of groundwater market structures through integrated policies that ensure equitable resource allocation and promote sustainable agricultural practices. It examines the interconnections between groundwater markets and agricultural resilience, highlighting priority areas for policy intervention. These areas include managing groundwater sustainably, protecting ecosystem services, and ensuring water equity.

The rapid global depletion of groundwater, primarily due to agricultural irrigation, is worsened in areas with informal groundwater markets. Here, unregulated abstraction often outpaces natural recharge rates, leading to increased irrigation costs, degraded water quality, and ecological impacts such as land subsidence (Sears et al., 2018).

In response, recent governance models emphasize legal frameworks, monitoring, decentralized water management, and the enforcement of regulations, often focusing on supply-side technological solutions (Closas & Villholth, 2019). However, fragmented governance and traditional socio-economic barriers impede sustainability, particularly in countries like India, China, and Pakistan (Schwartz et al., 2020; Razzaq et al., 2022b).

As climate change intensifies hydrological variability, transforming groundwater market structures becomes imperative. This paper aims to provide an overview of the current state of groundwater markets, highlight sustainability challenges, and propose policy frameworks for effective groundwater management. By integrating recent research and trends, this paper offers a comprehensive perspective on developing eco-sustainable, and resilient agricultural systems.

11.2 Overview of Groundwater Markets

Groundwater markets can be of two types: formal water markets and informal water markets. Formal water markets, typically prevalent in developed countries, involve trading water rights within a framework of rules set by states or federal governments. These markets allow trading over large areas, where water rights or the right to use a set amount of water are exchanged during a certain period. Conversely, informal water markets, often found in developing countries, operate without such structured regulations and formal recognition of water rights.

Groundwater markets are evolving rapidly, particularly in response to increasing water and land scarcity. For instance, northern China has witnessed the privatization of tubewells and a surge in groundwater market activities (Zhang et al., 2008). These markets vary significantly across regions, reflecting diverse institutional and socio-economic contexts.

In developed countries like Australia and the United States, groundwater markets have been shaped by well-established institutions and regulations, coupled with high-quality economic and scientific research (Wheeler et al., 2016). Conversely, in South Asian countries such as India and Pakistan, groundwater markets often function as informal "spot markets". Here, water is sold from private wells, mostly operating outside formal regulatory frameworks (Meinzen-Dick, 1998). These markets are shaped by hydro-geological factors, irrigation infrastructure, historical precedents, and social norms rather than direct policy interventions (Dubash, 2000).

An example of significant institutional challenges in groundwater trading can be seen in the Gnangara groundwater system of Western Australia. This system is characterized by conditional property rights, complex regulatory approval processes, inadequate infrastructure, and limited access to price information, which collectively hinder effective groundwater trading (Skurray et al., 2013). Moreover, recent trends indicate a global explosion in groundwater use, leading to severe overdraft and a resurgence of the 'resource-state', highlighting the need for recentralization of policies and institutions (Birkenholtz, 2015). The features of different water markets and influencing factors are shows in Fig. 11.1.

In India, the fragmentation of land holdings and specific governmental policies are primary factors driving groundwater exploitation, especially in northwestern regions (Jeet, 2015). Similarly, informal groundwater markets in regions like Assam ensure water access for farmers with limited capabilities to invest in technologies, albeit with the risk of aquifer overexploitation (Tamuli & Dutta, 2016).

These examples illustrate that groundwater markets are shaped by a variety of factors, including local contexts, legal frameworks, socio-economic dynamics, and ecological constraints. As such, the governance and structure of these markets are continuously evolving. This evolution reflects the diverse challenges and opportunities inherent in managing this essential resource in a sustainable and equitable manner.

11.3 Challenges in Sustainability of Groundwater Markets

The sustainability of groundwater resources remains a major global challenge, with over-exploitation leading to depletion and salinization, impacting both human and ecological systems. Key challenges include managing overdraft, increasing recharge, efficiently allocating pumping, and balancing the needs of people and nature (Matsumoto et al., 2019).

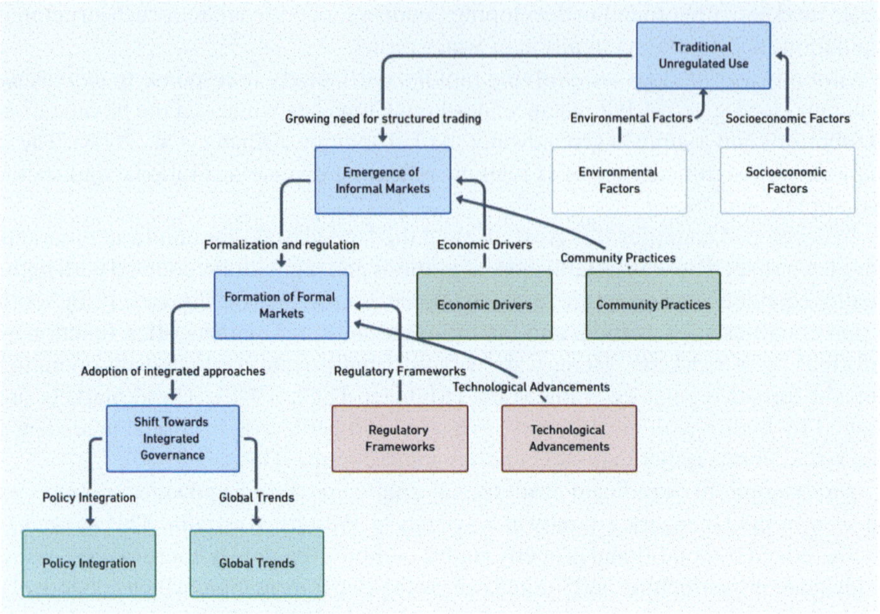

Fig. 11.1 Evolution of groundwater markets

Groundwater markets face both physical and policy constraints. These challenges determine where groundwater markets can operate effectively and how they adapt to future pressures on water availability due to climate change (De Luca & Sinclair, 2020). Furthermore, the excessive leaching and runoff of agricultural chemicals are affecting groundwater quality, soil salinity, and air quality, presenting additional sustainability challenges (Ahuja et al., 2007).

Co-management of groundwater, involving various stakeholders, has emerged as a potential approach to address these sustainability challenges. However, its success requires specific user community characteristics, credible environmental threats, joint monitoring, transparency, and substantial funding (Molle & Closas, 2020). In addition, integrating groundwater into development planning can improve economic efficiency, social equity, and environmental sustainability in groundwater markets (Foster & Ait-Kadi, 2012).

Increasing groundwater withdrawal is creating limits to growth, affecting agricultural communities by reducing profitability and resource depletion. This calls for sustainable quantitative management of groundwater, implementing effective water policy instruments, and enabling sectoral policies to work in synergy with water policy objectives (Rouillard, 2020). Local groundwater sustainability agencies should assess conditions and educate well owners to reduce risks to their water supplies for sustainable groundwater markets (El-naggar, 2015).

The above analysis shows that the sustainability challenges in groundwater markets are multifaceted, involving ecological, economic, social, and policy dimensions. Addressing these challenges requires integrated, multi-stakeholder approaches that consider both local contexts and broader policy frameworks.

11.4 Groundwater Markets and Social Equity

Groundwater markets have significant implications for social equity and water access (Bajaj et al., 2022). They can improve the equity of water access by enabling the transfer of water from large to small farmers, thereby increasing water use efficiency and productivity (Razzaq et al., 2019; 2022a). These markets provide non-well owners, particularly resource-poor farmers, equitable access to irrigation water, enhancing agricultural production and incomes (Kolvalli & Chicoine, 1989).

However, the equity implications of groundwater markets, including the distribution of market benefits and equitable access to water and rights, can be complex and multifaceted. While groundwater markets generate substantial net benefits and enable more efficient water use across sectors, they also raise concerns about environmental impacts and long-term sustainability (Ayres et al., 2019). Rapid urbanization and land rights issues in regions like Guwahati city, India, lead to increased groundwater extraction over safe yield, resulting in reduced accessibility and increased water affordability issues (Hazarika & Nitivattananon, 2016).

Declining groundwater quality can exacerbate inequity in terms of cost and volume of consumption between different income groups. This situation highlights the need for environmental and social priorities to extend beyond piped water to include groundwater resources (Kooy et al. 2016). Inequity issues can also arise from the potential exploitation of buyers, particularly small and marginal farmers, leading to unsustainable and inequitable use of groundwater resources (Sharma & Sharma, 2006).

While groundwater markets can offer benefits in terms of improved water access and agricultural productivity, they also pose challenges related to social equity, equitable water access, sustainability, and environmental impacts. Addressing these challenges requires a comprehensive understanding of the local socio-economic contexts and the development of policies that balance efficiency gains with equity and sustainability considerations.

11.5 Governance Frameworks in Groundwater Management

Effective groundwater management in the context of groundwater markets requires integrated policy frameworks that consider both environmental and socioeconomic systems. An integrated quantification framework and a revised policy-making procedure are essential to address the complex relationship between these systems (Yao et al., 2019).

Key components of better groundwater governance include improved legal frameworks, monitoring and control of access and abstraction, decentralized water management, enforcement of regulations, and supply-side technological solutions (Closas & Villholth, 2019). For effective governance, it is crucial to have high-quality legal provisions that are fair, equitable, coherent, and enforceable, coupled with well-sequenced transition processes, operational capacity, social consensus, stakeholder participation, and supportive socio-economic trends (Mechlem, 2012).

Institutional impediments, such as conditional property rights, regulatory approval, lack of facilitating infrastructure, and limited price information, can significantly hinder the development of effective groundwater markets (Skurray et al., 2013). In India, participatory forms of groundwater management using 'aquifer-based, common pool resource' approaches are being incorporated into practices and policies, emphasizing community involvement and sustainable usage (Kulkarni et al. 2015).

A smart spot market for groundwater can also be a promising tool to reduce transaction costs, users' risk, and increase the reliability of environmental flows, thereby contributing to more sustainable groundwater management (Raffensperger et al., 2009). In addition, establishing safeguard zones in groundwater bodies for drinking water through a three-phase methodology involving hydrogeological criteria, pressure evaluation, and wellhead protection areas can enhance groundwater quality and sustainability (Martínez-Navarrete et al., 2011).

The management of groundwater resources through markets requires innovative policy frameworks that are adaptable, participatory, and integrated, ensuring sustainable use from both environmental and socio-economic perspectives. These frameworks should address legal, institutional, technological, and community-based aspects of groundwater management to achieve long-term sustainability and equity. Table compares different water markets and their features across different regions (Table 11.1).

Table 11.1 Comparative analysis of groundwater management across different region

Region	Market type	Governance framework	Sustainability challenges	Climate change adaptation strategies	Equity implications
Murray-Darling Basin	Formal & integrated	Strong legal and institutional frameworks; adaptive management	Water scarcity; over-extraction	Water banks; managed aquifer recharge	Equitable water distribution; participation of indigenous communities
Upper Guadiana Basin	Informal & integrated	Community-based management; participatory decision-making	Aquifer over-exploitation; ecological degradation	Crop pattern adaptation; efficient irrigation systems	Access for small-scale farmers; social consensus
South Asia (General)	Informal & fragmented	Largely unregulated; local norms and customs	Over-exploitation; water quality issues	Adoption of drought-tolerant crops; groundwater recharge initiatives	Inequity in access; exploitation of resource-poor farmers
Western United States	Formal & market-driven	Well-defined water rights; market-based allocation	Groundwater depletion; land subsidence	Flexible water trading; conservation incentives	Varied impact on small vs. large farmers; market access challenges
East Anglia, UK	Integrated & adaptive	Horizontal and vertical integration; adaptive frameworks	Climate variability; demand management	Adaptive cropping patterns; local water management strategies	Community involvement; equitable resource use
China	Varied & evolving	State-regulated with increasing market mechanisms	Rapid industrialization and urbanization pressures; pollution	Implementation of large-scale water transfer projects; promoting water-saving technologies	Balancing economic growth with sustainable water use; regional disparities

11.6 Adaptation to Climate Change Through Groundwater Markets

Mitigating climate change necessitates a unified global effort, transcending individual continents and regions. Groundwater markets offer potential benefits for climate change adaptation by providing flexibility in allocating water resources amidst growing variability and uncertainty of supply. They buffer vulnerabilities to droughts and rainfall shocks, sustaining irrigation and rural livelihoods during water stress, thus contributing to resilience in agricultural systems (Foster and Perry, 2010; Mukherji, 2007).

Strategic approaches like water banks, managed aquifer recharge, and adjustable pumping restrictions can enhance resilience to anticipated climate impacts. Active recharge of aquifers using excess wet season flows, for instance, provides storage to supplement irrigation in dry periods, demonstrating an adaptive response to climate variability (Jha et al., 2018; Dillon et al., 2019). Local management strategies, such as reducing irrigated agriculture and applying adaptive cropping patterns, play a crucial role in groundwater sustainability and climate change adaptation (Afruzi et al., 2021).

However, the realization of these adaptation benefits through groundwater markets requires proactive policies and collective action. Unregulated pumping for trading poses risks to long-term sustainability and exacerbates scarcity. Robust monitoring, cropping restrictions, and conservation incentives are needed to moderate demand and prevent overexploitation (Giordano, 2009; Foster and Garduño, 2013).

Investing in climate-resilient agriculture is critical for adaptation. The adoption of low water-intensive, heat and drought-tolerant crop varieties can reduce pressure on groundwater resources and diversify rural livelihoods, thus building adaptive capacity (Burney et al., 2010; Akhtar et al., 2022). Ultimately, climate adaptation via groundwater markets necessitates integrated water-energy-agriculture planning and policies, emphasizing the need for strengthening institutions and decision-making processes to manage uncertainty and balance sustainability, efficiency, and welfare outcomes (Shah, 2009).

Pursuing synergies across governance levels and stakeholders is imperative to secure water access for the most vulnerable in climate-impacted regions, ensuring that groundwater markets bolster resilience in the face of climate variability and change.

11.7 Integrated Governance for Sustainable Groundwater Markets-A Conclusion

The complexity of managing groundwater markets in the context of environmental sustainability, equitable access, and climate change adaptation demands a paradigm shift towards integrated governance. Groundwater markets have the potential to enhance resource sustainability and resilience, yet this potential can only be realized through holistic, multifaceted governance frameworks that effectively blend market dynamics with regulatory oversight.

Integrated governance requires a balance between facilitating market mechanisms and ensuring regulatory compliance to prevent over-exploitation. For formal groundwater markets, this includes enhancing market transparency, establishing clear water trading platforms, and ensuring stringent regulatory oversight. In informal markets, steps such as formalizing water rights, providing better access to market information, and supporting small-scale farmers' participation are critical.

The transition from informal to formal groundwater markets should be managed carefully, considering the benefits of improved resource management and the challenges of legal and institutional reforms. This balance is critical for long-term aquifer sustainability and social equity. Effective governance models must include robust legal and institutional frameworks, adaptive management strategies, and participatory decision-making processes. These elements are key to responding to the dynamic relationship of climate change, socio-economic needs, and environmental preservation.

The importance of integrated governance is highlighted in successful examples such as the Murray-Darling Basin, where an integrated water management approach has proven efficient and flexible in dealing with water availability during both wet and dry periods (Ross, 2012). Similarly, in the Upper Guadiana Basin in central Spain, integrated catchment management has been identified as a key solution for sustainable groundwater management in areas of over-exploitation (Bromley et al., 2001).

Investing in strategies like managed aquifer recharge, water banks, and adaptive agricultural practices is also important. These strategies not only buffer against water scarcity but also aid in transitioning toward more formalized and sustainable groundwater market systems.

The path towards sustainable groundwater markets is paved with the principles of integrated governance, encompassing legal, economic, environmental, and social dimensions. It necessitates a concerted effort from policymakers, stakeholders, and communities to develop governance models that are equitable, efficient, and environmentally sustainable. This approach will not only safeguard groundwater resources but also ensure their judicious use for future generations, aligning with the broader goals of sustainable development and climate change adaptation.

References

Afruzi A, Abyaneh H, Abdolabadi H (2021) Local strategies to manage groundwater depletion under climate change scenarios—a case study: Hamedan-Bahar Plain (Iran). Arab J Geosci 14:1–18

Ahuja L, Andales A, Ma L, Saseendran S (2007) Whole-system integration and modeling essential to agricultural science and technology for the 21st century. J Crop Improv 19:103–173

Akhtar R, Masud MM, Rahman MK (2022) Farmers' livelihood and adaptive capacity in the face of climate vulnerability. Int J Soc Econ 49(5):669–684

Ayres A, Meng K, Plantinga A (2019) Do environmental markets improve on open access? evidence from California groundwater rights. J Polit Econ 129:2817–2860

Bajaj A, Singh SP, Nayak D (2022) Impact of water markets on equity and efficiency in irrigation water use: a systematic review and meta-analysis. Agric Water Manag 259:107182

Birkenholtz TL (2015) Recentralizing groundwater governmentality: rendering groundwater and its users visible and governable. Wires Water 2:21–30

Bromley J, Cruces J, Acreman M, Martinez L, Llamas M (2001) Problems of sustainable groundwater management in an area of over-exploitation: the upper Guadiana catchment, central Spain. Int J Water Resour Dev 17:379–396

Burney J, Woltering L, Burke M, Naylor R, Pasternak D (2010) Solar-powered drip irrigation enhances food security in the Sudano-Sahel. Proc Natl Acad Sci 107(5):1848–1853

Closas A, Villholth K (2019) Groundwater governance: addressing core concepts and challenges. Wiley Interdisciplinary Reviews: Water 7

De Luca J, Sinclair D (2020) Development of groundwater markets in Australia: insights from Victoria in the Murray Darling Basin. In: Rinaudo JD, Holley C, Barnett S, Montginoul M (eds) Sustainable groundwater management, global issues in water policy, vol 24. Springer, Cham, pp 21–37

Dillon P, Stuyfzand P, Grischek T, Lluria M, Pyne RDG, Jain RC, Sapiano M (2019) Sixty years of global progress in managed aquifer recharge. Hydrogeol J 27(1):1–30

Dubash N (2000) Ecologically and socially embedded exchange: "Gujarat model" of water markets. Econ Pol Wkly 35:1376–1385

El-naggar A (2015) A survey of the quality of water drawn from nine domestic. Int J Res 2:618–630

Foster S, Ait-Kadi M (2012) Integrated water resources management (IWRM): how does groundwater fit in? Hydrogeol J 20:415–418

Foster S, Garduño H (2013) Groundwater resource governance in the global context. Hydrogeol J 21(5):949–963

Foster SSD, Perry CJ (2010) Improving groundwater resource accounting in irrigated areas: a prerequisite for promoting sustainable use. Hydrogeol J 18(2):291–294

Giordano M (2009) Global groundwater? Issues and solutions. Annu Rev Environ Resour 34(1):153–178

Hazarika N, Nitivattananon V (2016) Strategic assessment of groundwater resource exploitation using DPSIR framework in Guwahati city, India. Habitat Int 51:79–89

Holman I, Trawick P (2011) Developing adaptive capacity within groundwater abstraction management systems. J Environ Manage 92(6):1542–1549

Jeet I (2015) Political ecology of groundwater depletion in Northwestern India. In: Dutt A, Noble A, Costa F, Thakur S, Thakur R, Sharma H (eds) Spatial diversity and dynamics in resources and urban development. Springer, Dordrecht

Jha CK, Gupta V, Chattopadhyay U, Amarayil Sreeraman B (2018) Migration as adaptation strategy to cope with climate change. Int J Clim Chang Strat Manag 10(1):121–141

Kolvalli S, Chicoine D (1989) Groundwater markets in Gujarat, India. Int J Water Resour Dev 5:38–44

Kooy M, Walter C, Prabaharyaka I (2016) Inclusive development of urban water services in Jakarta: the role of groundwater. Habitat Int 73:109–118

Kulkarni H, Shah M, Shankar P (2015) Shaping the contours of groundwater governance in India. J Hydrol: Regnal Stud 4:172–192

Martínez-Navarrete C, Jiménez-Madrid A, Sánchez-Navarro I, Carrasco-Cantos F, Moreno-Merino L (2011) Conceptual framework for protecting groundwater quality. Int J Water Resour Dev 27:227–243

Matsumoto S, Rohde M, Heard S (2019) Policy note: economic tools to achieve groundwater sustainability for nature: two experimental case studies from California. Water Economics and Policy

Mechlem K (2012) Groundwater governance: a global framework for country action-thematic paper 6: legal and institutional frameworks (June 1, 2012). GEF: Groundwater Governance: A Global Framework for Country Action, GEF ID 3726.
Meinzen-Dick RS (1998) Groundwater markets in Pakistan: institutional development and productivity impacts. In: Easter KW, Rosegrant MW, Dinar A (eds) Markets for water. Natural Resource Management and Policy, vol 15. Springer, Boston, MA (1998).
Molle F, Closas A (2020) Comanagement of groundwater: a review. Wires Water 7(1):e1394
Mukherji A (2007) The energy-irrigation nexus and its impact on groundwater markets in eastern Indo-Gangetic basin: evidence from West Bengal, India. Energy Policy 35(12):6413–6430
Raffensperger J, Milke M, Read E (2009) A deterministic smart market model for groundwater. Oper Res 57:1333–1346
Razzaq A, Qing P, Naseer M, Abid M, Anwar M, Javed I (2019) Can the informal groundwater markets improve water use efficiency and equity? evidence from a semi-arid region of Pakistan. Sci Total Environ 666:849–857
Razzaq A, Xiao M, Zhou Y, Liu H, Abbas A, Liang W, Naseer MAUR (2022a) Impact of participation in groundwater market on farmland, income, and water access: evidence from Pakistan. Water 14(12):1832
Razzaq A, Liu H, Zhou Y, Xiao M, Qing P (2022b) The competitiveness, bargaining power, and contract choice in agricultural water markets in Pakistan: implications for price discrimination and environmental sustainability. Front Environ Sci 10
Ross A (2012) Easy to say, hard to do: integrated surface water and groundwater management in the Murray-Darling Basin. Water Policy 14:709–724
Rouillard J (2020) Tracing the impact of agricultural policies on irrigation water demand and groundwater extraction in France. In: Rinaudo JD, Holley C, Barnett S, Montginoul M (eds) Sustainable groundwater management, global issues in water policy, vol 24. Springer, Cham
Schwartz FW, Liu G, Yu Z (2020) Hess opinions: the myth of groundwater sustainability in Asia. Hydrol Earth Syst Sci 24(1):489–500
Sears L, Lim D, Lawell C-YCL (2018) The economics of agricultural groundwater management institutions: the case of California. Water Economics and Policy 04(03):1850003
Shah T (2009) Taming the anarchy: groundwater governance in South Asia. Resources for the Future; Colombo, Washington, DC, USA. International Water Management Institute (IWMI) Sri Lanka, p 310
Sharma P, Sharma R (2006) Factors determining farmers' decision for buying irrigation water: study of groundwater markets in Rajasthan. Agric Econ Res Rev 19:39–56
Skurray J, Pandit R, Pannell D (2013) Institutional impediments to groundwater trading: the case of the Gnangara groundwater system of Western Australia. J Environ Planning Manage 56:1046–1072
Tamuli J, Dutta M (2016) Factors influencing reliability of groundwater markets in less water scarce regions: a case of Assam in Eastern India. Rev Dev Chang 21:66–92
Wheeler SA, Schoengold K, Bjornlund H (2016) Lessons to be learned from groundwater trading in Australia and the United States. In: Jakeman AJ, Barreteau O, Hunt RJ, Rinaudo JD, Ross A (eds) Integrated groundwater management. Springer, Cham (2016)
Yao Y, Zheng C, Andrews C, He X, Zhang A, Liu J (2019) Integration of groundwater into China's south-north water transfer strategy. Sci Total Environ 658:550–557
Zhang L, Wang J, Huang J, Rozelle S (2008) Development of groundwater markets in China: a glimpse into progress to date. World Dev 36:706–726

Chapter 12
The State of Energy in Sub-Saharan Africa and the Role of International Cooperation for Small Hydropower Development

Young Cui, Daniel Adu, Abdul-Nazif Mahmud, and Emmanuel Baffour Gyau

Abstract Africa is one of the continents globally with a large energy potential to be developed, particularly small hydropower resources. Hitherto, it is the region with low access to energy. This has crippled the region economically, socially, and slowed down its development. This study provides an overview of the state of hydropower potential in sub-Saharan Africa, particularly small hydropower, and the needs for its development to overcome the energy deficit in the region. The study also discusses an alternative for small hydropower development using pumps as turbines. Through the study, it has been revealed that the huge available rivers in the region carry the potential for small hydropower development using a pump as a turbine or pump-turbine, which is cheaper with low installation costs. It is, therefore, required by various governments and energy sector leaders to put appropriate measures and policies in place for its development. The development of small hydropower tends to be one of the likely solutions to overcome high energy demand for rural communities within sub-Saharan Africa. The study concludes on the importance of designing small hydropower plants for sub-Saharan Africa to achieve the UN sustainable development goal by 2030. This study provides useful information to overcome the energy crisis in Africa, particularly sub-Saharan Africa, where access to energy is a bigger challenge.

Keywords Energy generation · Sub-Saharan Africa · Sustainable development goal · International cooperation · Hydro potentials

Y. Cui · A.-N. Mahmud
Overseas Education College, Jiangsu University, Zhenjiang 212013, China

D. Adu (✉)
School of Management Science, Jiangsu University, Zhenjiang 212013, China
e-mail: adudaniel39@yahoo.com

School of Energy, University of Energy and Natural Resources, Sunyani, Ghana

E. B. Gyau
School of Finance and Economics, Jiangsu University, Zhenjiang 212013, China

H. Xu (ed.), *The 6th International Symposium on Water Resource and Environmental Management*, Environmental Science and Engineering,
https://doi.org/10.1007/978-3-031-55989-1_12

12.1 Introduction

There is a potential and substantial demand for the establishment of small-scale hydroelectric power stations in nations across sub-Saharan Africa (SSA). The purpose of this research is to aid in formulating a strategy for the advancement of small hydropower in rural regions of SSA. This report also discusses the potentials, challenges, and possible solutions. The economic landscape of the region is notably underdeveloped, marked by a scarcity of local industries and severely inadequate infrastructure, particularly in terms of poorly maintained roads. Numerous parts of the country lack access to electricity, with some regions relying on outdated, large hydropower facilities supplemented by costly gas and diesel power generation. A deficiency in energy is increasingly becoming a significant economic and social constraint in SSA. Small hydropower development potential in the region is very great, but most of these sites are still lying there undeveloped. African nations within the sub-Saharan region, fostering the growth of their emerging businesses and industries, are faced with difficulties in managing their electricity predicament. Electricity demand is increasing daily at a faster rate, whereas the generation has not increased at the same rate. This is attributed to population growth. Research has shown that several billions of people live in the region without access to clean and modern energy. This has contributed greatly to a lack of employment and an increase in mortality rate, particularly in mothers and children, as they depend on other means such as coal, firewood, and charcoal for cooking, which has harmful effects on them. Although there is high potential for hydropower resources, particularly small hydropower, there is a significant gap between electricity demand and supply in the Region. Figure 12.1 shows the aggregate capacity of small hydropower compared to the capacity implemented within the area. Although several rivers and potentials exist for small hydropower development in SSA, the development has not been easy due to challenges faced by governments and the energy sector ministries of the various countries in the region. These challenges include small hydropower nonexistence of adequate infrastructure, absence of technological know-how, and lack of funding. Several governments in SSA require support from advanced and technologically developed regions in planning and designing ways for improving small hydropower use and operation within the region to provide clean and modern energy to the people to decrease the high energy demand situation in the region, predominantly those in the rural communities. It is imperative to provide training for individuals in the SSA region who lack expertise in emerging technologies, particularly in the domain of small hydropower development. Comprehensive initiatives encompassing planning, construction, operation, and maintenance of power systems, along with dedicated funding for research projects, should be carefully orchestrated to prevent mismanagement.

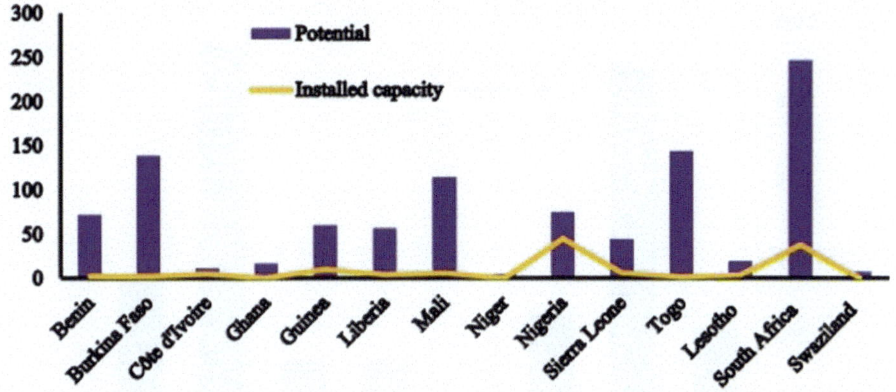

Fig. 12.1 Total small hydropower potential compared to the installed capacity in Africa (Liu et al. 2013)

12.2 Small Hydropower

Small hydropower (SHP) systems represent renewable energy technology designed for power generation. These systems are categorized into Pico, micro, mini, and small systems based on their installed capacity. There is no universally accepted definition for small hydropower, but a general distinction between 'large' and 'small' hydropower is based on installations with a capacity of up to 10 MW (IRENA 2015). Many countries in SSA and Europe adhere to the 10 MW limit when defining small hydropower, while Mozambique uses a 25 MW limit. SHP systems can be either reservoir or run-of-river types. Run-of-river types lack water storage for power generation; instead, water is diverted from the main river using a weir. Run-of-river systems are particularly suitable for small SHP capacities as they reduce the investment cost per kW of installation. Unlike other energy generation technologies like diesel generators, the maintenance cost of SHP plants is comparatively small, providing a long-term and reliable solution. A well-installed SHP plant can have a lifespan of 50 years or more without significant new investments in parts and replacements (although the average considered for investment purposes is around 30 years) (International Hydropower Association country profile statistics and database, 2021). Despite its significant potential, SHP technology is not widely adopted in the SSA region. A projection estimates a small hydropower potential of 21,800 MW from plants ranging from 1–10 MW, identified across 5383 sites in sub-Saharan Africa. The central part of the region, with contributions from South Africa, the DRC, and Sudan, holds about one-third of the identified potential (Duarte et al. 2010; Zhang et al. 2018). Figure 12.1 illustrates the total small hydropower potential compared to the installed capacity in sub-Saharan Africa.

Depending on the scale of hydropower potential, construction becomes more manageable and can be completed within a span of four to six years. Many experts attribute the challenges faced by large hydropower projects in Africa to construction

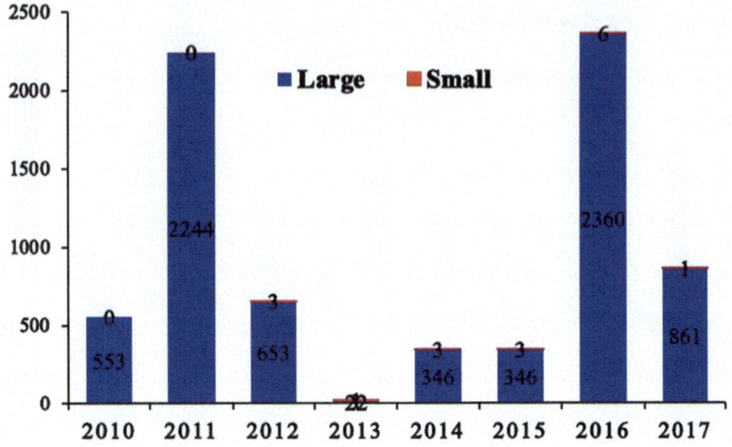

Fig. 12.2 Comparisons of small and large hydropower in selected Sub-Saharan African countries

delays, leading to setbacks in the hydropower industry on the continent. The Mozambique government has placed a strong emphasis on the sector, aiming to secure approximately $3 billion for the construction of the Mphanda Nkuwa Hydroelectric plant. This project is poised to transform Mozambique's social and economic landscape while addressing power deficits in the Southern African Development Community, boasting a potential annual generation of about 8,600 GW. The Mphanda Nkuwa project holds considerable promise, and while large-scale hydropower projects are being emphasized, the construction of small hydropower grid technologies proves highly suitable for providing electricity to rural communities in the region (Jonker 2012; IRENA 2015). Figure 12.2 presents a comparative analysis of small and large hydropower generation in selected SSA countries.

12.2.1 Literature on the State of Energy in Africa and the Need to Develop Available Potentials

Reliable electricity supply serves as a key driver for economic growth, particularly in SSA, where sluggish growth and insufficiencies prevail due to power shortages. It plays a pivotal role in expanding power generation capacity, mitigating energy deficits, and fostering economic development. Augmenting power generation substantially benefits the manufacturing and service sectors, contributing to an enhancement in living standards. Consistent electricity also positively impacts productivity at both individual and commercial levels. On an individual level, it facilitates better healthcare, access to quality education, and time savings through the use of appliances. Commercially, it enhances business capabilities and reduces operational costs. Contrary to the common belief that China's involvement in Africa is primarily centered around the resources sector, reports indicate that infrastructure projects constitute less than 10% of China's engagement in the region (Min Conf Water for Agriculture and Energy in Africa, 2008).

SSA witnessed a 45% growth in energy demand from 2000 to 2012, with Nigeria and South Africa leading as the largest energy demand centers, accounting for over 40% of the total demand. Despite hosting 13% of the global population, the region contributes only 4% to global energy demand. Although there has been significant economic growth in the past 15 years, energy growth has not kept pace with rising demand, straining existing resources. The growth of energy production, particularly in the power sector, has lagged, posing a significant challenge to the economic growth of the continent (International Hydropower Association, 2016). Access to electricity remains a critical issue, hindering economic growth in the region. Despite the abundant potential for sustainable energy sources, there is a persistent power supply shortage, affecting over 635 million people, mainly in rural areas. The electrification rate is less than one-third of the population, and although progress has been made, the number of people without electricity continues to rise due to rapid population growth. The EIA New Policies Scenario Report, aligned with the World Energy Outlook, predicts that the demand for electricity in Sub-Saharan Africa will triple by 2040, surpassing 1,300 TWh, based on existing and proposed government policies and measures. Addressing the energy deficit and improving infrastructure are crucial for sustained economic development in the region (EIA New Policies Scenario Report).

12.2.2 Overview of SSA's Small Hydropower Situation

The Democratic Republic of the Congo possesses a significant portion of Africa's potential resources. The Congo River, notably the Grand Inga site, stands as the continent's largest and most potent location, connecting to the Atlantic Ocean and extending to ten other nations within 150 km from its origin in the DRC. The Inga Falls site marks the highest elevation point of the river. Presently, it hosts 40 hydropower projects, nine of which are situated in the DRC, including the substantial Inga I (354 MW) and Inga II (1,424 MW). However, these projects are not operating at full capacity due to insufficient maintenance, leading to unstable production. Once fully realized, Grand Inga will become the world's largest hydropower project, boasting 40 GW capacity—twice that of China's Three Gorges Dam. This mega project has the potential to provide cost-effective energy solutions for sub-Saharan Africa, meeting 40% of the continent's electricity demand. Additionally, the Zambezi River serves as a crucial source of hydropower, and Zambia occupies a substantial portion of the river basin—41%, surpassing the other seven riparian countries. Zambia has achieved a hydropower installed capacity of 2,257 MW, representing a remarkable 94% of the country's total energy mix, with the upgraded Lunzua station (14.8 MW) contributing significantly in 2014. Despite this achievement, Zambia faces a growing demand for electricity, escalating at a rate of 150 to 200 MW annually. The country grapples with the challenge of harnessing over 6,000 MW of untapped hydropower resources. Table 12.1 provides an overview of installed capacity and hydrogenation in 2015 for selected SSA countries. Figure 12.3 illustrates the renewable potentials in the region by source, with hydropower consistently leading, offering both high potential and cost-effectiveness to address the region's increasing energy demands.

A study was conducted to investigate the position of power supply and the effects of inadequate production and distribution on society in SSA, as shown in Table 12.1. The study reveals that the population without access to energy will increase to about 645 million in 2030 (IEA 2022). Small hydropower was therefore recommended to be adopted to overcome this energy crisis (Zhang, et al. 2018; Korkovelos et al. 2018). In contrast, another researcher carried out a technical assessment research on small hydropower, considering the head ranges between 0.01–10 MW in SSA (IEA 2019a). The cost of installing hydropower plants increased from USD 1,232 per kW in 2010 to USD 1,492 per kW in 2018 (IEA 2019b).

Table 12.1 Hydropower installed capacity and generation in selected sub-Saharan African countries (Williams and Tabakov 2019)

Country	Area	Population	Installed hydro. Capacity (MW)	Hydropower generation (GWh)	Year
Ethiopia	1,000,000 sq km	96,959,000	2,552	9,000	2015
Guinea	245,836 km^2	12,276,000	368	500	2015
Mozambique	801,590 km^2	27,216,000	2,187	12,000	2015
Mexico	1,943,50sqkm	125,386,000	12,435	30,127	2015

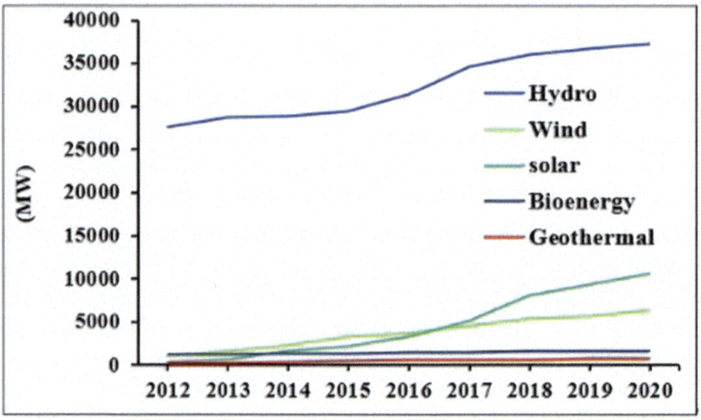

Fig. 12.3 Potential of available hydro production capacity in Sub-Saharan Africa

12.3 The Necessity for Developing Small Hydro Potential in SSA

A continuous and reliable source of electricity is crucial for fostering economic growth, particularly in SSA, where sluggish and insufficient development is attributed to power shortages. Using a pump as a turbine (PAT) will help to increase power generation capacity by reducing general energy shortages to achieve economic growth. The increase in power generation has become a favorable factor in both the manufacturing and service industries, in addition to increasing the living standards of the residents. Consistent electricity supply also has a positive influence on productivity, quality healthcare, and access to quality education. SSA's energy demand increased by 45% between 2000 and 2012, with Nigeria and South Africa as the most abundant energy demand centers, constituting over 40% of the overall energy consumption. Despite hosting 13% of the global population, the region only contributes to 4% of the global demand for modern energy services. Even though there is an upward trend, access to modern energy services in the Region remains notably restricted (Janneh, 2011). SSA GDP in the past 15 years has doubled. There was significantly faster growth just before the growth rate surged from an average of 5% per year during the late 1990s to an average of 5% per year between 2008 and 2015. However, this growth decelerated to 3.5% in 2015, with an expected increase to 4% in 2016 (NEPAD et al. 2015; World Energy Outlook 2014). The growth of energy is not increasing compared with the higher demand in developing regions, putting a lot of pressure on the old existing resources in the regions. There was an increase in the gross domestic product (GDP) between 2001 and 2005 in SSA, which grew by more than half (4.5%) of the various countries' annual growth rate, whereas power generation grew at 1.2% (IMF 2015). Inadequate electricity access is contributing significantly to the decline of economic growth in the region. Even though the region has great potential and much more productive sustainable energy sources, there is still a problem of a severe deficit in electricity provision. Over 635 million individuals within the region lack access to electricity, which constitutes less than one-third of the population's electrification rate (IMF 2016). Despite improvements in overall access to electricity in SSA, the population without electricity continues to increase due to rapid population growth. Nonetheless, recent enhancements in electricity infrastructure and increased investments in the power sector have expanded access to electricity for a larger portion of the region's population (International Hydropower Association country profile statistics and database, 2017; Foster and Briceno-Garmendia 2010). According to the EIA New Policies Scenario Report cited in the World Energy Outlook, the demand for electricity in SSA is anticipated to triple by 2040, surpassing 1,300 TWh, considering current and proposed government policies and initiatives (IEA (International Energy Agency 2015). Figure 12.4 illustrates the renewable energy potential categorized by source in SSA.

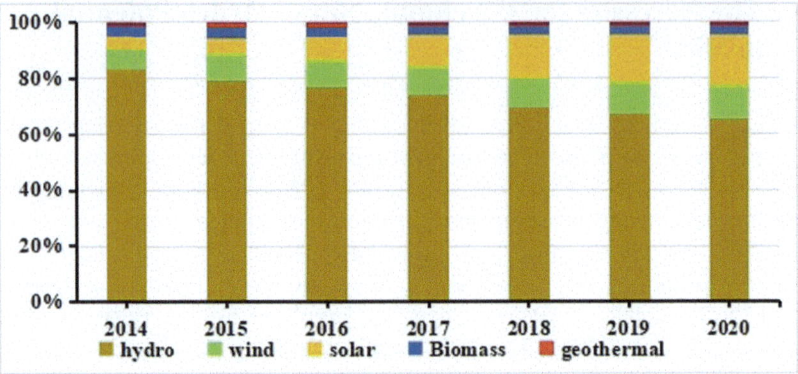

Fig. 12.4 Renewable energy potentials in sub-Saharan Africa by energy source

12.4 Conclusion

Energy remains one of the major pillars of the development of a country. However, energy access remains a bigger challenge in SSA, which has slowed down development in the region. This study discusses the state of energy in sub-Saharan Africa and suggests solutions for developing the huge available small hydropower potentials in the region, addressing the pressing energy demand issue within the area. The essence of developing these potentials, such as the provision of portable and accessible healthcare, education, the creation of jobs, and aiding in economic development, has also been discussed. The increase in power generation has become a favorable factor in the production and service industries, in addition to improving the living standards of the residents. Consistent electricity supply also has a positive influence on productivity. Thus, for individuals, the use of better-quality healthcare, and access to quality education have all been discussed. The study yields significant insights into strategies essential for harnessing untapped small hydropower sites in SSA. Embracing the utilization of pumps operating in reverse mode as turbines for small hydropower generation emerges as a pivotal approach to mitigate the prevailing energy crisis and foster the economic development of sub-Saharan Africa. This research contributes valuable knowledge to the limited information available on small hydropower technology in the region.

Acknowledgements The authors express their gratitude to the General Program of The Chinese Society of Academic Degrees and Graduate Education Research Fund for providing financial support under grant number 2020MSA350. Additionally, appreciation is extended to The Chinese Society of Educational Development Strategy Research Fund, acknowledging grant number SRB202131, and the Major Program of The Chinese Society of Academic Degrees and Graduate Education Research Fund, with grant numbers 2020ZAC11.

References

Adu D, Zhang J, et al (2018) The prospects & submissions for small hydropower development in sub-Saharan Africa. IOP Conf Ser: EarthEnviron Sci 163:012022

Duarte M, Nagarajan S, Brixiova Z (2010) Financing of sustainable energy solutions. Published by AFBD, Ivory Coast, AfBD Committee of Ten Policy Brief

Foster V, Briceno-Garmendia C (eds) (2010) Africa's infrastructure: a time for transformation. Agency Françoise De-Development, Washington DC

IEA (International Energy Agency (2015) World energy outlook. OECD/IEA, Paris

IEA, Africa energy outlook 2019. Paris. https://www.iea.org/reports/africa-energy-outlook

IEA, World Energy Outlook 2019 Electricity Access Database. www.worldenergyoutlook.org. Accessed 12 February 2021

IEA, World Energy Outlook. http://www.worldenergyoutlook.org/resources/energydevelopment/energyaccessdatabase/2017OECD/IEA. Accessed 10 March 2022

IMF (2015) Regional Economic Outlook: Sub-Saharan Africa, IMF. www.imf.org/external/pubs/ft/reo/2015/afr/eng/pdf/sreo0415.pdf

IMF (International Monetary Fund) (2016) World economic outlook update

International Hydropower Association country profile statistics and database. https://www.Hydropower.org/country-profiles. Accessed 10 March 2017

International Hydropower Association country profile statistics and database. https://www.hydropower.org/country-profiles. Accessed 10 March 2021

IRENA (2015) Renewable capacity statistics. United Arab Emirates, Abu Dhabi

Janneh A (2011) UN under-secretary-general of ECA, Speech

Jonker Klunne W (2012) Current status and future developments of small and micro hydro in southern Africa

Korkovelos A, Mentis D, Siyal SH, Christopher A, Holger R, Morgan B, Mark H, Hylke B (2018) A geospatial assessment of small-scale hydropower potential in Sub-Saharan Africa. Energies 11:3100

Liu H, Masera D, Esser L (ed) (2013) World small hydropower development report United Nations industrial development organization. Int Cent Small Hydro Power (2013)

NEPAD (New Partnership for Africa's Development) (2015) Infrastructure development within the context of Africa's co-operation with new and emerging development partners. NEPAD, Johannesburg, www.nepad.org/fr/Infrastructure-Development (2015).

Williams SE, Tabakov PY (2019) Power supply and the role hydropower plays in Sub-Saharan Africa's modern energy system and socioeconomic wellbeing. Int J Energy Econ Policy 9:347–363

World Energy Outlook 2014 Factsheet Energy in sub-Saharan Africa today

Chapter 13
Research and Application of the Value Accounting System for Aquatic Ecosystem Products

Nengjun Mao, Hongjie Yu, Hongxian Li, Xinxin Li, and Kaiwei Chen

Abstract The Aquatic ecosystem is an essential component of terrestrial ecosystems. It not only provide freshwater resources, fish products, hydropower, and other basic materials and energy sources for humans but also offer services such as flood control, climate regulation, and improving the quality of life for surrounding residents. Assessing the contributions of aquatic ecosystems to human well-being has been a focal point in international aquatic ecology research. Based on domestic and international research findings, this article enhances the inventory of aquatic ecosystem products, proposes a flood storage value accounting method based on flood control volume, and constructs a comprehensive accounting system for aquatic ecosystem product values. Furthermore, a case study was conducted, showing that the total value of aquatic ecosystem products in Jiande City in 2021 amounted to 10.691 billion RMB, accounting for 24.83% of the annual GDP of the region. The aquatic ecosystem plays a significant role in supporting the socioeconomic development of the city.

Keywords Aquatic ecosystem products · Value accounting system · Aquatic product inventory · Flood defense value · Aquatic value conversion

Scientifically and reasonably evaluating the value contained in green mountains and clear waters is a key foundation for establishing an ecological product value realization mechanism. In 2022, the China Development and Reform Commission and the Bureau of Statistics jointly issued the "Accounting Standard for Total Value of Ecological Products (Trial)" (Development and Reform Basic (2022) No. 481) Barrett (2000) which provides specific ecological product value accounting methods for forests, grasslands, farmlands, wetlands, deserts, cities, and more Fu (2009). The aquatic ecosystem is vital components of terrestrial ecosystems. It not only provide

N. Mao (✉) · H. Yu · H. Li · X. Li · K. Chen
Zhejiang Water Resources and Hydropower Survey and Design Institute Co., Ltd., Hangzhou 310002, Zhejiang, China
e-mail: 351299309@qq.com

H. Xu (ed.), *The 6th International Symposium on Water Resource and Environmental Management*, Environmental Science and Engineering,
https://doi.org/10.1007/978-3-031-55989-1_13

freshwater resources, fish products, hydropower, and other basic materials and energy sources for humans but also offer services like flood control, climate regulation, and improving the quality of life for surrounding residents Gao (2019). Assessing the contributions of aquatic ecosystems to human well-being has been a focal point in international aquatic ecology research (Hu 2022; Huang 2000; Lin 2023). Due to the diversity of aquatic ecosystem services, unclear beneficiaries, and inconsistent conceptual understandings, the scientific and standardized accounting of aquatic ecosystem product values remains a challenge (Liu 2013; National development and reform commission 2022; Ouyang 1999).

The accounting of aquatic ecosystem values holds significant importance for water resources management, the advancement of major water projects, and the transformation of aquatic ecosystem product values Shu (2016). Building upon the latest domestic and international research findings, this article further refines the inventory of aquatic ecosystem products, sum up value accounting methods for products like flood control, and carries out case studies in Jiande city.

13.1 The Aquatic Ecosystem Value Accounting System

13.1.1 Compilation of Aquatic Ecosystem Product Lists

Ecological products are defined as the final products or services provided for human well-being through biological production and their joint action with human labor, while maintaining ecological security and ecological regulatory functions Wang (2013). According to current national and Zhejiang land standards, aquatic ecosystem products can be classified into three major categories: supply products, regulatory services, and, cultural services (Table 13.1).

(1) Supply Products: These are materials and resources obtained from natural aquatic ecosystems, such as freshwater, fisheries products, hydroelectric power, shipping, and unconventional products like sand and gravel, without damaging the stability and integrity of the natural ecosystems.
(2) Regulatory Services: These are benefits provided by natural ecosystems to improve human living conditions and the environment, including flood regulation, water purification, carbon sequestration, oxygen production, and climate regulation.
(3) Cultural Services: These are non-material benefits provided by natural ecosystems and their coexisting historical and cultural heritage to support knowledge acquisition, leisure, and entertainment for humans. Aquatic ecosystems not only serve as attractions for water-related activities and aquatic tourism but also provide recreational areas for local residents, contributing significantly to leisure and entertainment.

Table 13.1 Aquatic ecosystem product list

Categories	Subjects	Ecological products
Supply products	Direct utilization supply products	Water resources
		Fishery products
		Shipping
		Sand and gravel
	Converted utilization supply products	Hydropower
Regulation services	Runoff regulation	Flood Defense
	Environmental climate regulation	Temperature regulation
		Humidity regulation
		Water environment purification
		Carbon sequestration
		Oxygen release
Cultural services	Tourism and recreation	Ecotourism
		Leisure and entertainment

13.1.2 Valuation of Supply Products

Water Resource Products. The supply of water is mainly categorized into three types: domestic, industrial, and agricultural. Water quantity is used as the functional quantity indicator for water resource products, and it is priced using the market value method (Zeng 2014; Zhong 2005). For high-quality development, premium water (high-quality industrial water), the market water price is used as the basis, combined with factors like economic scale and water brand empowerment index to determine the price.

$$V_{szy} = \sum_{i=1}^{n} W_i \times P_i \times \alpha_i \tag{13.1}$$

where: V_{szy} is the total value of water resources products, W_i is the supply volume, p_i is the reference price, and α_i is the premium index.

Fisheries and Hydropower Products. For fishery products, output is used as the functional quantity index of fishery products, and the market value method is used to determine the price of fishery products. The value of fishery products is calculated by multiplying the output of various fishery products by their market prices. Fishery products cultured in artificial fish ponds are not included in this accounting. Hydropower products use the quantity of electricity produced as the functional indicator of hydropower, use the market value method to determine electricity prices, and calculate the value of hydropower by multiplying power generation by the electricity price.

Shipping and Sand-Gravel Products. Taking shipping volume as the functional indicator of shipping, the market value method is used to determine the shipping price, and the shipping value is calculated by multiplying the shipping volume by the shipping price. The sand mining volume is used as the functional indicator of sand and gravel, the market value method is used to determine the sand and gravel price, and the sand and gravel value is calculated by multiplying the sand mining volume by the sand and gravel price.

13.1.3 Valuation of Regulatory Services

Flood Defense. About flood defense, the most commonly used method at present, is calculated based on water retention in lakes and reservoirs and is priced using market prices Barrett (2000). Using the replacement cost method, According to the "Technical Specifications for Accounting Gross Ecosystem Product (GEP) Terrestrial Ecosystems", the cost of unit storage capacity in 2015 are calculated to be 25.89 RMB/(m^3·a).

Water Environment Purification. Water environment purification is based on the purification quantity of pollutants, and it is priced using the replacement cost method Barrett (2000). Using the replacement cost method, According to the "Technical Specifications for Accounting Gross Ecosystem Product (GEP) Terrestrial Ecosystems", the cost of purification for COD is 8000 RMB/t, for ammonia nitrogen is 9573 RMB/t, and for total phosphorus is 10000 RMB/t.

Temperature and Humidity Regulation. Temperature regulation is based on the energy absorbed through water evaporation and transpiration from vegetation. It is priced using the replacement cost method Barrett (2000). Humidity regulation is calculated based on the water volume evaporated and transpired. The price is determined using the replacement cost method, taking the energy consumed by humidifiers to convert water into steam as the reference value.

Carbon Sequestration and Oxygen Release Regulation. Carbon sequestration is based on the quantity of carbon dioxide fixed, and it is priced using the market value method, and its value is 23.72 RMB/t recommended in the Zhejiang land area specification Barrett (2000). Oxygen release is based on the quantity of oxygen released and is priced using the replacement cost method, and its value is 1,200 RMB/t recommended in the Zhejiang land area specification Barrett (2000).

13.1.4 Valuation of Cultural Services

Ecotourism. The value of ecotourism is calculated using the travel cost method, taking the number of visitors to waterfront areas as the functional quantity and multiplying it by the travel cost per person.

Leisure and Entertainment. The value of leisure and entertainment is calculated based on the number of residents visiting waterfront spaces for leisure and entertainment. It uses the travel cost method, considering factors such as travel expenses and time value.

13.2 Jiande's Water Ecosystem Value Accounting in 2021

13.2.1 Overview

Jiande City is located in the southwest of Hangzhou. In 2021, the registered population of the city reached 507,510, with the total GDP reaching 43.06 billion RMB. Water is the soul of Jiande. According to the latest water area survey results, the total water area in the city is 87.02 square kilometers. The city is blessed with two national-level large reservoirs, Xin'an River and Fuchun River. The water in the Xin'an River meets the national primary drinking water standard, maintaining a temperature of 14–17 °C throughout the year. Additionally, the water from Xin'an River actively supports the healthy and thriving development of Jiande's water industry, such as the brand enhancement for Nongfu Spring, Zhonghe, and Tsingtao Beer (Fig 13.1).

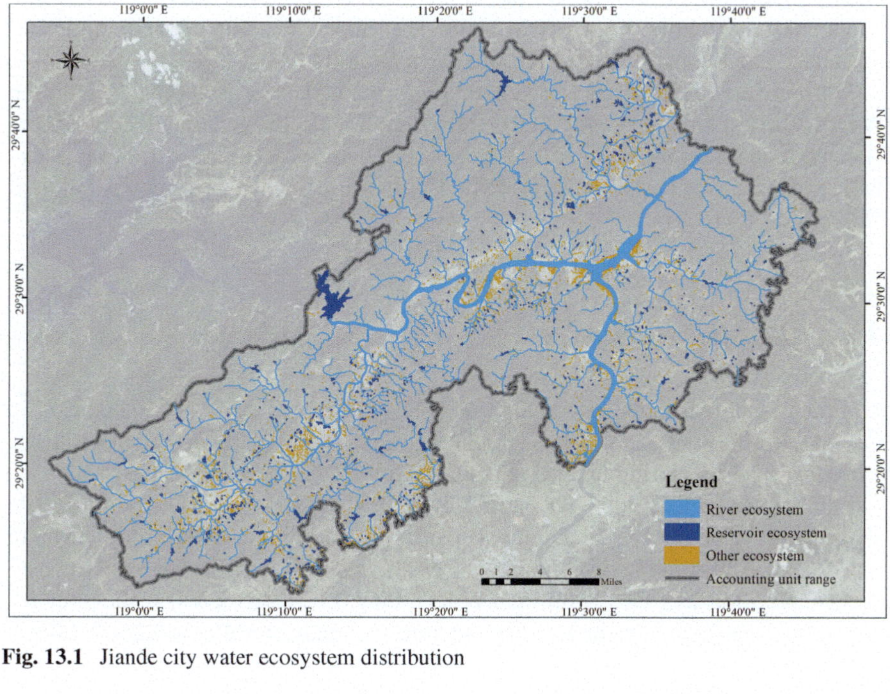

Fig. 13.1 Jiande city water ecosystem distribution

13.2.2 Supply Products

Water resources: In 2021, the domestic water consumption in Jiande City was 49 million cubic meters, agricultural water consumption was 105 million cubic meters, and general industrial water consumption was 30 million cubic meters. The high-end industrial water consumption, such as Nongfu Spring, was approximately 10 million cubic meters. The prices were set at 2.2 RMB/m3 for domestic use, 3.8 RMB/m3 for general industrial use, and 0.5 RMB/m3 for agricultural use. According to the research on water brand enhancement index, the premium index for high-end industrial water is 40. The total value of water resources in Jiande's water ecosystem in 2021 was 1.794 billion RMB.

Fishery products: Excluding the artificial pond breeding, the annual fishery product output in Jiande City in 2021 was 3026 tons, with a value of approximately 45 million RMB.

Navigation: Passenger transport value is calculated within the scope of ecological tourism. Here, only the value of cargo transportation is considered. In 2021, there were a total of 6 million tons of cargo transportation in Jiande, including four cargo terminals such as Shili Bu. The average value of cargo transportation was 6 RMB/ton, resulting in a navigation value of 36 million RMB.

Sand and gravel: In 2021, the total legal excavation of sand and gravel in Jiande City was 1,255,588 tons, including 25,146 tons of fine sand and 1,230,442 tons of pebbles. According to the market price, the price of fine sand was 150 RMB/ton, and the price of pebbles was 58 RMB/ton. Therefore, the value of sand and gravel in Jiande's water ecosystem in 2021 was 75 million RMB.

Hydropower: The hydropower generation in Jiande City was about 1.855 billion kW·h in 2021, of which the Xin'an River hydropower station generated 1.824 billion kW·h, and the total power generation from small hydropower was 31 million kW·h. The electricity price was set at 0.54 RMB/kW·h, resulting in a hydropower value of 1.002 billion RMB.

13.2.3 Regulatory Services

According to statistics, the effective flood control storage capacity of Jiande City in 2021 44.71 million m^3, the cost of unit storage capacity in 2021 are calculated to be 38 RMB /(m^3·a), resulting in a flood control value of 1.699 billion RMB.

Temperature regulation: With a water area of 81.87 square kilometers in Jiande City, the average evaporation depth was 2.17 mm, and the number of days with an average temperature equal to or higher than 26 °C was 107 days. The heat absorbed by water evaporation was 11.934 billion kW·h. Taking 0.18 RMB/kW·h as the price for temperature regulation products, the temperature regulation value was 2.148 billion RMB.

Humidity regulation: The number of days with a daily average humidity equal to or less than 60% was 28 days, so the water volume evaporated in terms of humidity regulation products was 4.9744 million cubic meters. The price for humidity regulation products was set at 64.8 RMB/m3. Therefore, the humidity regulation value in Jiande's water ecosystem in 2021 was 0.322 billion RMB.

Water environment purification: In 2021, the total COD emissions from all sewage treatment plants in Jiande was 239.35 tons, and ammonia nitrogen emissions were 4.14 tons. The price for COD purification was set at 8,000 RMB/ton, and the price for ammonia nitrogen purification was set at 9,573 RMB/ton. Thus, the water environment purification value in Jiande's water ecosystem in 2021 was 20 million RMB.

Carbon sequestration and oxygen release: Using the MODIS NPP dataset, the NEP value was calculated, taking into account the total NPP within the accounting area. The direct carbon fixation by vegetation (including corresponding soil) was 7,600 tons. Jiande's hydropower generation was about 1.855 billion kW·h in 2021, resulting in a reduction of 1,595,300 tons of carbon dioxide emissions. The price for carbon sequestration was based on the market carbon trading price of 23.72 RMB/ton, resulting in a carbon sequestration value of 4 million RMB. In 2021, Jiande's water ecosystem released 5,527 tons of oxygen. The price for oxygen release was set at 1,200 RMB/ton, resulting in an oxygen release value of 7 million RMB.

13.2.4 Cultural Services

Ecotourism: Major waterside scenic areas in Jiande City include Dreamy Xin'an River and Lingsi Cave, among others, which received a total of 1.38 million visitors in 2021. The cost of travel per person was approximately 1,847 RMB. Therefore, the value of ecotourism in Jiande's water ecosystem in 2021 was 2.549 billion RMB.

Leisure and entertainment: In 2021, the permanent residents in Jiande City numbered 444,000, most of whom were within a 15-min drive of waterside areas. Considering that residents engage in leisure and entertainment activities in waterside areas once a week, with an average duration of 2.42 h each time, the time value was determined based on the 2021 Zhejiang minimum hourly wage rate (18 RMB). The value of leisure and entertainment in Jiande's water ecosystem in 2021 was 1.008 billion RMB.

13.3 Summary

The total value of Jiande City's water ecosystem products in 2021 was 10.691 billion RMB. Among these, regulatory services had the highest value at 4.182 billion RMB, accounting for 39.11%, while supply products had a value of 2.952 billion RMB, accounting for 27.61%, and cultural services had a value of 3.557 billion

Table 13.2 2021 Jiande city water ecosystem product value accounting results

Categories	Subjects	Ecological products	Worth (100 millions)
Supply products	Direct use of supply products	Water resources	17.94
		Fishery products	0.45
		Shipping	0.36
		Sand and gravel	0.75
	Conversion and utilization of supply products	Hydropower	10.02
Regulation services	Runoff regulation	Flood defense	16.99
	Environmental climate regulation	Temperature regulation	21.48
		Humidity regulation	3.22
		Water environment purification	0.02
		Carbon sequestration & release oxygen	0.11
Cultural services	Tourism and leisure	Eco-tourism	25.49
		Leisure and entertainment	10.08
Total			106.91

RMB, accounting for 33.27%. In 2021, Jiande City's regional GDP was 43.06 billion RMB, and the value of water ecosystem products represented 24.83% of the GDP. Water is the driving force for Jiande's high-quality development, and the water ecosystem plays a significant role in supporting the city's socio-economic development (Table 13.2).

13.4 Conclusion and Outlook

This paper, summed up a list of water ecosystem products and existing value accounting methods for water products, and conducted a case study using Hangzhou's Jiande City as an example. Water is the driving force for Jiande's high-quality development, and the water ecosystem plays a significant role in supporting the city's socio-economic development.

Knowing the value of water ecosystem products is only the first step. The ultimate goal is to promote the realization of the value of water ecosystem products and effectively incentivize the construction of ecological civilization.

For future research on water ecosystem GEP and its applications, the following suggestions are made:

1. Emphasize the improvement of the value increment evaluation system for water resource projects before and after the construction of water projects. This promotes the exploration of the multidimensional value of water projects, which is conducive to addressing issues such as investment and financing difficulties for water projects.
2. Further research and improve the standardized accounting process for water ecosystem GEP, focusing on specific research areas such as data automation and standardization of statistical results.

References

Barrett J (2000) Component ecological footprint: developing sustainable scenarios. Impact Assess Proj Apprais, 19(2) pp. 107–118

Fu BJ, Zhou GY, Bai YF (2009) Main terrestrial ecosystem service functions and ecological security in China. Adv Earth Sci, (6) 6. (In Chinese)

Gao MX (2019) Connotation, accounting framework and implementation conditions of ecosystem gross product: Design and demonstration from statistical perspective. Acta Ecol Sin, 40(2). (In Chinese)

Hu XY (2022) Gross ecosystem product calculation in the new era: review on the theory and method of gross ecosystem product calculation. Wetl Sci Manag, 18(1)1. (In Chinese)

Huang YJ, Chen XJ (2000) Analysis of the current situation and trend of the demand management of water resources in our country. Adv Water Sci 11(2):215–220 (In Chinese)

Lin YQ, Chen M, Zhang JY (2023) Calculation system and application of gross product of water ecosystem. Water Resour Prot, 02 (13)1–13. (In Chinese)

Liu YL (2013) China's first Gross ecosystem product (GEP) evaluation project starts. For Man, (3)1. (In Chinese)

National development and reform commission, national bureau of statistics. Standard for Accounting the Total Value of Ecological Products (Trial): Basis of Development and Reform [2022] No. 481 (2022). (In Chinese)

Ouyang ZY, Wang XK, Miao H (1999) A preliminary study on the service function and eco-economic value of terrestrial ecosystem in China. Acta Ecol Sin 19(5):7 (In Chinese)

Shu QK, Gao YN, Liu YZ, Wang Y, Bao GY (2016) Calculating ecological value of land use in the coastal areas of jiangsu province. J Geo-Inf Sci, 18(6)

Wang XH (2013) Connotation of ecological products, necessity and measures of production. Glob Mark Inf Rev, 000(012)50, 99. (In Chinese)

Zeng HF (2004) Study on forest ecological value Accounting—Forest ecological value accounting and forestry sustainable development. Probl For Econ

Zhong QL, Xue CL (2005) Chinese water resource price study (Two articles)—Chinese water price control study. Gansu Soc Sci, (3) 4. (In Chinese)

Chapter 14
Degradation of Enrofloxacin Antibiotic by Ultraviolet Activated Persulfate Advanced Oxidation Process: Inorganic Anions Effects and Mechanisms

Juxiang Chen, Kai Wang, Wanying Xie, Yuqiong Gao, Zhenliang Liao, and Naiyun Gao

Abstract Enrofloxacin (ENR) is continually detected in natural water environment, becoming one threaten if treating untimely. Performance and mechanism of ENR degradation by ultraviolet (UV)/persulfate (PDS) technology was systematically investigated. Degradation efficiency by UV/PDS was 5.4 times of UV alone, due to the higher rate constant (k_{obs}) of 5.848 × 10^{-2} min^{-1}. Quencher experiments showed, $\cdot SO_4^-$ and ·OH were the major free radicals generated by UV/PDS, and the corresponding contribution rate on degrading ENR was 65.2 and 17.2%. Results of electrical energy per order (EE/O) demonstrated, UV/PDS was less energy consumption and more economical. Moreover, when high concentration of Cl^- (10 mM) or low concentration of NO_3^- (1 mM) co-existed with ENR, the degradation capability by UV/PDS process was slightly promoted, while the negatively effect of CO_3^{2-} was obvious. Using density functional theory (DFT) and Fukui functions to analyze the possible active sites of ENR, and the results illustrated that C_{28} and C_{30} on piperazine ring of ENR was attacked by UV/PDS process, according to the effect of substitution and de-alkylation.

Keywords ENR antibiotic · UV/PDS process · Degradation efficiency · Sulfate free radicals · Degradation pathway

---

J. Chen · K. Wang · W. Xie (✉)
College of Civil Engineering and Architecture, Xinjiang University, Urumqi, Xinjiang 830017, China
e-mail: xwy@xju.edu.cn

Y. Gao
School of Environment and Architecture, Shanghai University of Technology, Shanghai 200093, China

Z. Liao · N. Gao
State Key Laboratory of Pollution Control and Resources Reuse, College of Environmental Science and Engineering, Tongji University, Shanghai 200092, China

H. Xu (ed.), *The 6th International Symposium on Water Resource and Environmental Management*, Environmental Science and Engineering,
https://doi.org/10.1007/978-3-031-55989-1_14

14.1 Introduction

Fluoroquinolone antibiotics (FQs) as one broad-spectrum antibiotic, have the characteristics of tiny side-effect, applied in many antibacterial fields as a result (He et al. 2016). The FQs mainly include ENR, ciprofloxacin (CIP), norfloxacin (NOR), ofloxacin, and so forth (Van et al. 2014). As one typical FQs, ENR is widely used in oral medicine and intravenous preparations (Xu et al. 2016), due to its highly effective tissue permeability. Besides, ENR has spectral activity against a large number bacterium of pathogenic gram-negative and gram-positive, and is also used for veterinary treatment of animals and livestock infections in animal husbandry (Guinea et al. 2010). ENR is hard to be completely transformed in mammals but easily discharged into water, meanwhile, overuse of ENR leads to the frequently detected (Bound and Voulvoulis 2006). In water environment, the toxic effects of residual ENR on vertebrates and invertebrates consist of central nervous system excitation, convulsions, reproductive dysfunction and gastrointestinal disorders (Anderson 1992). And, if residual ENR enters into human body, food poisoning and allergic reaction probably occurs which seriously threaten human health (Zou et al. 2011). Moreover, the residual antibiotics also directly affect the growth of bacterial communities and enzyme activities, and thus affect the entire ecosystem (Pallecchi et al. 2008). Residual antibiotics also possibly induce the development of quinolone-resistant pathogens (Janecko et al. 2016) posing potential menace to humans and other organisms. In China, ENR antibiotic has been frequently detected in water environment, and the observed concentration up to 0.50–4.24 μg/L (Wei et al. 2012). At the same time, serious concern has been expressed for the ENR migration into water environment to form residual antibiotic (Wang et al. 2019), which presenting severe health risks and water environmental issues and should be solved urgently.

In recent years, the treatment process of ENR antibiotic has aroused widespread attention, like oxidants or advanced oxidation processes (AOPs) which containing UV (Xiang et al. 2022), PDS (Jiang et al. 2016), potassium permanganate ($KMnO_4$) (Xu et al. 2016), chlorine (Dodd et al. 2005), ozones (Li et al. 2013), UV-activated PDS (UV/PDS process) (Ou et al. 2017), thermal activated persulfate, and so forth. If only using UV, PDS, or $KMnO_4$, the C/C_0 of ENR nearly around 90% after 90 (Xiang et al. 2022), 200 (Jiang et al. 2016), and 120 min (Xu et al. 2016), illustrating that degradation efficiency was not satisfying. Chlorine (Dodd et al. 2005) and chlorine dioxide (Wang et al. 2010) showed fast degradation rate of ENR but meanwhile with some by-products, and the high costs of ozone might be one obstacle to degrade ENR in actual application (Guinea et al. 2009). Among them, UV/PDS process exhibited some advantages during antibiotics degradation, like high efficiency, no secondary pollution and gentle reaction conditions (Guo et al. 2017). As one sulfate radical based AOPs, UV/PDS process is one promising method to guarantee degradation efficiency of treating antibiotics, as compared to activation by heat (Guo et al. 2016), transition metal ions or oxides (Tan et al. 2014; Fang et al. 2013). When using UV/PDS process to degrade ENR, the degradation paths have not been deeply understood. Some study (Guo et al. 2017) used the results of degradation products to illustrate that, the possible dominant path of ENR degradation was hydrogenation, dehydrogenation and ring cleavage reaction of alpha-C atom on the ethyl of piperazine group, as compared to the cracked C–F bond of ENR and ring-opening reaction and decarboxylation occurred on piperazine ring of ENR.

However, the effects of some co-exist ions on ENR degradation mechanisms are not well considered yet, which is not conducive to the development of ENR treatment technology (UV/PDS process), some anions or cations inevitably exist in natural water environment after all. Additionally, there also has been few articles deeply studying the ENR degradation paths by UV/PDS process, from the point view of combining structural site attack and intermediate products. In this study, selecting ENR as one typical FQs antibiotic, degradation performance by UV/PDS process was systematically optimized, and effect of some typical anions on degradation efficiency was analyzed. The treating performance of UV/PDS process was evaluated by degradation kinetics of ENR, depending on various parameters, such as, PDS dosage, ENR concentration, pH, inorganic anions (Cl^-, HCO_3^-, SO_4^{2-}, NO_3^-). In order to better analyze the relationship between anions and degradation efficiency of ENR, the result of RSM model was exhibited. Moreover, the possible degradation mechanism of ENR was analyzed by structural site attack, depending on the condensed Fukui function and electron density. In addition, the main products and degradation pathways of ENR by UV/PDS process was identified by HPLC-MS-MS.

14.2 Materials and Methods

14.2.1 Materials

Chemical agents used in this study were ENR ($C_{19}H_{22}FN_3O_3$), PDS ($Na_2S_2O_8$), sodium chloride (NaCl), sodium bicarbonate ($NaHCO_3$), potassium dihydrogen phosphate (KH_2PO_4), sodium thiosulfate ($Na_2S_2O_3$), disodium hydrogen phosphate (Na_2HPO_4), sodium dihydrogen phosphate (NaH_2PO_4), sodium sulphate (Na_2SO_4), tertbutyl alcohol ($C_4H_{10}O$), Sodium nitrate ($NaNO_3$), p-benzoquinone ($C_6H_4O_2$), sodium tetraborate decahydrate ($Na_2B_4O_7 \cdot 10H_2O$), sodium carbonate ($NaHCO_3$), methanol, acetonitrile, and formic acid. All reagents are of analytical grade or higher quality.

Using Purelab Flex system (Elga®Labwater, United Kingdom, resistivity >18.2 Ω m) to obtain ultrapure water and prepare reaction solutions. The stock solution of ENR and PDS were prepared by dissolving $C_{19}H_{22}FN_3O_3$ and $Na_2S_2O_8$ in ultrapure water, respectively, and then stored in amber glass bottles at 4 °C.

14.2.2 Experimental Procedures

The experimental procedures were described as following. Put ENR and PDS into the crystallizing dish, and then place it under the preheated UV lamp. After reaction of 30 min, took out the sample (1 mL) and then was quenched immediately by excessive $Na_2S_2O_3$. The pretreated sample was analyzed with calibration curve of ENR, which kept the experimental error was less than 5%. In addition, the reaction time was selected as 30 min, relying on the consideration of economic-feasibility of UV/PDS process. And, the highest PDS dosage was set at 20 mg/L, depending on the detection limit of liquid chromatography.

According to previous study (Buxton et al. 1988; Neta et al. 1979), the second-order reaction rates of methanol (MeOH) with •OH and $SO_4^{\cdot-}$ were $9.7 \times 10^8\ M^{-1}\ s^{-1}$ and $1.1 \times 10^7\ M^{-1}\ s^{-1}$ (Buxton et al. 1988), respectively. The second-order reaction rates of tert-butanol (TBA) with •OH and $SO_4^{\cdot-}$ were $6.0 \times 10^8\ M^{-1}\ s^{-1}$ and $4.0 \times 10^5\ M^{-1}\ s^{-1}$, respectively. Thus, MeOH was considered as an effective scavenger for •OH and $SO_4^{\cdot-}$, while TBA was regarded as an effective scavenger for •OH (Neta et al. 1979; Xie et al. 2015). Both of them cannot absorb 254 nm UV (Xiao et al. 2016). Besides, p-benzoquinone (BQ) was recognized as an effective scavenger for $O_2^{\cdot-}$. Therefore, the type of free radicals produced in the reaction system was judged by the inhibition of MeOH, TBA and BQ on the degradation of the UV/PDS process.

In the influence factor experiments, the parameter setting was given as following. Set PDS dosage at 1/5/10/20 mg/L, and ENR concentration at 1/5/10/15 mg/L, and solution pH at 3/5/7/9/11. To avoid the effect of some anions (e.g. CO_3^{2-}/HCO_3^-) on the oxidation process by sulfate radicals (Jiang et al. 2017; Song et al. 2021), all experiments were carried out in the buffer-free solution which mixing with organic compound ENR (25 ± 1°C). During experiments of inorganic anions, set Cl^- concentration at 0/1/5/10 mM, HCO_3^- and SO_4^{2-} both at 0/1/3/5 mM, NO_3^- at 0/1/5/10 mM. Besides, to adjust the ultrapure water added with different concentrations of ENR to the required pH (3/5/7/9/11) by adding NaOH (0.1 M) and H_2SO_4 (0.1 M).

14.2.3 Analytical Method

ENR concentration was determined by the ultra-high performance liquid chromatography with C18 column (250 nm × 4.6 nm × 5 μm, Waters), and the specific condition was described as following. The UV wavelength was 276 nm, the mobile phase was 20% acetonitrile and 80% water (0.1% formic acid), the flow rate was 0.15 mL/min, the column temperature was 30 °C, the injection volume was 10 μL, and detection time was 3 min.

The mass spectrometry conditions were given in detail. Using the electrospray ionization (ESI) ion source at positive ion mode, the mass scanning range of M/Z was 100–500. Fragments of the main products were dissociated by collision induction, and the collision energy varied by the products. The argon flow rate was 0.15 mL/

min, and pressure of collision chamber was 5×10^{-3} mbar, cycle time was 0.5 s, residence time was 10 min. Injection volume was 250 uL, injection flow rate was 5 uL/min, capillary voltage was 3500 V, sample cone voltage was 43 V, the source temperature was 200 °C, and the dissolvent gas flow rate was 650 L/h.

14.2.4 Theoretical Calculation

Energy consumption analysis. During ENR degradation, using electrical energy per order (EE/O) to evaluate the energy consumption by UV/PDS process. The EE/O was recommended by the International Union of Pure and Applied Chemistry (Bolton et al. 1996). And, EE/O means the energy consumed when the concentration of the target pollutant was reduced by one logarithmic concentration per cubic meter of water solution. The value of EE/O consists of electric energy cost and oxidant cost, which was showed in Eqs. 14.1–14.3 (Tian et al. 2020).

$$\mathrm{EE/O} = \mathrm{EE/O_{UV}} + \mathrm{EE/O_{oxidant}} \tag{14.1}$$

$$\mathrm{EE/O_{UV}} = (P \times t \times 1000)/(V \times \log(C_0/C_t)) \tag{14.2}$$

$$\mathrm{EE/O_{oxidant}} = \mathrm{Eq_{oxidant}} \times ([\text{oxidant}] \times 1000)/(\log(C_0/C_t)) \tag{14.3}$$

where, V and P meant volumes of the oxidation samples (L) and input power of the applied UV irradiation (kW), respectively; t represented the photoreaction time (h); C_0 and C_t were the initial and instantaneous concentration of ENR (mg/L), respectively; [oxidant] referred to the oxidant concentration (mol/L); $EqNa_2S_2O_8$ stood for the equivalent electrical energy (kW) converted to form 1 mol PDS ($EqNa_2S_2O_8 =$ 4.4124 kW hmol^{-1}). Depending to the current market situation, the average electricity cost was 0.416 yuan/kW h, and the average price of $Na_2S_2O_8$ was 1200 yuan/ton.

Calculation of first-order reaction kinetics. In all experiments, the degradation of ENR was fitted with a pseudo-first-order kinetic model, and the corresponding calculation of K_{obs} was in the Eq. 14.4. Where, t was the reaction time; C_0 and C_t indicated the concentration of reactant at time 0 and t, the concentration unit of ENR or PDS was mg L^{-1}, the concentration unit of inorganic ions was M; Kobs was the first order reaction rate constant, min^{-1}.

$$\ln\left(\frac{C_0}{C_t}\right) = \mathrm{Kobs} \cdot \mathrm{t} \tag{14.4}$$

Recognition of reactive sites. The B3LYP method and 6–311 g** basis-set were used to perform density functional theory (DFT) calculations. Firstly, through the optimized geometric structure to find the most stable configuration of ENR molecule. And then, using the optimized structure to calculate the single point energy, to obtain the HOMO and LUMO of ENR. The HOMO–LUMO energy gap value from the energy difference between HOMO and LUMO tracks (ΔE) was calculated according to the previous study (Xu et al. 2021). All DFT calculations were performed by Gauss 09 software. The ENR degradation mechanism of electrophilic attack, nucleophilic attack and free radical attack was explained by the condensed Fukui function, which based on the change of electron density. The calculation of Fukui index was shown as Eqs. 14.5–14.7 (Oláh et al. 2002).

$$f^{+} = Q_k(N+1) - Q_k(N) \tag{14.5}$$

$$f^{-} = Q_k(N) - Q_k(N-1) \tag{14.6}$$

$$f^{0} = [Q_k(N+1) - Q_k(N-1)]/2 \tag{14.7}$$

where, Q_k represented the electronic population of atoms in the molecule given by the density integral in the atomic region; N was the number of electrons; K represented an atom in the molecule; f^{+} value indicated the degree of vulnerability to electrophilic attack; f^{-} value indicated the degree of susceptibility to nucleophilic attack; f^{0} value signified the activity of the atom and the ease of being attacked by free radicals (Morrison 2018).

14.3 Results and Discussion

14.3.1 Performance of ENR Degradation by UV/PDS Process

The ENR degradation performance by using UV or PDS alone, and UV/PDS process was showed in Fig. 14.1a. It showed that the ENR degradation rate (equals to the decreased C_t/C_0 was only 7.4%, illustrating that ENR wasn't effectively degraded by PDS alone in this study which consisted with conclusion of other studies (Li et al. 2021; Liu et al. 2022; Ma et al. 2021). PDS exists in the form of $S_2O_8^{2-}$ in water and was stable at room temperature, leading to the modest reaction rate with ENR, even though its redox potential ($E_0 = 2.01$ V) (Kolthoff and Stenger 1947) is similar to ozone ($E_0 = 2.07$ V) (Li et al. 2013). And, when using UV irradiation alone, the decreased C_t/C_0 of ENR was also as low as 15.2%. For UV irradiation process, the degradation was caused by that the wavelength of UV light ($\lambda = 254$ nm) within the range of ENR absorbable wavelength ($\lambda_{max} = 266$ nm) (Babić et al. 2013). Some photolysis of partial ENR degradation might be realized by UV process, because free radical like $\cdot OH$ (Han et al. 2019) might oxidize ENR directly or indirectly. At the same time, the structural stability of ENR (Michael et al. 2013) resulted in the limited degradation rate (15.2%), which was consistent with the previous study (Dhaka et al. 2017). Compared to PDS or UV alone, the degradation rate of ENR was obviously improved to 82.2% by UV/PDS process, owing to the free radicals might be generated by this advanced oxidation process meanwhile showed the highest K_{obs} (Fig. 14.1b). As can be seen in Fig. 14.1b, the UV/PDS process without adding any radical scavengers was marked as control group, corresponding to the highest K_{obs} of 5.848×10^{-2} min^{-1}. After adding MeOH of 10 and 100 mM, the K_{obs} of UV/PDS process sharply decreased to 1.735×10^{-2} and 1.030×10^{-2} min^{-1}, respectively. The decreased K_{obs} of 70.3 and 82.4% illustrated that both $SO_4^{\cdot-}$ and $\cdot OH$ which generated by UV/PDS process were evidently inhibited by the radical scavenger of MeOH. When compared to control group, the K_{obs} slightly decreased to 5.471×10^{-2} and 4.843×10^{-2} min^{-1} after adding TBA of 10 and 100 mM. This indicated that the degradation effect of ENR by $\cdot SO_4^-$ was more obvious during the UV/PDS process when comparing to $\cdot OH$, due to the decrease degree of K_{obs} by TBA was less obvious than that by MeOH. In addition, the K_{obs} of UV/PDS process remained close to the control group after adding BQ, demonstrating that degradation effect of ENR by $\cdot O_2^-$ was weaker than $\cdot OH$ and $\cdot SO_4^-$ in this study, due to BQ majorly scavenged $\cdot O_2^-$ (Fouad et al. 2020).

Relying on the calculation by Eqs. 14.9–14.10 of MeOH and TBA (100 mM), contributions of $\cdot SO_4^-$ ($R_{\cdot SO_4^-}$) and $\cdot OH$ ($R_{\cdot OH}$) during the UV/PDS process to the ENR degradation were estimated to be 65.2% and 17.2%, respectively. This deduced that $SO_4^{\cdot-}$ and $\cdot OH$ was the primary and subordinate free radical during ENR degradation, which was consistent with the K_{obs} results in Fig. 14.1b. Some study (Zhang and Huang 2005) pointed that, the activation of PDS by UV irradiation was able to generate $SO_4^{\cdot-}$ which had a high oxidation–reduction potential ($E_0 = 2.60$ V) (Eq. 14.10). In addition, the $\cdot OH$ was simultaneously generated due to the reaction of $SO_4^{\cdot-}$ in aqueous solution (Eqs. 14.11–14.12) (Moussa and Serge 2014; Li et al.

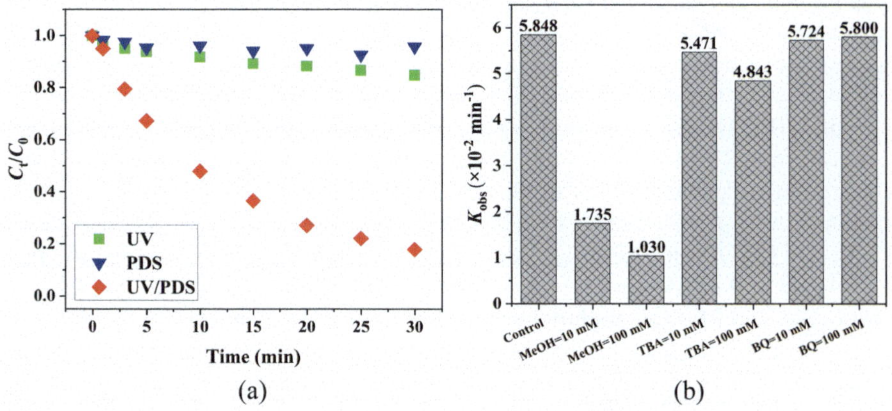

Fig. 14.1 Degradation of ENR by UV or PDS alone, and UV/PDS process (**a**); impacts of radical trapping agents on K_{obs} by UV/PDS process (**b**). Condition: [ENR] = 1 mg/L, [PDS] = 20 mg/L, [MeOH] = 10 or 100 mM, [TBA] = 10 or 100 mM, [BQ] = 10 or 100 mM

2016) to remove organic pollutants (Han et al. 2019). In this study, it was considerable that both $SO_4^{\bullet-}$ and •OH were responsible for the ENR degradation, due to the strong oxidizing properties of those two free radicals led to the destruction of ENR.

$$R_{\cdot OH} = (1 - (K_{obs}, TBA/K_{obs})) \times 100 \quad (14.8)$$

$$R_{\cdot SO_4^-} = (1 - (K_{obs}, MeOH/K_{obs})) - (1 - (K_{obs}, TBA/K_{obs})) \times 100 \quad (14.9)$$

$$S_2O_8^{2-} + UV \rightarrow 2SO_4^{\bullet-} \quad (14.10)$$

$$SO_4^{\bullet-} + H_2O \rightarrow \cdot OH + H^+ + SO_4^{2-} \quad (14.11)$$

$$SO_4^{\bullet-} + OH^- \rightarrow \cdot OH + SO_4^{2-} \quad (14.12)$$

14.3.2 Energy Consumption Analysis

To evaluate the applicable feasibility of UV/PDS process on ENR degradation, the energy consumption was exhibited in Table 14.1. When using PDS or UV alone to degrade ENR, the value of EE/O was 18.24 and 11.73 kW h m^{-3} $order^{-1}$, respectively, obviously higher than that of UV/PDS process. This indicated that the energy consumption of ENR degradation by UV/PDS process was smaller, indicating the corresponding characteristic of economical friendliness. When initial ENR concentration raised from 1 to 15 mg/L, both EE/O_{UV} and $EE/O_{oxidant}$ increased, the energy consumption ratio of UV ((EE/O_{UV})/(EE/O)) and PDS (($EE/O_{oxidant}$)/(EE/O)) maintained at 67% and 33%, respectively. This implied that UV irradiation of UV/PDS process was the major contributor to energy consumption, when ENR concentration increased. In addition, when PDS concentration raised from 1 to 10 mg/L, the (EE/O_{UV})/(EE/O) decreased from 98 to 80%, meanwhile the ($EE/O_{oxidant}$)/(EE/O) increased from 2 to 20%. At the same time, the EE/O presented the decrease tendency. These findings denoted that UV irradiation was still a major consumer of energy, even though $EE/O_{oxidant}$ increased at higher PDS dosage. Moreover, when pH raised from 3 to 9, both EE/O_{UV} and $EE/O_{oxidant}$ decreased, meanwhile the EE/O sharply dropped from 10.39 to 0.82 kW h m^{-3} $order^{-1}$. When pH further increased to 11, EE/O_{UV}, $EE/O_{oxidant}$, and EE/O all slightly increased. In the pH range of 3–11, the (EE/O_{UV})/(EE/O) and ($EE/O_{oxidant}$)/(EE/O) were still be 67% and 33%, respectively. It was difficult to promote UV/PDS process in multiple processing scenarios, due to the high operating cost (Chu et al. 2016).

Based on the above analysis, when the ENR concentration was 1 mg/L, cost of the degradation performance was decided by UV irradiation which acting as the major energy consumer, rather than PDS. And, two strategies like increasing PDS dosage

Table 14.1 Energy consumption calculation of UV/PDS process degradation of ENR under various conditions

$C_{(ENR)}$	$C_{(PDS)}$	UV intensity	pH	EE/O_{UV}	$EE/O_{oxidant}$	EE/O
(mg/L)		(mW cm^{-2})	/	(kW h m^{-3} $order^{-1}$)		
1	20	0	7	0.00	18.24	18.24
1	0	2.35	7	11.73	0	11.73
1	20	2.35	7	0.98	0.49	1.47
5	20	2.35	7	1.35	0.67	2.02
10	20	2.35	7	1.94	0.96	2.90
15	20	2.35	7	2.59	1.28	3.87
1	1	2.35	7	3.52	0.09	3.61
1	5	2.35	7	2.30	0.29	2.59
1	10	2.35	7	1.41	0.35	1.76
1	20	2.35	3	6.95	3.45	10.39
1	20	2.35	5	2.80	1.39	4.19
1	20	2.35	9	0.55	0.27	0.82
1	20	2.35	11	0.81	0.40	1.21

or adjusting pH to 9 were proposed to bring down the energy consumption, which helped to improve the practicability of the UV/PDS process.

14.3.3 *Effects of Inorganic Anions on ENR Degradation Efficiency*

Inorganic anions generally exist in surface water and groundwater (Jutarvutikul et al. 2021), it was necessary to estimate the influence of typical inorganic ions on the ENR degradation by UV/PDS process, which was analyzed in this section.

Effects of four inorganic anions (Cl^-, HCO_3^-, SO_4^{2-}, NO_3^-) on ENR degradation by UV/PDS process were shown in Fig. 14.2. As Fig. 14.2a exhibited, when Cl^- concentration increased from 0 to 5 mM, the ENR degradation rate decreased from 82.2 to 75.0%. And, when Cl^- further increased to 10 mM, the decreased C_t/C_0 was slightly promoted to 86.0%. The Cl^- might react with $\cdot SO_4^-$ to generate ·Cl and even $\cdot Cl_2^-$ (Eqs. 14.13–14.14), then were further oxidized to produce Cl_2 (Eqs. 14.15–14.16) (Xiao et al. 2016), due to they had relative lower E_0 ($E_0(\cdot Cl/\cdot Cl_2^-) = 2.41$ V, $E_0(Cl_2/Cl^-) = 1.36$ V) than that of $\cdot SO_4^-$ ($E_0(\cdot SO_4^-/SO_4^{2-}) = 2.6$ V). When Cl^- at low or medium concentration, the generated ·Cl and even $\cdot Cl_2^-$ possibly snatched $\cdot SO_4^-$ from ENR, resulting in the lower degradation rate. When at high concentration, the Cl^- might transform to HClO (Eq. 14.17) which acting as oxidant (Yang et al. 2014) to further participate the ENR degradation.

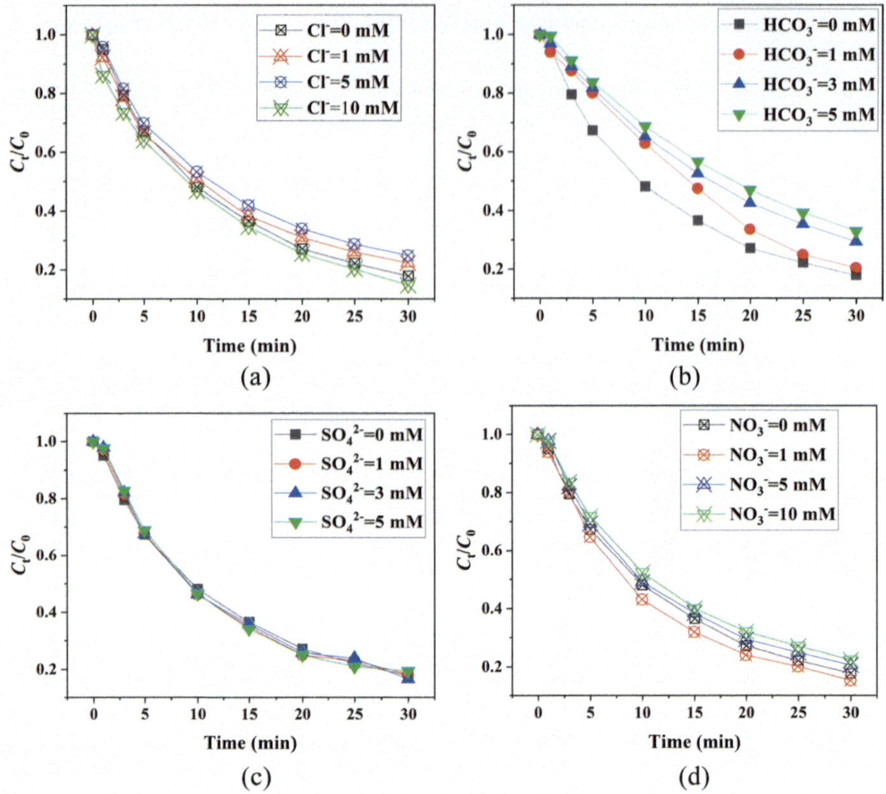

Fig. 14.2 Effect of Cl^- (**a**), HCO_3^- (**b**), SO_4^{2-} (**c**) and NO_3^- (**d**) on ENR degradation by UV/PDS process. Condition: pH = 7, [PDS] = 20 mg/L, [ENR] = 1 mg/L

$$\cdot SO_4^- + Cl^- \leftrightarrow SO_4^{2-} + \cdot Cl \tag{14.13}$$

$$\cdot Cl + Cl^- \rightarrow \cdot Cl_2^- \tag{14.14}$$

$$\cdot Cl_2^- + \cdot Cl_2^- \rightarrow Cl_2 + 2Cl^- \tag{14.15}$$

$$\cdot Cl_2^- + \cdot Cl \rightarrow Cl_2 + Cl^- \tag{14.16}$$

$$Cl_2 + H_2O \rightarrow Cl^- + HClO + H^+ \tag{14.17}$$

Effect of HCO_3^- concentration on ENR degradation by UV/PDS process was exhibited in Fig. 14.2b. As HCO_3^- increased from 0 to 5 mM, the ENR degradation ate obviously decreased from 82.15 to 66.85%, due to HCO_3^- or CO_3^{2-} (Eq. 14.18) were commonly recognized as the scavengers for free radicals like $\cdot SO_4^-$ and $\cdot OH$.

When HCO_3^- or CO_3^{2-} was abundant, they could be reacted with $\cdot OH$ (Eqs. 14.19–14.20) or $\cdot SO_4^-$ (Eqs. 14.21–14.22) to generate $\cdot CO_3^-$ (Yang et al. 2014; Ji et al. 2015). The HCO_3^- was easily to react with $\cdot SO_4^-$ and $\cdot OH$, owning to the high K_{obs} of 9.1 $\times 10^6$ and 8.5 $\times 10^6$ M^{-1} s^{-1}. However, $\cdot CO_3^-$ had selectivity on the degradation of organic pollutants, which was not beneficial to guarantee the ENR degradation rate.

$$HCO_3^- \rightarrow H^+ + CO_3^{2-} \tag{14.18}$$

$$\cdot OH + CO_3^{2-} \rightarrow \cdot CO_3^- + OH^- \tag{14.19}$$

$$\cdot OH + HCO_3^- \rightarrow \cdot CO_3^- + H_2O \tag{14.20}$$

$$HCO_3^- + \cdot SO_4^- \rightarrow SO_4^{2-} + H^+ + \cdot CO_3^- \tag{14.21}$$

$$CO_3^{2-} + \cdot SO_4^- \rightarrow SO_4^{2-} + \cdot CO_3^- \tag{14.22}$$

Effect of SO_4^{2-} on ENR degradation rate by UV/PDS process was described in Fig. 14.2c. When SO_4^{2-} increased from 0 to 5 mM, the degradation rate of ENR maintained around 82%, illustrating the influence was negligible. After adding SO_4^{2-}, the reaction between SO_4^{2-} and $\cdot OH$ occurred slowly (Eq. 14.23) (Ji et al. 2015), due to the K_{obs} was only 3.5 $\times 10^5$ M^{-1} s^{-1}. At the same time, SO_4^{2-} barely reacted with $\cdot SO_4^-$. These illustrated the negligible influence of SO_4^{2-} on the concentration of $\cdot OH$ and $\cdot SO_4^-$, meanwhile the ENR degradation rate.

$$\cdot OH + SO_4^{2-} \rightarrow OH^- + \cdot SO_4^- \tag{14.23}$$

Effect of NO_3^- on ENR degradation was drawn in Fig. 14.2d. After adding 1 mM NO_3^-, the degradation rate of ENR slightly increased from 82.2 to 84.5%. When NO_3^- increased to 5 and 10 mM, the degradation rate decreased to 79.3% and 77.5%, respectively. Under UV irradiation, the NO_3^- might generate $\cdot O^-$ and NO_2^- (Eq. 14.24), the $\cdot O^-$ then reacted with H2O to produce $\cdot OH$ (Eq. 14.25). Moreover, the $\cdot SO_4^-$ in UV/PDS process might react with NO_3^- to form $\cdot NO_3^-$ (Eq. 14.26). When at low NO_3^- concentration, both $\cdot OH$ and $\cdot NO_3^-$ ($E_0 = 2.30$ V) were responsible for the ENR degradation. When NO_3^- concentration increased to 5–10 mM, NO_3^- possibly consumed $\cdot SO_4^-$ to decrease ENR degradation, which was consistent with the decreased K_{obs}.

$$NO_3^- + UV \rightarrow \cdot O^- + NO_2^- \tag{14.24}$$

$$H_2O + \cdot O^- \rightarrow OH^- + \cdot OH \tag{14.25}$$

$$NO_3^- + \cdot SO_4^- \rightarrow \cdot NO_3^- + SO_4^{2-} \tag{14.26}$$

In brief, inorganic ions like Cl^-, HCO_3^- and NO_3^-, indeed affected the ENR degradation rate by UV/PDS process, while the effect of SO_4^{2-} was negligible. When high concentration of Cl^- (10 mM) or low concentration of NO_3^- (1 mM) co-existed with ENR, the degradation capability by UV/PDS process was slightly promoted, owning to some intermediates showing strong oxidizing property were produced, like Cl_2, HClO, or $\cdot NO_3^-$.

14.3.4 Degradation Mechanism

Before analyzing the specific intermediates, the optimized molecular structure and the reactive sites of ENR was analyzed briefly in this section, by the results of HOMO and LUMO distribution and the DFT calculation. After high-precision single-point energy calculation, the ENR structure was optimized to be stable and have the lowest energy (Fig. 14.3a). Based on this optimized structure of ENR, the highest occupied molecular orbital and the lowest unoccupied molecular orbital was created and displayed in Fig. 14.3b, c, corresponding to the HOMO and LUMO structure. As HOMO structure showed, only the piperazine ring of ENR was occupied by a bigger electron cloud. This speculated that the piperazine ring of ENR was more prone to lose electrons, and more susceptible attacked by free radicals generated by UV/PDS process, such as $\cdot OH$ and $\cdot SO_4^-$. And, this speculation was consistent with other study (Jiang et al. 2016). Except for piperazine ring, another two rings of ENR molecular were filled by a bigger electronic clouds in the LUMO structure, such as, the ring including C_1–C_6, or including C_3, C_4, C_8, C_{11}, C_{10} and C_{14}. This was benefit to improve their propensity to take electrons. Besides, the electron affinity energy was related to the LUMO energy level of the molecular (Zhan et al. 2003), the lower orbital energy of LUMO structure made it simpler for the molecular to obtain electrons (Asghar et al. 2017). In addition, the HOMO–LUMO energy gap, ΔE, was another important quantum chemical parameter which favors to predict the chemical reactivity and polarizability of chemical species participating in chemical reactions (Asghar et al. 2017). The ΔE was calculated as high as 4.24 eV, indicating that the molecular structure of ENR was stable.

Furthermore, the condensed Fukui function including f^+, f^- and f^0, was calculated and illustrated in Table 14.2. Specifically, using f^+ to explain the difficulty of nucleophilic attack, and using f^- to explain the difficulty of electrophilic attack, as well as using f^0 to explain the difficulty of free radical attack (Morrison 2018). The higher value of Fukui function, the higher reactivity of the corresponding site (Liu et al. 2021). The f^+ value of C_{10}, C_5, C_2, C_6, C_8 and C_3 was higher than 0.058, demonstrating these sites of ENR molecule were more easily to be attacked by nucleophilic reactions. The f^- value of O_{13}, N_{41} and N_{40}, was higher than 0.094, reflecting these sites of ENR molecule were most facilely to be attacked by electrophilic reactions. Meanwhile, the sites of O_{24} and C_{11} also showed some activity to

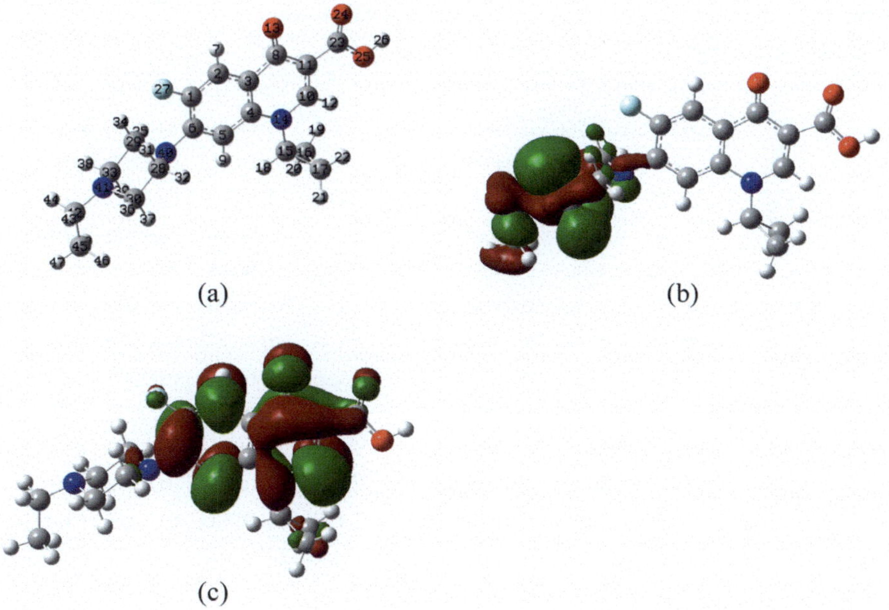

Fig. 14.3 For ENR molecular, the optimized structure (**a**); the HOMO structure (**b**); and the LUMO structure (**c**)

electrophilic reactions, due to their f^- values were also higher than 0.049. Similarly, depending on the relative higher f^0 values of O_{13}, N_{41}, C_{10}, N_{40}, C_{11}, revealing that these sites of ENR molecule were more likely assaulted by free radicals.

The degradation mechanism of ENR by UV/PDS process was comprehensively analyzed by DFT calculation, Fukui index, and intermediate products. According to the DFT calculation results, the piperazine ring of ENR was more likely to lose electrons and to be attacked by free radicals. And, some sites of ENR were vulnerable to be attacked by free radicals like $\cdot SO_4^-$ and $\cdot OH$, including O_{13}, N_{41}, C_{10}, N_{40}, and C_{11}, depending on the value of Fukui functions (f^+, f^- and f^0). Analysis of intermediate products showed that ENR was degraded by substitution and de-alkylation effect, on account of the C_{28}, C_{30}, $C_{42,}$ and C_{45} of ENR were probably attacked by UV/PDS process. According to this results, the possibility of C_{28} and C_{30} being attacked was consistent with the results of DFT calculation that piperazine ring was easily to be attacked. Compared to the intermediate products analysis, the active sites analysis by Fukui functions showed certain limitations. This was resulted from the fact that, active sites were calculated depending on the optimized ENR structure which obtained from DFT calculation result, and the oxidation conditions of UV/PDS process was not considered meanwhile, leading to the analysis difference between the pathways and active sites. In brief, the combination of active sites analysis and intermediate products was instructional to expound the degradation mechanism of ENR by UV/PDS process.

Table 14.2 Values of f^+, f^-, f^0 and each atom in ENR molecule

Number	Atom	f^+	f^-	f^0	Number	Atom	f^+	f^-	f^0
1	C_1	0.04995	0.0312	0.04058	14	C_{17}	0.01484	0.00845	0.01164
2	C_2	0.06287	0.02766	0.04527	15	C_{23}	0.04469	0.01490	0.02980
3	C_3	0.05800	0.03799	0.04799	16	O_{24}	0.03587	0.06478	0.05032
4	C_4	0.05375	0.0245	0.03913	17	O_{25}	0.01579	0.01967	0.01773
5	C_5	0.06389	0.02988	0.04689	18	F_{27}	0.00943	0.01190	0.01067
6	C_6	0.06276	0.03063	0.04669	19	C_{28}	0.00700	0.01891	0.01295
7	C_8	0.05854	0.02638	0.04246	20	C_{29}	0.00700	0.01891	0.01295
8	C_{10}	0.09695	0.01818	0.05757	21	C_{30}	0.00490	0.02039	0.01265
9	C_{11}	0.05197	0.04949	0.05073	22	C_{33}	0.00490	0.02039	0.01265
10	O_{13}	0.04720	0.13353	0.09037	23	N_{40}	0.00862	0.09478	0.05170
11	N_{14}	0.04248	0.03490	0.03869	24	N_{41}	0.00143	0.12712	0.06428
12	C_{15}	0.01410	0.00984	0.01197	25	C_{42}	0.00422	0.01699	0.01060
13	C_{16}	0.01483	0.00844	0.01164	26	C_{45}	0.00533	0.01439	0.00986

14.4 Conclusion

In this study, ENR was selected as one typical fluoroquinolone medicine, its degradation performance and mechanism by UV/PDS process was comprehensively investigated. The results showed that ENR degradation rate by UV/PDS process was 82.2% with the highest K_{obs} of 5.848 × 10^{-2} min^{-1}, due to the generation of free radicals like $\cdot SO_4^-$ ($R_{\cdot SO_4^-}$ = 65.2%) and $\cdot OH$ ($R_{\cdot OH}$ = 17.2) to accelerate degradation efficiency. Energy consumption analysis demonstrated that UV irradiation was the major contributor to energy consumption when using UV/PDS process to degrade ENR, and it was implemental to decrease energy costs by increasing PDS dosage or maintaining pH to 9. When Cl^- (10 mM) or NO_3^- (1 mM) co-existed with ENR, degradation capability of UV/PDS process was slightly promoted, owning to some intermediates (Cl_2, HClO, or $\cdot NO_3^-$) showed strong oxidizing property. According to the RSM results, the quadratic model displayed that ENR degradation rate by UV/PDS process could be guaranteed at 74.8% with less negatively influenced, when Cl^- (7 mM), HCO_3^- (2 mM), and NO_3^- (3 mM) presented. Besides, the results of DFT calculation illustrated that that the piperazine ring of ENR was more prone to lose electrons and more susceptible attacked by free radicals.

Acknowledgements This work was funded by the National Natural Science Foundation of China (52060023 and 51768067). The authors also are deeply grateful to Changyuan Water Research Institute (Urumqi, China) for providing research facilities.

References

Anderson G (1992) Quinolone antimicrobial agents. Annu Rev Med 43:465–486

Asghar A, Abdul RAA, Daud WRA (2017) Reactivity, stability, and thermodynamic feasibility of H_2O_2/H_2O at graphite cathode: application of quantum chemical calculations in MFCs. Environ Prog Sustain Energy 37(4):1291–1304

Babić S, Periša M, Škorić I (2013) Photolytic degradation of norfloxacin, enrofloxacin and ciprofloxacin in various aqueous media. Chemosphere 91(11):1635–1642

Bolton JR, Bircher KG, Tumas W, Tolman CA (1996) Figures-of-berit for the technical development and application of advanced oxidation processes. J Adv Oxid Technol 1:13–17

Bound JP, Voulvoulis N (2006) Predicted and measured concentrations for selected pharmaceuticals in UK rivers: implications for risk assessment. Water Res 40(15):2885–2892

Buxton GV, Greenstock CL, Helman WP, Ross A (1988) Critical review of rate constants for reactions of hydrated electrons, hydrogen atoms and hydroxyl radicals ($\bullet OH/\bullet O^-$) in aqueous solution. J Phys Chem Ref Data 17:513–886

Chu W, Chu T, Du E, Yang D, Guo Y, Gao N (2016) Increased formation of halomethanes during chlorination of chloramphenicol in drinking water by UV irradiation, persulfate oxidation, and combined UV/persulfate pre-treatments. Ecotoxicol Environ Saf 124:147–154

Dhaka S, Kumar R, Khan MA, Paeng K-J, Kurade MB, Kim S-J, Jeon B-H (2017) Aqueous phase degradation of methyl paraben using UV-activated persulfate method. Chem Eng J 321:11–19

Dodd MC, Shah AD, von Gunten U, Huang CH (2005) Interactions of fluoroquinolone antibacterial agents with aqueous chlorine: reaction kinetics, mechanisms, and transformation pathways. Environ Sci Technol 39(18):7065–7076

Fang G-D, Dionysiou DD, Al-Abed SR, Zhou D-M (2013) Superoxide radical driving the activation of persulfate by magnetite nanoparticles: implications for the degradation of PCBs. Appl Catal B Environ Int J Devot Catal Sci Appl 129:325–332

Fouad K, Alalm MG, Bassyouni M, Saleh MY (2020) A novel photocatalytic reactor for the extended reuse of W-TiO_2 in the degradation of sulfamethazine. Chemosphere 257:127270

Guinea E, Brillas E, Centellas F, Cañizares P, Rodrigo MA, Sáez C (2009) Oxidation of enrofloxacin with conductive-diamond electrochemical oxidation, ozonation and Fenton oxidation. A comparison. Water Res 43(8):2131–2138

Guinea E, Garrido JA, Rodríguez RM, Cabot PL, Arias C, Centellas F, Brillas E (2010) Degradation of the fluoroquinolone enrofloxacin by electrochemical advanced oxidation processes based on hydrogen peroxide electrogeneration. Electrochim Acta 55(66):2101–2115

Guo H, Gao N, Yang Y, Zhang Y (2016) Kinetics and transformation pathways on oxidation of fluoroquinolones with thermally activated persulfate. Chem Eng J 292:82–91

Guo H, Ke T, Gao N, Liu Y, Cheng X (2017) Enhanced degradation of aqueous norfloxacin and enrofloxacin by UV-activated persulfate: kinetics, pathways and deactivation. Chem Eng J 316:471–480

Han R, Wu Q, Lin C, Zhang L, Zhai Z, Sun P, Fang Y, Wu J, Liu H (2019) Degradation of UV filter ethyl 4-aminobenzoate (Et-PABA) using a UV-activated persulfate oxidation process. Appl Sci 9(14):2873

He X, Deng M, Wang Q, Yang Y, Yang Y, Nie X (2016) Residues and health risk assessment of quinolones and sulfonamides in cultured fish from Pearl River Delta, China. Aquaculture 458:38–46

Janecko N, Pokludova L, Blahová J, Svobodova Z, Literák I (2016) Implications of fluoroquinolone contamination for the aquatic environment—a review. Environ Toxicol Chem 35:2647–2656

Ji Y, Dong C, Kong D, Lu J, Zhou Q (2015) Heat-activated persulfate oxidation of atrazine: implications for remediation of groundwater contaminated by herbicides. Chem Eng J 263:45–54

Jiang C, Ji Y, Shi Y, Chen J, Cai T (2016) Sulfate radical-based oxidation of fluoroquinolone antibiotics: kinetics, mechanisms and effects of natural water matrices. Water Res 106:507–517

Jiang M, Lu J, Ji Y, Kong D (2017) Bicarbonate-activated persulfate oxidation of acetaminophen. Water Res 116:324–331

Jutarvutikul K, Sakulthaew C, Chokejaroenrat C, Pattanateeradetch A, Imman S, Suriyachai N, Satapanajaru T, Kreetachat T (2021) Practical use of response surface methodology for optimization of veterinary antibiotic removal using UV/H_2O_2 process. Aquacult Eng 94:102174

Kolthoff IM, Stenger VA (1947) Titration methods: acid-base, precipitation, and complex-formation reactions. 52

Li Y, Zhang F, Liang X, Yediler A (2013) Chemical and toxicological evaluation of an emerging pollutant (enrofloxacin) by catalytic wet air oxidation and ozonation in aqueous solution. Chemosphere 90:284–291

Li Y, Liu LD, Liu L, Liu Y, Zhang HW, Han X (2016) Efficient oxidation of phenol by persulfate using manganite as a catalyst. J Mol Catal A: Chem 411:264–271

Li H, Yang YL, Li X, Fan XY, Wang N (2021) Enhanced degradation mechanism of sulfamethazine by vacuum ultraviolet/persulfate. J Environ Chem Eng 9:106489

Liu Y, Qu G, Sun Q, Jia H, Wang T, Zhu L (2021) Endogenously activated persulfate by non-thermal plasma for Cu(II)-EDTA decomplexation: synergistic effect and mechanisms. Chem Eng J 406:126774

Liu Z, Lan H, Wang Y, Zhang J, Qin J, Zhang R et al (2022) Highly efficient degradation of bisphenol A with persulfate activated by vacuum-ultraviolet/ultraviolet light (VUV/UV): experiments and theoretical calculations. Chem Eng J 429:132485

Ma X, Tang L, Deng J, Liu Z, Li X, Wang P, Li Q (2021) Removal of saccharin by UV/persulfate process: degradation kinetics, mechanism and DBPs formation. J Photochem Photobiol A Chem 420:113482

Michael I, Rizzo L, McArdell CS, Manaia CM, Merlin C, Schwartz T, Dagot C, Fatta-Kassinos D (2013) Urban wastewater treatment plants as hotspots for the release of antibiotics in the environment: a review. Water Res 47(3):957–995

Morrison RC (2018) Fukui functions for the temporary anion resonance states of Be^-, Mg^-, and Ca^-. Acta Phys Chim Sin 34:263–269

Moussa M-A, Serge C (2014) Ciprofloxacin oxidation by UV-C activated peroxymonosulfate in wastewater. J Hazard Mater 265:41–46

Neta P, Huie RE, Ross A (1979) Rate constants for reactions of inorganic radicals in aqueous solution. J Phys Chem Ref Data 17:1027–1284

Oláh J, Alsenoy CV, Sannigrahi AB (2002) Condensed Fukui functions derived from stockholder charges: assessment of their performance as local reactivity descriptors. J Phys Chem A Mol Spectrosc Kinet Environ Gen theory 106(15):3885–3890

Ou H, Liu J, Ye J, Wang L, Gao N, Ke J (2017) Degradation of tris(2-chloroethyl) phosphate by ultraviolet-persulfate: kinetics, pathway and intermediate impact on proteome of Escherichia coli. Chem Eng J 308:386–395 (2017)

Pallecchi L, Bartoloni A, Paradisi F, Rossolini GM (2008) Antibiotic resistance in the absence of antimicrobial use: mechanisms and implications. Expert Rev Anti Infect Ther 6:725–732

Song W, Li J, Fu C, Wang Z, Zhou Y, Zhang X, Yang J, Wang K, Liu Y, Song Q (2021) Establishment of sulfate radical advanced oxidation process based on Fe^{2+}/O_2/dithionite for organic contaminants degradation. Chem Eng J 410:128204

Tan C, Gao N, Deng Y, Deng J, Zhou S, Li J, Xin X (2014) Radical induced degradation of acetaminophen with Fe_3O_4 magnetic nanoparticles as heterogeneous activator of peroxymonosulfate. J Hazard Mater 276:452–460

Tian S, Zhang C, Huang D, Wang R, Zeng G, Yan M, Xiong W, Zhou C, Cheng M, Xue W, Yang Y, Wang W (2020) Recent progress in sustainable technologies for adsorptive and reactive removal of sulfonamides. Chem Eng J 389:123423

Van DX, Dewulf J, Van LH, Demeestere K (2014) Fluoroquinolone antibiotics: an emerging class of environmental micropollutants. Sci Total Environ 500–501:250–269

Wang P, He YL, Huang CH (2010) Oxidation of fluoroquinolone antibiotics and structurally related amines by chlorine dioxide: reaction kinetics, product and pathway evaluation. Water Res 44(20):5989–5998

Wang G, Zhang Q, Li J, Chen X, Lang Q, Kuang S (2019) Combined effects of erythromycin and enrofloxacin on antioxidant enzymes and photosynthesis-related gene transcription in Chlorella vulgaris. Aquat Toxicol 212:138–145
Wei R, Ge F, Chen M, Wang R (2012) Occurrence of ciprofloxacin, enrofloxacin, and florfenicol in animal wastewater and water resources. J Environ Qual 41(5):1481–1486
Xiang Y, Lu X, Liu Y, Yu C, Yang H, Gao N, Chu W, Zhang Y (2022) Influence of chemical speciation on enrofloxacin degradation by UV irradiation: kinetics, mechanism and disinfection by-products formation. Chemosphere 286(Part 1):131559
Xiao Y, Zhang L, Zhang W, Lim KY, Webster RD, Lim TT, Webster RD, Lim TT (2016) Comparative evaluation of iodoacids removal by UV/persulfate and UV/H_2O_2 processes. Water Res 102:629–639
Xie P, Ma J, Liu W, Zou J, Yue S, Li X, Wiesner RM, Fang J (2015) Removal of 2-MIB and geosmin using UV/persulfate: contributions of hydroxyl and sulfate radicals. Water Res 69:223–233
Xu Y, Liu S, Guo F, Zhang B (2016) Evaluation of the oxidation of enrofloxacin by permanganate and the antimicrobial activity of the products. Chemosphere 144:113–121
Xu Y, Chu Q, Chen D, Fuentes A (2021) HOMO-LUMO gaps and molecular structures of polycyclic aromatic hydrocarbons in soot formation. Front Mech Eng 7:1–10
Yang Y, Pignatello J, Ma J, Mitch W (2014) Comparison of halide impacts on the efficiency of contaminant degradation by sulfate and hydroxyl radical-based advanced oxidation processes (AOPs). Environ Sci Technol 48:2344–2351
Zhan CG, Nichols JA, Dixon DA (2003) Ionization potential, electron affinity, electronegativity, hardness, and electron excitation energy: molecular properties from density functional theory orbital energies. J Phys Chem A 107:4184–4419
Zhang H, Huang C-H (2005) Oxidative transformation of fluoroquinolone antibacterial agents and structurally related amines by manganese oxide. Environ Sci Technol 39:4474–4483
Zou S, Xu W, Zhang R, Tang J, Chen Y, Zhang G (2011) Occurrence and distribution of antibiotics in coastal water of the Bohai Bay, China: impacts of river discharge and aquaculture activities. Environ Pollut 159(10):2913–2920

Chapter 15
Runoff and Sediment Transport Simulation in the Jinghe River Basin Based on the Combination of SWAT and CA-Markov Models

Lixia Wang, Xu Luo, Fangxia Wang, Jiawei Zhang, and Zhao Liu

Abstract Hydrological process is an important indicator for the health evaluation of the hydrological ecosystem in a basin. The aims of this study were to explore the impact of land use change on the runoff and sediment transport in the Jinghe River Basin and provide a new method for the quantitative prediction of runoff and sediment transport based on the combination of SWAT and CA-Markov Models. The simulated results indicated that the distribution of runoff and sediment transport differs greatly at the spatio-temporal scales, and the runoff is more sensitive to external changes. From 2005 to 2025, with the continuous changes in land use, the runoff of the basin showed a trend from increase then to decrease, while the amount of sediment transport showed a trend of continuous increasing.

Keywords SWAT model · Hydrological ecology · Runoff · Sediment transport · Jinghe River Basin

15.1 Introduction

The water cycle that connects the Earth's biosphere, atmosphere and hydrosphere is the energy connection between the various layers on Earth and is the most active factor in environmental evolution (Yang et al. 2018). With the drastic climate change and rapid population growth, unprecedented challenges have been encountered in water resource management and planning. Knowledge of changes in the surface runoff and

L. Wang (✉) · X. Luo · J. Zhang · Z. Liu
Chang'an University, Shaanxi Xi'an 710064, China
e-mail: zylxwang@chd.edu.cn

F. Wang
Qinzhou Highway Section, Tianshui Highway Development Center, Gansu Tianshui 741000, China
e-mail: zylxwang@chd.edu.cn

H. Xu (ed.), *The 6th International Symposium on Water Resource and Environmental Management*, Environmental Science and Engineering,
https://doi.org/10.1007/978-3-031-55989-1_15

sediment transport conditions requires an analysis of land use. Land use changes have altered the characteristics of the underlying surface, affected various hydrological processes in the basin, and led to changes in hydrological factors such as near-surface evapotranspiration, interception, and infiltration, which will interfere with the process of runoff and sediment transport. Due to eco-environmental problems such as soil erosion, land degradation, and water environment deterioration, research on the hydrological response of land use changes has significantly increased (Yang et al. 2013). The watershed hydrological model is a comprehensive system that integrates mathematical models, computer technology and system analysis. It is the most effective means to analyze and explore the complex hydrological cycle process, and provides an effective tool for solving the actual watershed hydrological problems (Dong et al. 2012; Xu and Cheng 2010).

The soil and water assessment tool (SWAT) model is a physical process-based distributed watershed hydrological model developed by the United States Department of Agriculture. It has a strong physical foundation and can effectively simulate the transportation and migration processes of watershed runoff, sediment, nutrients, etc. under different climate, land use, and agricultural management measures (Havrylenko et al. 2016; Lai et al. 2012). The SWAT model is suitable for complex and large watersheds, can be utilized in areas where there is deficient data, and applied widely in the cold regions of North America and Canada (Neitsch et al. 2011). It was introduced in China relatively late, however, understanding of the model has significantly improved, and the model has been widely used by researchers in China (Xie et al. 2018). Also, the applications of the model mainly focused on non-point source pollution research, sediment/flow simulation and model input parameter uncertainty. Feng et al. (2010) built a SWAT distributed hydrological model in the Wuyuer River Basin and applied it to study the runoff change in the future climate change scenarios, the results show that under future climate change scenarios, the runoff will continue to decrease over time. The impact of different sub-basin divisions on the output of the SWAT model is due to the combined effects of the model's spatial parameter set and the model structure mismatch. Foreign scholars have also studied the relationship between land use change and hydrological effects by utilizing the SWAT model. Githui et al. (2009) used the SWAT model to study the impact of climate and land cover changes on the hydrological effects of western Kenya. Stehr A's (2010) research showed that runoff decreases with increasing agricultural land area and decreasing forestland area. The study by Ciampalini et al. (2012) showed that severe soil erosion has occurred in the Aksum region of Ethiopia due to the conversion of large areas of arable land into pastures in order to meet the economic development need. Recently, the researchers have paid more attention to the combination of the SWAT model and other models. Wei et al. (2019) simulated nitrate (NO_3) fate and transport in a watershed system based on a model that couples the SWAT model with the groundwater flow model MODFLOW and the groundwater solute reactive transport model RT3D. Sriworamas et al. (2020) assessed the changes in hydrologic responses of the Ubolratana Reservoir in northeastern Thailand due to increasing forest areas. Assessment results indicate that increasing forest areas cause a decrease of runoff, peak flow, and hence, inflow volume into the reservoir.

In summary, the current simulation studies based on the SWAT model are mostly long-term research on the model fitting accuracy and the temporal and spatial characteristics of runoff and sediment transport. However, as a basis for decision-making, we may need to pay more attention to possible future changes in runoff and sediment transport, and then consider the future health of the hydrological ecosystem in a basin under the existing land use patterns and planning models. In this paper, we provide a new method for simulating and predicting the runoff and sediment transport based on the combination of the SWAT and the cellular automata-(CA) Markov models in order to reveal the hydrological ecological effects of land use change. Also, the driving mechanisms of different land use types on runoff and sediment transport have been studied.

15.2 Study Area

The Jinghe River originates from the eastern foothills of the Liupan Mountain in the Ningxia Hui Autonomous Region and flows from west to east through the Shaanxi and Gansu provinces and Ningxia Hui Autonomous Region. The Jinghe River Basin (106°14′-108°42′E, 34°46′37°19′N) is located in the hinterland of the Loess Plateau with an area of 45,421 km^2. The basin has the higher topography in the northwest and lower in the southeast where the soil erosion is severe. It is located in a temperate semi-arid and semi-humid continental monsoon climate zone with frequent meteorological disasters, and the vegetation in the basin is poor. Due to regional development and population growth, a series of ecological and environmental problems have emerged, such as vegetation degradation, land desertification, soil erosion, seasonal rivers, river shrinkage, and water quality declining. Therefore, using the SWAT model to study the hydrological effects can assist in understanding the characteristics and evolution of runoff and sediment transport, and provide some theoretical decisions for basin planning and soil erosion control, which has great significance to its ecological construction and sustainable development.

15.3 Data Sources and Processing

15.3.1 Spatial Data Sources and Processing

The spatial data utilized in this study mainly include land use, DEM, and soil data.

Land use data in 2005 and 2015 were obtained from the Resource and Environmental Science Data Center of the Chinese Academy of Sciences (http://www.resdc.cn/Default.aspx). The land use data were classified based on the Chinese land resource classification standard, which includes six classifications: cultivated farmland, forestland, grassland, water, urban, and bare land.

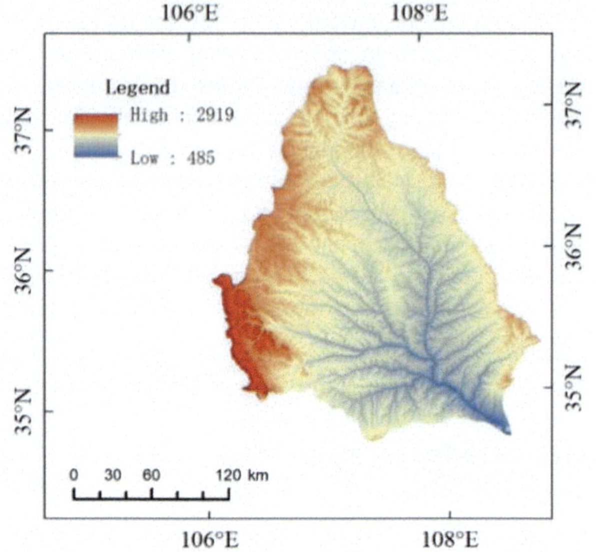

Fig. 15.1 DEM of the Jinghe River Basin

The second version of the ASTER GDEM data were used from the geospatial data cloud (http://www.gscloud.cn/). The spatial resolution of the data is 30 m, and there are 12 sceneries involved in the Jinghe River Basin. After data acquisition, the images were stitched, cropped, and projected in ArcGIS (Fig. 15.1).

The soil data were obtained from the Resource and Environmental Science Data Center of the Chinese Academy of Sciences (http://www.resdc.cn/Default.aspx). Using the traditional "soil occurrence classification" system, the soil in the study area was divided into 16 types, which mainly included loessian soil, cinnamon soil, sticky black loam soil and new soil, etc. (Fig. 15.2). The parameters in the soil database need to be recalculated, including soil depth, soil moisture density, effective water holding capacity, organic carbon content, saturated hydraulic conductivity coefficient and soil erosion force factor. HRU) water cycle.

15.3.2 Attribute Data Sources and Processing

The attribute data selected in this study included meteorological and hydrological data. Meteorological data collected from four meteorological stations (Changwu, Huanxian, Kongtong, and Xifeng) were obtained from the National Meteorological Science Data Center (http://data.cma.cn/site/index.html). The daily precipitation, temperature, sunshine, wind speed, and relative humidity were obtained from 2005 to 2010.

The hydrological data was obtained from the Yellow River Basin Hydrological Yearbook of the People's Republic of China, which included monthly runoff and

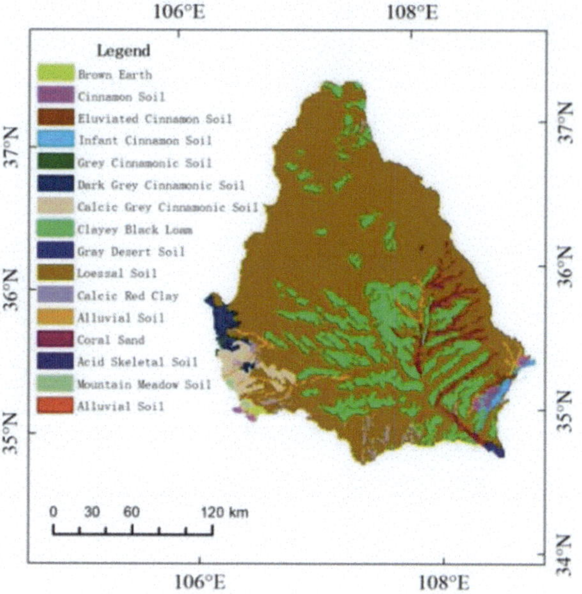

Fig. 15.2 Soil types in the Jinghe River Basi

sediment transport of Zhangjiashan hydrological station in the Jinghe River Basin from 2006 to 2010. The hydrological data was mainly used for model parameter calibration and verification.

15.4 Research Method

15.4.1 SWAT Model Construction

Sub-basin and HRU Division. The geomorphology of the study area was analyzed based on the watershed's DEM to obtain the flow direction of one-unit grid in the SWAT model. Then the river network was extracted, and the sub-basins were divided. In this study, 45 sub-basins were defined according to the DEM (Fig. 15.3).

Based on land use type, soil type and slope level, the sub-basin was divided into areas with unique combinations of land use, soil and slope type. The model calculates the runoff of each HRU first, then by the individual sub-basin, and finally, it calculates the runoff from the sub-basin to the basin's outlet. To ensure the accuracy of the model, the thresholds used in this study were 10% of the soil type, 15% of the land use type and 10% of the slope type, respectively, which generated 193 HRUs.

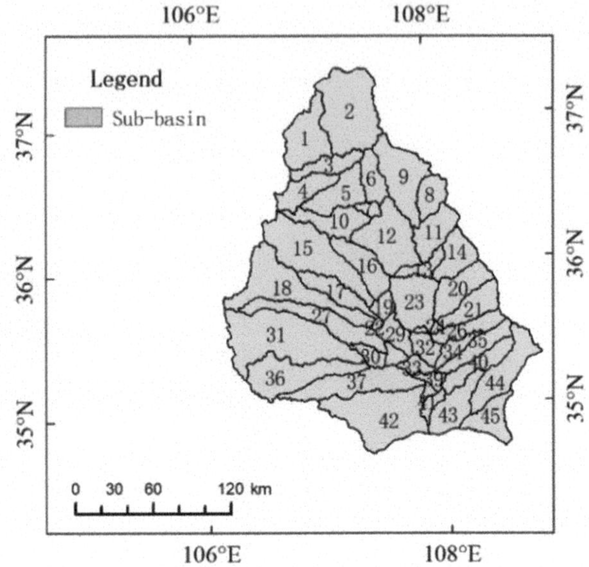

Fig. 15.3 Sub-basins of the Jinghe River Basin

Weather Generator Database Building. The weather generator, average precipitation, and maximum and minimum temperatures in the meteorological database must be entered into the SWAT model. For data such as solar radiation, average wind speed, and average relative humidity, the actual meteorological data can be input into the model. The main function of the weather generator is to generate climate data and supply missing data (Pang et al. 2007). The required parameters of the weather generator can be calculated using the SWAT weather software. A suitable weather generator database was established for the study area by manually entering five meteorological data types into a Microsoft Excel spreadsheet, which was stored in a.txt format according to the model requirements.

15.4.2 Parameter Calibration and Verification

This paper uses automatic calibration combined with manual fine-tuning to calibrate and verify the model parameters. The automatic calibration uses the Sufi-2 algorithm of the SWAT-CUP software to perform sensitivity analysis and calibration on the relevant parameters of the simulated water and sediment transport. The principle of the Sufi-2 algorithm is to first determine the maximum value range of each parameter, and perform multiple iterations on the selected parameters within this range, so as to select a new value range according to the results of each iteration until the requirements are met (Ji 2020). The parameter uncertainty of the algorithm considers sources of uncertainty, such as precipitation, actual data errors, etc., and the obtained results are more accurate. By adopting the global sensitivity method provided in SWAT-CUP for analysis, and combining with the actual situation of the basin, 14 highly sensitive parameters (SCS runoff curve coefficient, evaporation of the soil compensation coefficient, soil depth, alpha of the base flow, maximum water storage of the canopy, saturated wet density of the surface soil, effective hydraulic conductivity of the river channel, alpha of the base flow in bank regulation and storage, average slope length, average slope, soil absorption and compensation factor, biological mixing efficiency, plant ET curve coefficient, and groundwater re-evaporation coefficient) were selected for calibration and verification purposes. For this study, the periods of 2005, 2006–2008, and 2009–2010 were selected as the warm-up, calibration, and verification periods, respectively.

According to the official SWAT model calibration techniques and methods, first, the SCS runoff curve coefficient is adjusted until the surface runoff meets the requirements. If the surface runoff does not meet the requirements after adjusting the SCS runoff curve coefficient value, the soil depth or the soil evaporation compensation factor are adjusted. Next, after adjusting and calibrating the surface runoff, the simulated and measured values of the base flow are compared. Finally, the previous operations are repeated until the base flow value reaches an acceptable level (Liu 2014). Therefore, the continued adjustment of the parameters affecting the sediment transport is necessary to set the parameter calibration range to achieve parameter calibration (Table 15.1).

After calibrating the SWAT model parameters, it is necessary to analyze the model's applicability. Generally, two indicators are used to evaluate the applicability, the correlation coefficient (R^2), and the efficiency coefficient (Nash–Sutcliffe coefficient, NSE), which can be calculated using the SWAT-CUP software. R^2 is calculated as follows:

$$R^2 = \frac{\left[\sum_{i=1}^{n}\left(Q_{m\cdot i} - \overline{Q}_m\right)\left(Q_{s\cdot i} - \overline{Q}_s\right)\right]^2}{\sum_{i=1}^{n}\left(Q_{m\cdot i} - \overline{Q}_m\right)^2\left(Q_{s\cdot i} - \overline{Q}_s\right)^2} \tag{15.1}$$

where n is the number of observed values, $Q_{m\cdot i}$ is the observed value, $\overline{Q}_m$ is the average of the observed values, $Q_{s\cdot i}$ is the simulated value, and $\overline{Q}_s$ is the average of the simulated values.

NSE is calculated as follows:

$$\mathrm{NSE} = 1 - \frac{\sum_{i=1}^{n}\left(Q_{o\cdot i} - Q_{p\cdot i}\right)}{\sum_{i=1}^{n}\left(Q_{o\cdot i} - Q_{\mathrm{avg}}\right)} \tag{15.2}$$

where n is the number of observed values, $Q_{o\cdot i}$ is the observed value, $Q_{p\cdot i}$ is the simulated value, and Q_{avg} is the average value of the observed values.

The evaluation results of the simulated runoff and sediment transport in the Jinghe River Basin are presented in Table 15.2. When $R^2 = 1$ the simulated value is very consistent with the observed value. However, when $R^2 < 1$, the closer the value is to 1, the more accurate the simulation results. For NSE, the closer NSE is to 1, the greater the similarity between the simulated value and the observed value. When

Table 15.1 Parameter Calibration Results

Objects to Adjust	Number	Method	Parameter Name	Parameter Definition	Best Value
Runoff	1	v	CN2	**SCS runoff curve Coefficient**	50.000
	2	v	SOL_Z	Soil depth	2.000
	3	v	CANMX	Maximum water storage of the Canopy	13.000
	4	v	ESCO	Evaporation of the soil Compensation Coefficient	0.040
	5	v	SLSUBBSN	Average slope length	10.000
	6	v	CNCOEF	Plant ET curve coefficient	0.700
	7	v	GW_REVAP	Groundwater Re-evaporation Coefficient	0.200
	8	v	ALPHA_BNK	Alpha of the base flow in Bank Regulation and Storage	0.060
	9	v	HRU_SLP	Average slope	1.000
	10	v	BIOMIX	Biological mixing Efficiency	1.000
	11	v	EPCO	Soil absorption and Compensation Factor	0.590
	12	v	SOL_BD	Saturated wet density of the Surface Soil	2.500
	13	v	ALPHA_BF	Alpha of the base flow	0.600
	14	v	CH_K2	Effective hydraulic Conductivity of the River Channel	1.600
Sediment transport	1	v	CN2	**SCS runoff curve Coefficient**	200.000
	2	v	SOL_Z	Soil depth	200.000
	3	v	SPCON	Linear parameter for Calculating the Newest Maximum Amount of Sediment in the Channel Sediment Calculation	0.004

(continued)

Table 15.1 (continued)

Objects to Adjust	Number	Method	Parameter Name	Parameter Definition	Best Value
	4	v	SPEXP	Index parameter for Calculating the Maximum Newest Added Sediment Volume in the River Sediment Calculation	1.500
	5	v	OV-N	Manning coefficient of the Slope	30.000

Table 15.2 Evaluation index results of runoff and sediment transport in Jinghe River Basin

Hydrological Data	Calibration Period		Verification Period	
	R^2	NSE	R^2	NSE
Runoff	0.86	0.70	0.76	0.71
Sediment Transport	0.89	0.86	0.82	0.67

NSE < 0, the simulation result is unacceptable, and when NSE <0.54, the simulation result is acceptable. If $0.54 \leq$ NSE < 0.65, this indicates that the model highly fits the observation, and when NSE ≥ 0.65, the simulation results are fairly good (Zhou et al. 2020).

15.4.3 CA-Markov Model

CA is a local grid dynamic model with discrete time, space, and state (Zheng et al. 2010) that can simulate the spatiotemporal evolution of complex systems and has been widely utilized in fields such as land change, population migration, and urbanization. (Hu et al. 2018; Karimi et al. 2018). The CA model can be expressed as follows:

$$S_{ij}^{t+1} = f_n\left(S_{ij}^{t}\right) \tag{15.3}$$

where S_{ij} is the state of the cell ij; f is the conversion function that defines the conversion rule of the cell from time t to $t + 1$, and n is the neighborhood of the cell and belongs to the input variable f. The cell transformation in the local space can be realized by applying the defined conversion rules.

The Markov Model is a special random motion process with no after effect, which is applied in the simulation and prediction of many geographic phenomena (Chuang et al. 2011; Culik et al. 1990; Durmusoglu and Tanriover 2017). The basic principle of the method is to use the experience transfer probability between discrete states in the observation system to determine the changing trend of each state of the system, so as to predict the future state. In the Markov model, the mutual influence and interaction between land use types, in other words, the process of land use types in a specific area from time t to t + 1, can be directly reflected in the state transition probability matrix P, the process can be described by the following formula:

$$P = \left(P_{ij}\right) = \begin{bmatrix} p_{11} & p_{12} & \cdots & p_{1n} \\ p_{21} & p_{22} & \cdots & p_{2n} \\ \vdots & \vdots & \cdots & \vdots \\ p_{n1} & p_{n2} & \cdots & p_{nn} \end{bmatrix} \tag{15.4}$$

In the formula, P_{ij} represents the transition probability from level i to level j in a period, ranging from 0 to 1, and $\sum p_i=1$. According to the Markov process principle, the transition probability matrix in n periods is obtained by multiplying P_n. Specifically, if S_0 represents the initial land use type distribution, the land use type distribution after n periods can be determined as:

$$S_n = S_0 \times P_n \tag{15.5}$$

The Markov model cannot predict the spatial variables of land use changes; however, the CA model has a strong ability to simulate the spatio-temporal evolution of complex spatial systems (Hou et al. 2004). The CA-Markov module in the IDRISI 17.0 software combines the two models to simulate and predict the spatiotemporal pattern of dynamic land use type changes.

According to the transition probability matrix of land use from 2005 to 2015, the number of cycles is given, the land use types distribution in 2025 is calculated, as shown in Fig. 15.4, and the area and change rate of land use type are shown in Table 15.3. The results show that farmland and urban land areas gradually increase, particularly for urban area whose change is large, and other land types have decreased, of which bare land decreased the most. Therefore, in the future, it is necessary to protect grassland and forestland areas, moderately increase urban land, and promote the conversion of farmland to forestland and grassland.

15.5 Results and Analysis

15.5.1 Time Distribution of Simulation Results

As shown in Fig. 15.5, the simulated runoff value fits the observed value. The runoff is unevenly distributed during the year, and the runoff varies significantly between floods and low flow events. The peak runoff mainly occurs from July to September every year, indicating that precipitation has a significant effect and the SWAT model can accurately simulate the runoff. The precipitation amounts in July and August are

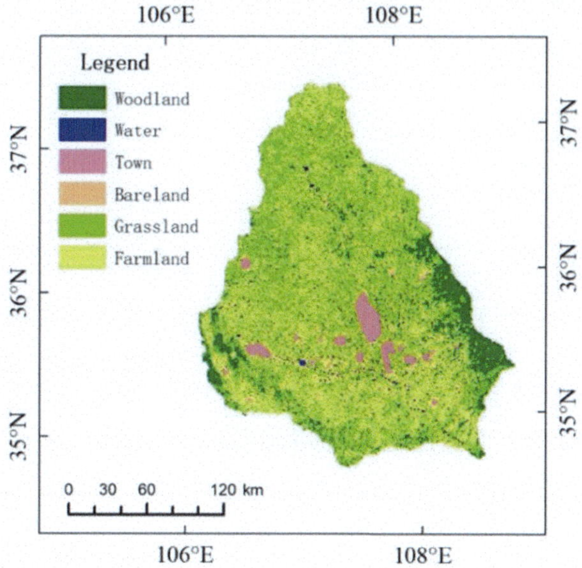

Fig. 15.4 Land use prediction in 2025

Table 15.3 Area and change rate of land use in the Jinghe River Basin in 2005 and 2025

Types of land use	Year		
	2005 (km^2)	2025 (km^2)	Area and change rate from 2005 to 2025 (%)
Forestland	4925.86	4847.48	−1.59
Water	202.40	192.14	−5.07
Construction land	1058.20	1428.26	34.97
Bare land	181.34	99.60	−45.08
Grassland	17,332.37	16,587.67	−4.30
Farmland	17,277.12	17,822.14	3.15

the highest in the Jinghe River Basin. Due to the steep river basin slope and poor vegetative coverage, the floods caused by heavy rain events have characteristics that are more obvious and the floods rise and fall rapidly.

Generally, the runoff stabilized from November 2007 to May 2010, and the peak runoff occurred around September 2010. This is due to the 2010 rainstorm weather process in the Qianhe River Basin, located in the upper reaches of the Jinghe River Basin, which caused a flood at the Weihe and Jinghe River confluence that was observed at the Lintong hydrological station on the Weihe River.

As shown in Fig. 15.6, the simulated sediment transport value is consistent with the observed value. The water and soil loss in the Jinghe River Basin is severe, and the amount of sediment transport is large. The distribution is uneven during the year, and the sediment transport peak, like the peak runoff, occurs mainly in the flood season.

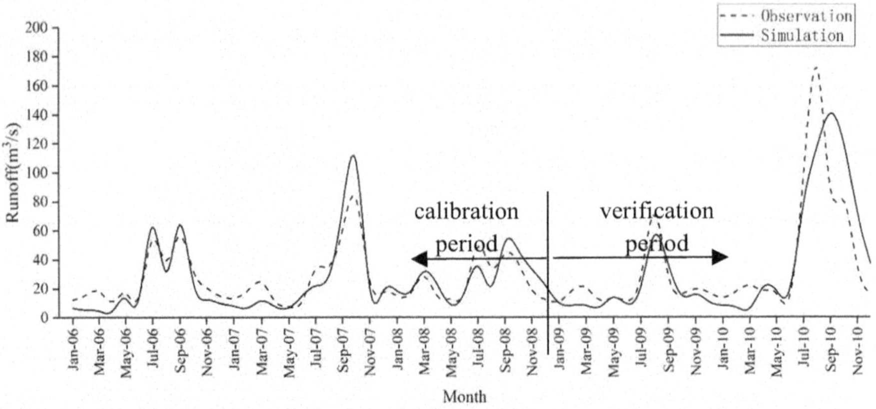

Fig. 15.5 Observed and simulated monthly average runoff in the Jinghe River Basin

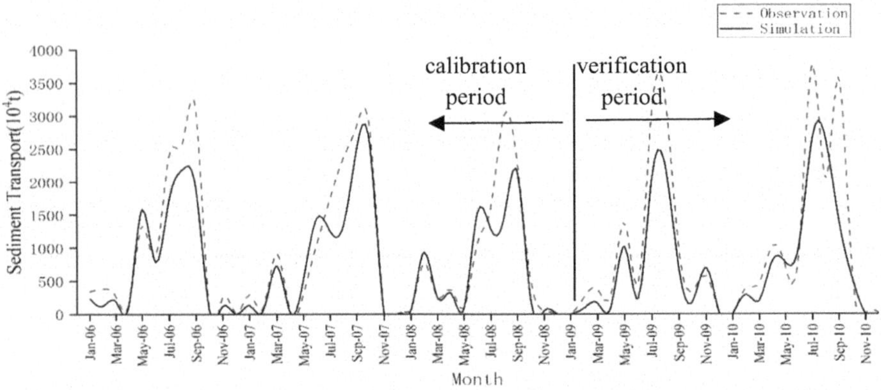

Fig. 15.6 Observed and simulated monthly average sediment transport

15.5.2 Spatial Distribution of Simulation Results

The distribution of annual average runoff and sediment transport in each sub-basin was obtained according to the simulation data statistics, as shown in Figs. 15.7 and 15.8.

Taking the Qingyang, Yuluoping and Zhangjiashan hydrological stations as nodes, the Jinghe River is divided into upper, middle and lower reaches, respectively. The runoff from the Qingyang Station upstream to the Zhangjiashan Station downstream shows a significant increasing trend. The gully region is more than 90 km long and runs from the Tangqu section in Changwu County to the Zaofantou section in Binxian County. The vally region, which is more than 120 km long, runs from the Zaofantou section to the Zhangjiashan Mountain. The area approximately 60 km

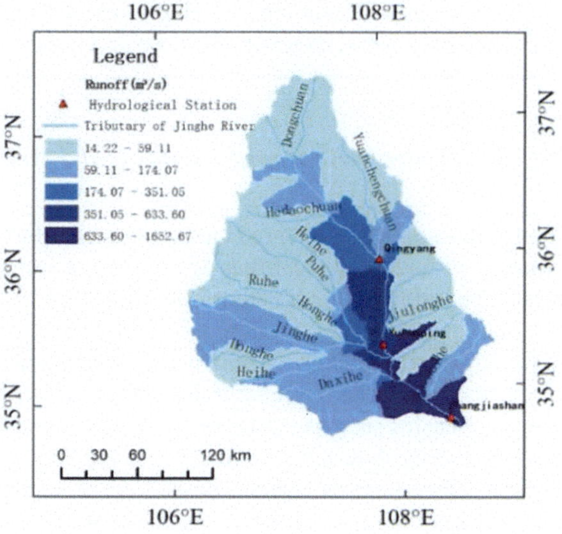

Fig. 15.7 Spatial distribution of runoff

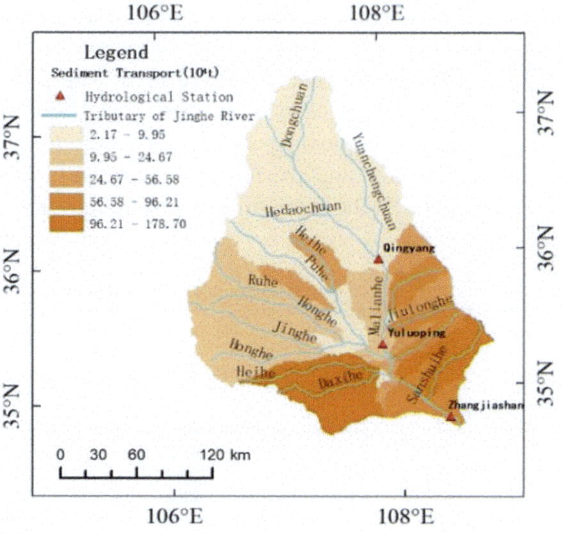

Fig. 15.8 Spatial distribution of sediment transport

below Zhangjiashan Mountain is flat, open and fertile with a slow water flow that is conducive to irrigation (Zhao 2010).

The spatial distributions of the sediment transport and runoff are similar and show a gradually increasing trend from upstream to downstream. The largest sediment transport amount is mainly concentrated in the sub-basins such as the Daxihe and Sanshuihe River Basins located in the lower reaches of the Malian River. Because the soil in the upper Duanjing Village in Binxian County is Cretaceous red sandstone and Quaternary loess, its texture is loose and is easily transported.

15.5.3 Hydrological and Ecological Effect Predictions

The land use data in 2025 predicted by the CA-Markov model and the inter-annual variation of meteorological data were utilized in the established SWAT model to predict the runoff and sediment transport in 2025 (Figs. 15.9 and 15.10).

From the Qingyang station upstream, both predictions show a significant increasing trend, reaching the maximums in the middle and lower reaches. In contrast to previous years, the sediment transport amount in 2025 is no longer concentrated in the sub-basins such as the Daxihe and Sanshuihe River Basins. It is more evenly distributed in the downstream watershed of the Yuluoping hydrological station, which shows that forestland expansion and ecological environment improvement are the key factors controlling river sediment production. Afforestation engineering and ecological restoration are two of the most effective ways to control the water and soil loss in the Jinghe River Basin. Planting trees and grass can conserve water resources, improve soil, increase ground cover to prevent water erosion, reduce water and soil loss, and has obvious positive effects on the ecological environment.

By simulating the runoff and sediment transport under the distribution of land use types in different years, it is found that under relatively stable climatic conditions, the continuous expansion of construction land destroys the terrain, reduces soil water retention and erosion resistance, and leads to a gradual increase in sediment transport. At the same time, forest and grass vegetation can reduce the kinetic

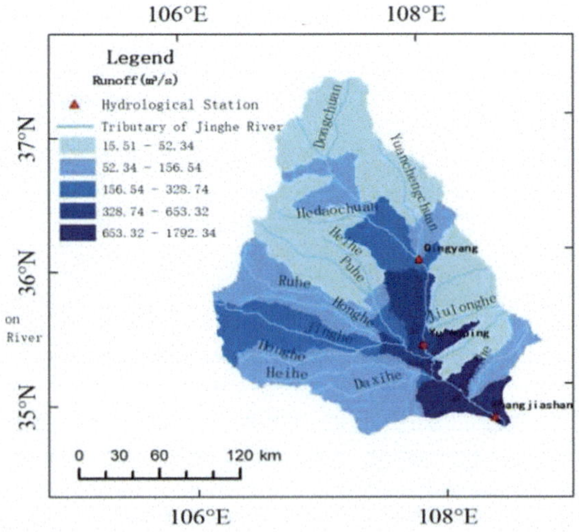

Fig. 15.9 Spatial distribution of runoff in 2025

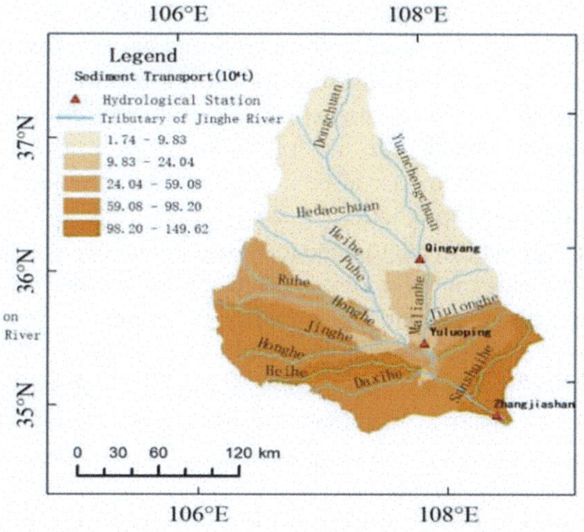

Fig. 15.10 Spatial distribution of sediment transport in 2025

energy of precipitation, block surface runoff, increase soil permeability, and reduce soil erosion. Therefore, the reduction of the area of forestland and grassland reduces the amount of runoff and increases the amount of sediment transport. In addition, although the area of farmland is declining, the soil structure of farmland is destroyed during repeated cultivation, soil water permeability becomes worse, and it is sensitive to rainwater, which is easy to cause soil erosion.

15.6 Discussion and Conclusion

15.6.1 Discussion

The SWAT model is widely applied for the assessment of soil–water resources and the relative fields. The scholars have achieved good research results utilizing it. The Jinghe River Basin has always been one of the key areas for soil erosion control in the Loess Plateau. Since the 1960s, large-scale soil and water conservation work has been performed, including warping dam and terrace installation and afforestation, and as a result, since 1995, the soil and water conservation areas have increased substantially. Soil and water conservation measures can effectively block water and soil transportation, conserve water resources, increase vegetation coverage, and improve the ecological environment and basic conditions of agricultural production. The SWAT model was used to simulate runoff and sediment transport in this area because it has important theoretical significance for exploring the hydrological and ecological effects in the context of climate models and land use changes. At present, many scholars have started with the fundamental issue of the influencing factors and mechanism of runoff and sediment change and have carried out research in the Jinghe river basin (Yan et al. 2017, Liu et al. 2019, Ma et al. 2019). However, these studies still face many problems, such as the uncertainty of climatic changes and the complexity of human activities. In the follow-up research, the scenario model theory and the CA-Markov method can be combined to carry out multi-scenario model research simulation and prediction.

It should be noted that the most significant feature of the SWAT model is the inclusion of a large number of parameters; however, there is a certain correlation between some parameters, and when applied in the model, the input of one can affect another parameter, causing the model results to differ. Second, there are some uncertainties in the model structure. For example, during the construction of the hydrological model, relevant empirical formulas are utilized to describe the hydrological cycle of the basin (Li et al. 2017) and many data that have uncertainties are input into the model, such as the uncertainty in the DEM spatial resolution and remote sensing image processing, and the discontinuity of the individual meteorological data. Therefore, accurate adjustment of the parameter input into the model based on a comprehensive analysis and consideration of the role of multiple factors also needs to be further explored.

15.6.2 Conclusion

Runoff and sediment transport are important indicators for characterizing hydrological and ecological effects and are of great significance for exploring watershed hydrological characteristics. In this study, the SWAT-CUP software was used for parameter calibration and verification to explore the spatio-temporal distributions of the runoff and sediment transport. Then, the CA-Markov model was used to predict the land use distribution, and the results were entered into the SWAT model. Finally, based on the predicted land use distribution in 2025, the spatio- temporal runoff and sediment transport distributions were predicted. The conclusions are as follows:

(1) The SWAT model was used to simulate the monthly average runoff and sediment transport in the Jinghe River Basin. In the calibration period, the R2 and NSE of the runoff were 0.86 and 0.70, respectively, and in the verification period, R2 and NSE were 0.76 and 0.71, respectively. During the calibration period, the R2 and NSE of sediment transport were 0.89 and 0.86, respectively, and during the verification period, R2 and NSE are 0.82 and 0.67, respectively. That indicates the SWAT model has good applicability in the Jinghe River Basin.
(2) The interannual variability of the runoff is characterized by a gradual peak in late 2007, followed by a stable period through May 2010, and then increasing its peak runoff value around September 2010, which was caused by the heavy rain weather occurrence in the summer of 2010. The annual runoff variation is characterized by a large difference between flood and low flow events, and

the peak value generally occurs from July to September each year, indicating that precipitation has a significant impact on runoff, which is attributed to two main causes. The water and soil erosion is severe, and the sediment transport amount is large. There is minimal interannual variation in sediment transport and it is unevenly distributed throughout the year. The peak sediment transport value mainly occurs in the flood season, and the high sediment transport lasts for a longer period.

(3) On the spatial scale, The runoff from the Qingyang station upstream to the Zhangjiashan station downstream showed a significant increasing trend. The spatial distribution characteristics of the sediment transport are roughly similar to those of the runoff and the sediment amount transported in the surrounding watershed downstream of the Malian River is the greatest. The areas with large sediment transport amounts are mainly concentrated in the sub-basins affected by the tributaries because the sediment amount in the tributary is greater than the amount in the river. The main reason for this is that Dong Zhiyuan, the largest loess plateau in China, which has easily eroded soil, is located between the Malian River and the Pu River; therefore, the sediment amount is significantly higher.

(4) The predicted average monthly runoff and sediment transport in the Jinghe River Basin in 2025 are basically similar to those in previous years; however, the annual average runoff and sediment transport are predicted to decrease and increase significantly, respectively, compared with the actual values in 2010. This might be due to the predicted decrease in grassland and forestland and the increase in urban area and farmland. The spatial distribution of the runoff and sediment transport is predicted to be similar to the previous years included in this study.

References

Chuang CW, Chien CH, Lin CY (2011) Application of Markov-chain model for vegetation restoration assessment at landslide areas caused by a catastrophic earthquake in Central Taiwan. Ecol Model 222(3):835–845

Ciampalini R, Billi P, Ferrari G et al (2012) Soil erosion induced by land use changes as determined by plough marks and field evidence in the Aksum area (Ethiopia). Agric, Ecosyst & Environ 146(1):197–208

Culik II K, Hurd LP, Yu S (1990) Computation theoretic aspects of cellular automata. Physica D 45(1), 357–378

Dong LH, Li S, Xiong LH, Yu KX (2012) Research progress on the impact of climate change and human activities on hydrology. Adv Water Sci 23(02):278–285

Durmusoglu ZO, Tanriover AA (2017) Modelling land use/cover change in Lake Mogan and surroundings using CA-Markov chain analysis. J Environ Biol 38(5):981–989

Feng XQ, Yin XR, Zhang GX (2010) Hydrological response of climate change in Wuyuer River Basin based on SWAT model. Prog Geogr 29(7):827–832

Githui F, Gitau W, Mutua F et al (2009) Climate change impact on SWAT simulated streamflow in western Kenya. Int J Climatol 29(12):1823–1834

Hamel P, Valencia J, Schmitt R et al (2020) Modeling seasonal water yield for landscape management: applications in Peru and Myanmar. J Environ Manage 270:110792
Havrylenko SB, Bodoque JM, Srinivasan R, Zucarelli GV, Mercuri P (2016) Assessment of the soil water content in the Pampas region using SWAT. Catena 137, 298–309
Hou XY, Chang B, Yu XF (2004) Research on land use change of Hexi Corridor based on CA-Markov. Trans Chin Soc Agric Eng 20(5):286–291
Hu ZZ, Zhao ZL, Zhang GX (2018) Simulation of the spatial distribution pattern of Beijing population under the background of non-capital functional unraveling. J Geo-Inf Sci 20(2):71–82
Ji GX (2020) Geographical calculation of runoff changes and droughts and floods in the Yellow River Basin under future climate change. East China Normal University
Karimi H, Jafarnezhad J, Khaledi J et al (2018) Monitoring and prediction of land use/land cover changes using CA-Markov model:a case study of Ravansar County in Iran. Arab J Geosci 11(19):1–9
Lai GY, Wu DY, Zhong YX et al (2012) Development and application progress of SWAT model. J Hohai Univ (Nat Sci) 40(3), 243–251
Li YY, Chang JX, Jin WT, et al (2017) Runoff simulation in subzones of the Weihe River Basin based on the SWAT model. J Northwest & F Univ (Nat Sci) 45(04), 204–212
Liu W (2014) Hydrological simulation and runoff response analysis based on SWAT model. Northwest University
Liu F, Xiong W, Wang YH, Yu PT, Xu LH (2019) The response relationship between landscape pattern and runoff in Jinghe Basin based on LUCC. Arid Land Resour Environ 33(1), 137–142(2019)
Ma CH, Huang Q, Guo AJ (2019) Feature analysis and uncertainty evaluation of joint distribution of water and sediment in Jinghe River Basin. J Hydraul Eng 50(02):273–282
Neitsch SL, Arnold JG, Kiniry JR, Williams JR (2011) Soil & water assessment tool theoretical documentation version 2009. Texas Water Resources Institute, pp 1–647
Pang JP, Liu CM, Xu ZX (2007) Construction and verification of weather generator and database in SWAT model. Journal of China Hydrology 27(5):25–30
Stehr A, Aguayo M, Link O et al (2010) Modelling the hydrologic response of a mesoscale Andean watershed to changes in land use patterns for environmental planning. Hydrol Earth Syst Sci 14(10):1963–1977
Sriworamas K, Prasanchum H, Supakosol J (2020) The effect of forest rehabilitation on runoff and hydrological factors in the upstream area of the Ubolratana Reservoir in Thailand. J Water Clim Chang 11(04):1009–1020
Wei XL, Bailey RT, Records RM, Wible TC, Arabi M (2019) Comprehensive simulation of nitrate transport in coupled surface-subsurface hydrologic systems using the linked SWAT-MODFLOW-RT3D model. Environ Model & Softw 122, 104242
Xie ZL, Gao LM, Si JJ, et al (2018) Research and application progress of SWAT model. J Shandong Ind Technol (16), 211–235
Xu ZX, Cheng L (2010) Research and application progress of distributed hydrological model. J Hydraul Eng 41(09):1009–1017
Yan M, He L, Zheng Mg, Sun LY, Pei L, Wang SJ (2017) Research on the Scale Effect of High Sediment Flow in Jinghe River Basin. J Soil Water Conserv 31(04), 75–80
Yang DW, Li Z, Xu ZX et al (2018) Progress and prospect of hydrology research. Prog Geogr 37(1), 36–45
Yang YZ, Feng ZM, Liu D, Zhang WK (2013) Study on the effect of land-use change on the balance of water and soil resources——Taking the Xiliao River Basin as an example. J Nat Resour 28(03):437–449
Zhao YP (2010) Study on the historical changes and current situation of Jinghe River. Northwest University

Zheng QH, Luo GP, Zhu L, et al (2010) Landscape pattern prediction of Yili River Delta based on CA-Markov model. Chin J Appl Ecol 21(4), 873–882
Zhou Z, Wu JF, Yang Y, et al (2020) Surface runoff simulation of Beishan Reservoir Basin based on SWAT model. South-to-North Water Transf Water Sci & Technol 18(1), 66–73

Chapter 16
A Laboratory Experiment: Remediation of Metal-Contaminated Marine Sediments with Medium Coarse Sand, Fine Sand, Silty Fine Sand and Clay Caps

Jiachen Zeng, Yutuo Cheng, Bin Guo, He Bai, Yuchi Hao, and Hanghang Lv

Abstract In recent years, there has been an increasing demand for disposal of marine dredging materials. In suit capping (ISA) is an effective method for recycling these materials by using capping substances to keep pollutants in storage for an extended period. This research aims to evaluate the effectiveness of medium coarse sand, fine sand, silty fine sand, and clay proposed for capping marine sediments contaminated with trace heavy metals (including AS, Cd, Cr, and Cu) in marine sediments. The metal-contaminated sediment, capping substances and seawater are added to the experimental device successively and the effects of different cap substances and thicknesses on metal spillage within 48 h are monitored by ion chromatography. Results demonstrate that, due to the physical isolation, stabilization and chemical isolation of the capping method, the concentration of heavy metal ions reaches a stable state after 48 h of diffusion when the thickness of the capping material remains unchanged. The release barrier effect of 5 cm caps of the four tested materials on heavy metals can reach more than 90%. It's noted that the higher the water depth, the higher the concentration of heavy metal. Additionally, the diffusion effect of heavy metal ions in contaminated soil is positively correlated with the thickness of caps. Suitable particle size gradation makes the capping effect of clay the best among the four test materials: after 48 h, the blocking effect of 9 cm clay on the release of heavy metals can reach 99%.

Keywords Heavy metal · Dredged material · In suit capping · Metal overflow

Dredging materials refer to sediment excavated underwater by dredging in order to meet the construction needs during the implementation of port and waterway projects. It's the primary product of dredging process (Zhao et al. 2016). The utilization

J. Zeng · Y. Cheng · B. Guo (✉) · H. Bai · Y. Hao · H. Lv
CCCC National Engineering Research Center of Dredging Technology and Equipment Co., Ltd, Shanghai 200082, China
e-mail: guobing@cccc-drc.com

H. Xu (ed.), *The 6th International Symposium on Water Resource and Environmental Management*, Environmental Science and Engineering,
https://doi.org/10.1007/978-3-031-55989-1_16

method of dredged material depends on its physical and chemical properties. As a valuable resource with potential utilization value, dredged soil can be categorized into agricultural utilization, engineering utilization and environmental improvement utilization according to its regeneration value and utilization methods (Huang et al. 2019). Due to the high cost associated with large-scale disposal of marine dredged materials, common methods are land reclamation and offshore dumping (Wang et al. 2019). With the implementation of China's reclamation policy, offshore dumping has become the most important disposal method of dredging materials. The demand for dumping treatment of dredging materials has been increasing, resulting in a significant rise in the volume of dumped materials. According to China's "Marine Environmental Status Bulletin", the amount of dredging materials dumped in China's ocean has reached hundreds of millions of cubic meters per year over the past five years. Comparative analysis of the volume of dredging materials dumped in China and on the international market reveals that China accounts for approximately one-third of the global total, making it a major contributor to ocean dredging material waste (Cheng et al. 2023).

Heavy metals accumulated in dredged sediment represent persistent and highly toxic pollutants that pose significant ecological risks (Zhang et al. 2021). In China, the heavy metal pollution in the sediment of sea and lake cannot be ignored (Han et al. 2011), among which the types of heavy metals mainly include Cu, Pb, Hg and Cr (Ma et al. 2021a). Existing remediation techniques are economically impractical for dealing with large quantities of heavy metal-contaminated sediment. In situ capping (ISC), a cost-effective and long-lasting alternative to dredging, involves applying a protective layer of cleaning material on top of the contaminated sediment. The goal of the ISC is to isolate the sediment from the overlying water column and biota (Boyd et al. 1988), reduce the flux of pollutants into the bioactive portion of the sediment, create new habitats for aquatic organisms (Reible et al. 2010). Commonly used cap materials include granular substances like sand, stone, as well as impermeable or permeable materials such as synthetic geotextiles (Palermo 1998). However, in China, limited research has focused on the treatment of heavy metal pollutants in sediments and dredged materials using the capping method. This study highlights the urgency for research on the use of dredging-derived capping substances in situ for capping heavy metal sediments.

This study focuses on evaluating the barrier effect of different particle size capping materials, various thicknesses, and combinations of cap substances on pollutant diffusion under still water conditions. Cap material selection criteria include effectiveness in blocking pollutant diffusion, stability against erosion, and material availability. Four representative cap materials with different particle sizes were chosen for experimental investigation. In a specially designed capping material device, sediments, capping materials, and seawater were sequentially added and left undisturbed for 48 h. The inhibitory effect of the cap materials on heavy metal diffusion was explored considering factors such as cap species, thickness, and duration.

16.1 Materials and Methods

16.1.1 Main Instruments and Reagents

Inductively coupled plasma Atomic Emission Spectrometry Instrument (ICP-AES, Prodigy7, Leeman, USA); Milli-pore ultra-pure water device.

Copper sulfate ($CuSO_4 \cdot 5H_2O$), Mercury nitrate ($Hg(NO_3)_2 \cdot H_2O$), Chromium nitrate ($Cr(NO_3)_3 \cdot 9H_2O$), Arsenic oxide (As_2O_3) and other reagents used for trace metal analysis are of superior purity or above, and the solution is equipped with Milli-pore ultra-pure water.

All utensils should be soaked in deionized water overnight before use, washed and set aside.

16.1.2 The Selection of Cap Material

The selection of cap materials should follow the principles of large specific surface area, high stability, low cost, easy to obtain and simple pretreatment. In this experiment, medium coarse sand, fine sand, silty fine sand and clay (high silicon content/ high calcium content) are selected as capping substances. Medium coarse sand refers to sand with fineness modulus of 1.6–3.7; Fine sand refers to sand with fineness modulus of 1.6–2.2; Silty fine sand refers to sand with a particle size between 0.075 mm and 4.75 mm after processing and screening; Clay refers to clay with few sand grains and is viscous (Ma et al. 2021b). The particle size grading diagram of several cap substances is as follows (Fig. 16.1):

16.1.3 The Selection of Pollutant

The contents of major heavy metals and organic pollutants in dredges from Yangtze River Estuary are copper (50 mg/kg), cadmium (30–50 mg/kg), chromium (30–40 mg/kg) and lead (15–30 mg/kg). In this experiment, copper, mercury, chromium, arsenic and sulfide are selected as target pollutants for cap disposal research. In order to avoid chemical reactions with cap substances, $CuSO_4 \cdot 5H_2O$, $Hg(NO_3)_2 \cdot H_2O$, $Cr(NO_3)_3 \cdot 9H_2O$ and As_2O_3 dissolved in deionized water with concentrations of 500 mg/L, 200 mg/L, 100 mg/L and 50 mg/L are used to prepare pollution dredges. Place the dredge (about 700 g) in the device, slowly pour the metal contaminant mixture until the dredge is completely soaked, let it stand for 3 days.

16.1.4 Capping Test

The test device is shown in Fig. 16.2. The contaminated sediment, cap material and seawater are loaded in layers. The material of the test device is plexiglass with a thickness of 5 mm, and the size of the test device is 100 × 100 × 100 cm in length × width × height. The dredged sediment thickness is 25 cm; Cleaning and dredging material cover layer thickness is 5 cm, 7 cm and 9 cm.

There are three main samples involved in the test: seawater samples, cap materials and sediment samples. Seawater samples were taken from three water outlets in the seawater plate at different depths (three water samples) for index detection in Sect. 1.5; Cap material and sediment samples are sampled through homemade plexiglass tubes and corresponding supporting facilities.

The specific test design scheme is shown in Tables 16.1 and 16.2.

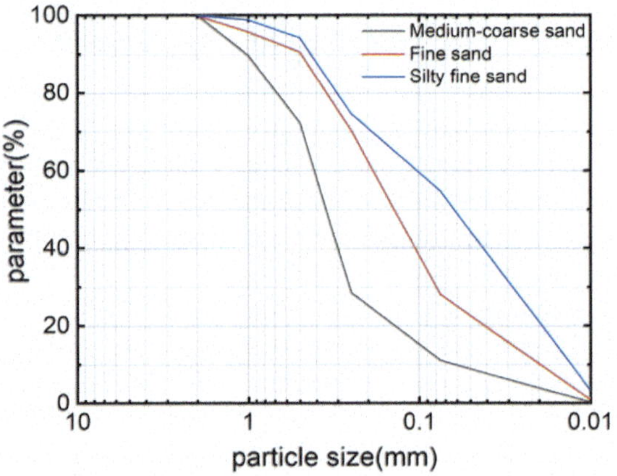

Fig. 16.1 Size gradation of cap materials

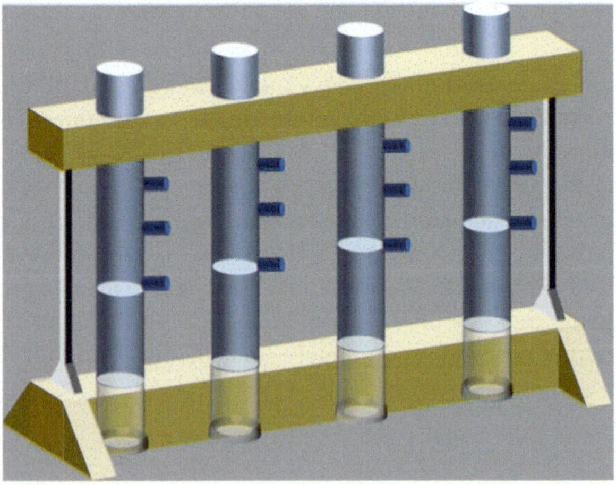

Fig. 16.2 Capping test device 3D model diagram (four reactors)

Table 16.1 Parameters of ocean dumping pollutant diffusion test $L_{16}(4^3)$

Number	Type of capping materials	Thickness of capping materials (cm)	Thickness of polluted sludge (cm)	Depth of water (cm)
1	Medium coarse sand	5	25	40
2	Silty fine sand			
3	Fine sand			
4	Medium coarse sand	7		
5	Silty fine sand			
6	Fine sand			
7	Medium coarse sand	9		
8	Silty fine sand			
9	Fine sand			
10	clay			

Table 16.2 Cotent of various heavy metal in artificially simulated contaminated soil

	Cr(mg/L)	Cu(mg/L)	Cd(mg/L)	Pb(mg/L)
Contaminated sludge 1	3.5802	11.5308	10.4865	11.7811
Contaminated sludge 2	2.6659	7.7328	7.4268	8.0904

16.1.5 Capping Materials Effect Characterization

The adsorption and isolation effect of porous cap material on pollutants (heavy metals, etc.) in dredged soil is explored by characterizing the contact layer between dredged soil and cap material. The solid phase pollutants are dried and digested, and the contents of copper, cadmium, chromium, and lead in the pollutants were determined by inductively coupled plasma atomic emission spectrometry (ICP-AES, Prodigy7, United States). Samples were collected from different diffusion times (6 h, 24 h, 48 h) and different sampling water depths (10cm, 20cm, 30cm), with a total of 90 detection groups. The blocking efficiency is calculated as follows:

$$\varepsilon = \frac{(C_0 - C_1)}{C_0} \times 100\%$$

C_0 is the concentration of heavy metals in the sediments at the beginning of the capping experiment;

C_1 is the heavy metal concentration (mol/L) in the water covered by the capping group at the end of the capping experiment.

16.2 Results and Discussion

16.2.1 Analysis of Diffusion Trend of Heavy Metals

Medium coarse sand as capping material. Figure 16.3 depicts the diffusion trend of four heavy metal ions (Cr, Cu, Cd, Pb) over a 0–48 h period when medium coarse sand as the capping material (at a thickness of 5 cm). Subfigures (a)–(c) respectively illustrate the relationship between the diffusion concentration of heavy metal ions in the aqueous solution and different water depths. The concentrations of the four heavy metal ions in the contaminated soil were found to stabilize under varying water depths after 48 h of being covered with a 5 cm layer of medium coarse sand. Regarding water depth, the concentration of diffused heavy metal ions followed the order: underwater 30 cm > underwater 20 cm > underwater 10 cm. After 48 h, the diffusion ratios for Cr, Cu, Cd, and Pb at a depth of 30 cm were 1.58%, 0.36%, 1.09%, and 0.63%, respectively.

Furthermore, Fig. 16.3d presents the variation in heavy metal ion concentrations within the solution under different cap material thicknesses (medium coarse sand

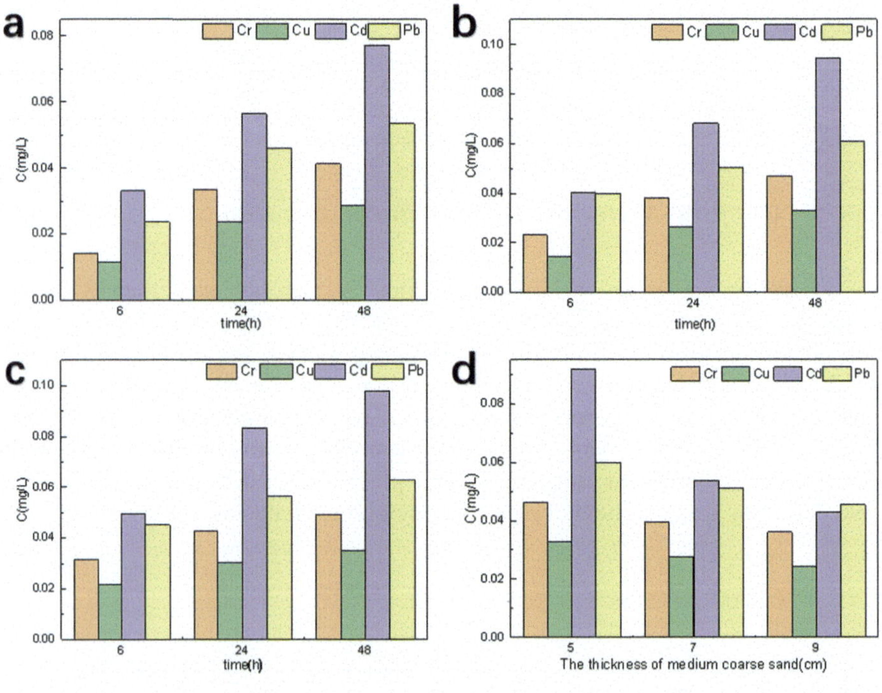

Fig. 16.3 Diffusion of heavy metal ions at **a** 10 cm **b** 20 cm **c** 30 cm under water when medium coarse sand is 5 cm. **d** Influence of thickness of medium coarse sand cap material on diffusion of heavy metal ions when capping time is 48 h

at 5 cm, 7 cm, and 9 cm) after a 48-h diffusion process. The results indicate a negative correlation between the concentration of heavy metal ions in the solution after 48 h and the thickness of the capping material. In other words, a thicker layer of fine sand provides a better inhibitory effect on the diffusion of heavy metal ions in contaminated soil.

Fine sand as capping material. Figure 16.4a–c show the diffusion of four heavy metal ions (Cr, Cu, Cd. Pb) within 0–48 h when the capping material is fine sand (5 cm) respectively. After the four heavy metal ions in the contaminated soil were covered by 5 cm fine sand for 48 h, the ion concentration tended to be stable under different water depths, and the metal ion concentration increased with the increase of water depth. After 48 h, the diffusion ratios of Cr, Cu, Cd and Pb at 30 cm underwater were 1.37%, 0.43%, 0.58% and 0.52%.

Figure 16.4d shows the change of heavy metal ion concentration in solution under different cap material thickens (fine sand 5cm, 7cm, 9cm) after 48h diffusion process. The results show that the concentration of heavy metal ions in 48 h solution is negatively correlated with the thickness of fine sand. The thicker the fine sand, the better the inhibition effect on the diffusion of heavy metal ions in contaminated soil.

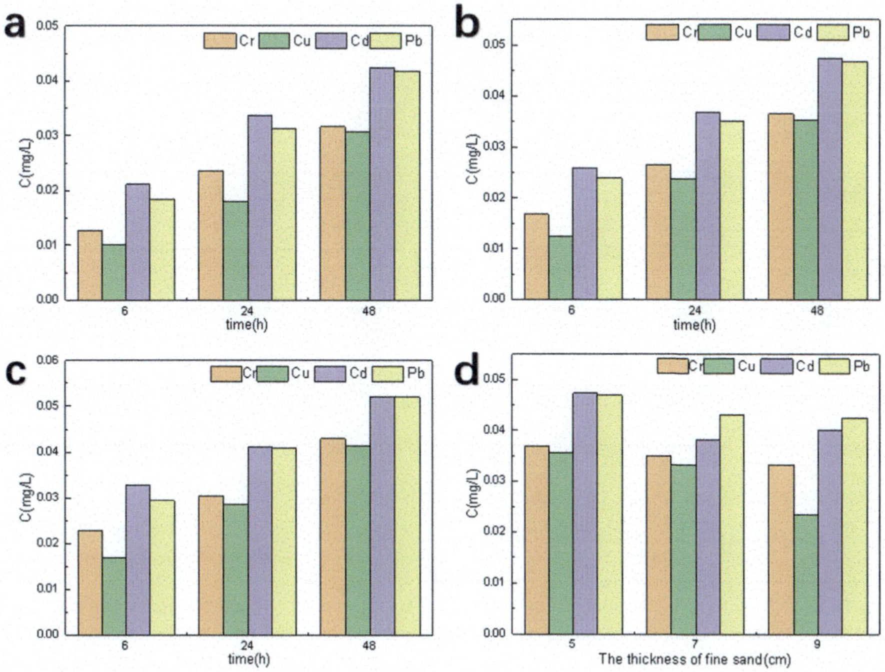

Fig. 16.4 Diffusion of heavy metal ions at **a** 10 cm **b** 20 cm **c** 30 cm under water when fine sand is 5 cm. **d** Influence of thickness of fine sand cap material on diffusion of heavy metal ions when capping time is 48 h

Silty fine sand as capping material. Figure 16.5a–c respectively show the diffusion of four heavy metal ions (Cr, Cu, Cd, Pb) within 0–48 h when the capping material is silty fine sand (5 cm). After being covered by 5 cm fine sand for 48 h, the ion concentrations of the four heavy metal ions in the contaminated soil tended to be stable at different water depths, and the metal ion concentrations increased with the increase of water depth and the metal ion concentrations increased with the increase of water depth. After 48 h, the diffusion ratios of Cr, Cu, Cd and Pb at 30 cm underwater were 1.24%, 3.69%, 0.52% and 0.46%.

Figure 16.5d shows the change of heavy metal ion concentration in solution under different cap material thicknesses (fine sand 5 cm, 7 cm, 9 cm) after 48 h diffusion process. The results show that the concentration of heavy metal ions in 48 h solution is negatively correlated with the thickness of fine sand, and the thicker the thickness of fine sand, the better the inhibition effect on the diffusion of heavy metal ions in contaminated soil, which is consistent with the results of the previous two sections.

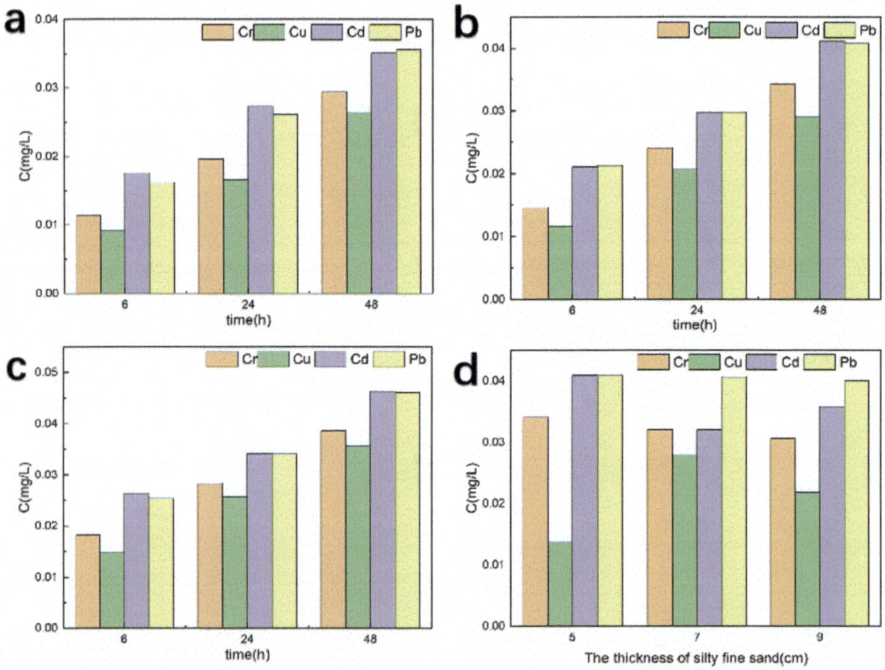

Fig. 16.5 Diffusion of heavy metal ions at **a** 10 cm **b** 20 cm **c** 30 cm under water when silty fine sand is 5 cm. **d** Influence of thickness of silty fine sand cap material on diffusion of heavy metal ions when capping time is 48 h

16.2.2 *Effect of Capping Species on Diffusion Concentration of Heavy Metal Ions*

The change of heavy metal ion diffusion in contaminated soil is similar to that of medium coarse sand at different time and thickness of capping material. When the thickness of capping material remains unchanged, the concentration of heavy metal ions gradually stabilizes after 48 h of diffusion, and the concentration of heavy metal ions is positively correlated with the water depth. The higher the underwater depth, the higher the concentration of heavy metal ions. Furthermore, the diffusion effect of heavy metal ions in contaminated soil is positively correlated with the capping thickness of fine sand and silty fine sand.

Figure 16.6 shows the residual ion concentration of heavy metal ions in the contaminated soil after 48 h diffusion when the cap thickness of medium coarse sand, fine sand, silty fine sand and clay is 9 cm. The concentration of the four heavy metal ions in the figure from high to low is: medium coarse sand > fine sand > fine sand > clayy soil. In terms of the effect of capping substances on inhibiting the diffusion of pollutants, clayy soil has the best effect. The barrier effect of clayy soil on heavy metals can reach 99% after 48h. In addition, the diffusion effect of heavy metal ions is related to the size gradation of the caps. Table 16.3 also shows the results of the diffusion of other types of capping substances to heavy metal ions.

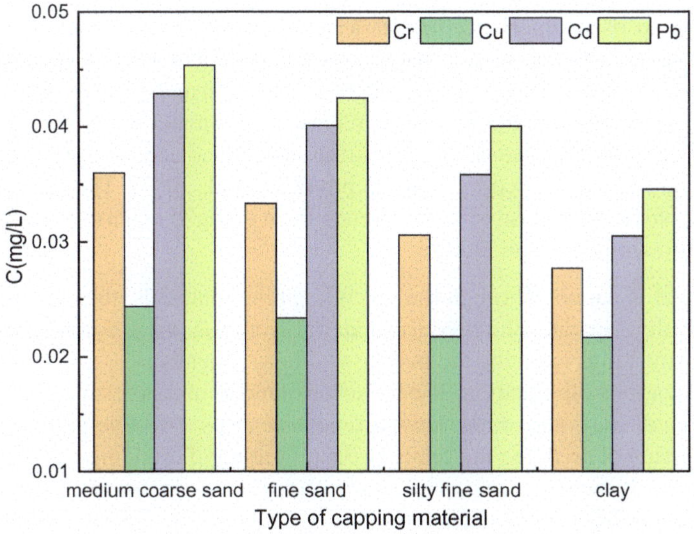

Fig. 16.6 Capping effect of different types of capping substances on heavy metal ions

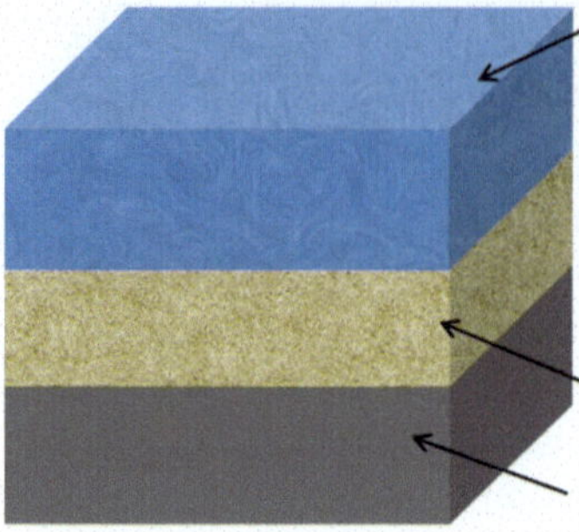

Fig. 16.7 Schematic diagram of capping materials (Giovanni et al. 2017)

Table 16.3 Diffusion ratio of heavy metal ions with different capping substances (9 cm)

	Cr	Cu	Cd	Pb
Medium coarse sand	0.0115	0.0025	0.0048	0.0046
Silty fine sand	0.0107	0.0024	0.0045	0.0043
Fine sand	0.0098	0.0023	0.0040	0.0040

16.2.3 Mechanism of Action of Heavy Metals in Dredged Materials by Capping

The utilization of sediment capping as a treatment method for dredged materials may not effectively reduce or eliminate heavy metal content. However, under specific conditions and suitable sites, capping can be a long-term and effective approach. The capping materials used in this study are mainly neutral materials such as sand, stone and clay, which have the advantages of affordability and accessibility (Zhang et al. 2016), but are easily limited by water depth (Eek et al. 2007). The mechanisms of capping to block the release of heavy metals from dredged materials can be divided into the following (Liu et al. 2001):

(1) Physical isolation: Through the use of capping materials, the capping method physically separates the dredged material from the water column and aquatic organisms, thus reducing the overflow of heavy metals.;
(2) Stabilization/solidification: The inclusion of medium coarse sand, fine sand,silty fine sand, clay and other low-particle size caps effectively inhibit the re-suspension of dredging materials or the migration of dredging materials caused by water disturbance, and achieve the effect of solidification of contaminated soil containing heavy metals;
(3) Chemical isolation: Capping method can effectively block the migration of dissolved pollutants from dredging materials to natural overlying water bodies, and the active capping materials contained in capping substances can also strengthen the chemisorption in the capping process to a certain extent, thus greatly improving the disposal efficiency (Zhang et al. 2016).

16.3 Conclusion

The effects of medium coarse sand, fine sand, silty fine sand and clay as cap materials on reducing the release and stabilization of heavy metals in marine contaminated sediments are evaluated. In several groups of experiments, the heavy metal content in water increase with the increase of water depth, decrease with the increase of the thickness of the capping material. Medium coarse sand, fine sand and fine sand of 5 cm can limit the overflow of heavy metals in contaminated soil to more than 90%. The cap layer effectively inhibits the release of the mobile heavy metals Cd, As, Cr and Cu. The inhibition effect of 9 cm medium coarse sand, fine sand, fine sand and clay on heavy metal overflow of contaminated soil within 48 h can reach 99%, and the particle size gradation of caps plays an important role in the process of heavy metal release. The mechanism of blocking heavy metals by caps only includes physical isolation, stabilization treatment and chemical isolation.

Although capping method is an effective method to interrupt the release of heavy metals, traditional materials are not effective in fixing heavy metals. Compound dredging materials can be studied to further stabilize them. In addition, subsequent experiments also need to further extend the experimental time to improve the accuracy of continuous extraction experiments, so as to analyze the efficiency of the capping material.

References

Boyd AS, Lee JF, Mortland MM (1988) Attenuating organic contaminant mobility by soil modification. Nature 333:345–347

Cheng HF, Liu J, Chen FK, Fu G, Wang ZZ, Xie YF (2023) Study on comprehensive utilization of dredged soil in yangtze estuary channel during the 14th five-year period. Resour Environ Yangtze Basin 32:331–338

Eek E, Godoy A, Aagaard P, Breedveld AD (2007) Experimental determination of efficiency of capping materials during consolidation of metal-contaminated dredged material. Chemosphere 69:719–728

FRTR. FRTR remediation technologies screening matrix and reference guide, version 4.

Giovanni L, Diego M, Giusy L, Marco G, Mauruzio C, Francesco A, Barbara C, Michele N (2017) Toxicity assessment within the application of in situ contaminated sediment remediation technologies: a review. Sci Total Environ 621:85–94

Han ZX, He GD, Li X, Cui YZ (2011) Safety evaluation of dredge mud dumping in Jiangsu sea area on marine environment. Environ Pollut Cont 33(3):11–14

Huang XF, Wang Z, Ye G, Ye GY, Li SQ, Li SY, Liu J (2019) Improvement of soilphysico-chemical properties by dredged sediments to promote the rapid colonization of Phragmites. Acta Sci Circum 39:4261–4268

Liu CH, Jay JA, Ika R, Shine JP, Ford TE (2001) Capping Efficiency for metal-contaminated marine sediment under conditions of submarine. Environ Sci Technol 35:2334–2340

Ma GX, Liu S, Liu HY, Huang LL, Ma T, Chen HD, Liu J (2021a) Geotextile tubes technology for dewatering and volume reduction of sediments polluted by heavy metals and its engineering case. Chin J Environ Eng 15:3773–3780

Ma GL, He X, Lu HM, Wu HR, Liu HL, Chu J, Xiao Y (2021b) Strength of biocemented coarse sand with kaolin micro-particle improved nucleation. Chin J Geotech Eng 43:290–299
Palermo MR (1998) Design considerations for in-situ capping of contaminated sediments. Water Sci Technol 37:315–321
Reible D, Hayes D, Lue-Hing C, Patterson J, Bhowmik N, Johnson M, Teal J (2010) Comparison of the long-term risks of removal and in situ management of contaminated sediments in the fox river. Soil Sediment Contam: Int J 12(3):325–344
Wang XM, Kong LN, Chen JY, Zhao DM, Chen H, Sun RJ, Yang WC, Han JB (2019) Distribution of butyltins at dredged material dumping sites around the coast of China and the potential ecological risk. Mar Pollut Bull 138:491–500
Zhang C, Zhu MY, Zeng GM, Yu ZG, Cui F, Yang ZZ, Shen LQ (2016) Active capping technology: a new environmental remediation of contaminated sediment. Environ Sci Pollut Res 23:4370–4386
Zhang YY, Labianca C, Chen L, Degisi S, Notarnicola M, Guo BL, Sun J, Ding SM, Wang L (2021) Sustainable ex-situ remediation of contaminated sediment: a review. Environ Pollut 117333:287
Zhao M, Liu H, Wang WJ, Tian T, Peng X, Wu Y (2016) Pollution situation of dredged sediment and treatment technology. J Salt Chem Ind 8:1–7

Chapter 17
Study on Water Supply Guarantee in Xi'an Urban Area Before the Operation of Hanjiang-to-Weihe River Valley Water Diversion Project

Kunming Lu, Jiawei Kou, Gaofeng Zhang, and Zhuoli Xu

Abstract In recent years, Xi'an urban population continues to grow, and the urban water supply projects have been basically full load. It is urgent to find new stable water sources to ensure water supply during the peak period. In order to solve the problem of water shortage in Xi'an 's economic and social development before the operation of Hanjiang-to-Weihe River Valley Water Diversion Project, based on the current situation and problems of urban water supply, analyzed the balance between water supply and demand in Xi'an urban area. By combing the available water sources, water transmission and distribution projects and water plants in the local and surrounding areas, formulated the annual water supply guarantee scheme from 2022 to 2027, and put forward the water supply safety guarantee measures. It is helpful to alleviate the increasingly prominent contradiction between water supply and demand in Xi'an, improve the safety of water supply in the city, and help the economic and social development of the city.

Keywords Hanjiang-to-Weihe River Valley Water Diversion Project · Xi'an · Water supply · Guarantee

17.1 Introduction

In recent years, with the economic and social development of Xi'an City by leaps and bounds, urban water supply demand has been rising year after year, the water supply capacity of the Heihe River, the Lijiahe River, and groundwater has basically reached full capacity, and the pressure of water supply has become increasingly severe.

K. Lu · J. Kou (✉) · G. Zhang
PowerChina Northwest Engineering Corporation Limited, Xi'an 710065, China
e-mail: koujiawei_hhu@163.com

Z. Xu
College of Electrical Engineering, Xi'an Jiaotong University, Xi'an 710049, China

H. Xu (ed.), *The 6th International Symposium on Water Resource and Environmental Management*, Environmental Science and Engineering,
https://doi.org/10.1007/978-3-031-55989-1_17

As a major inter-basin water transfer project in Shaanxi Province, the Hanjiang-to-Weihe River Valley Water Diversion Project is a strategic water source project to alleviate the water shortage in Guanzhong area in the future, and the Hanjiang-to-Weihe River Valley Water Diversion Project will supply 5×10^8 m^3 of water to Xi'an city per year before the completion of the South-to-North Water Diversion Middle Line Project, which will effectively solve the problem of water shortage for the economic and social development of Xi'an city (Wei 2016). However, the construction cycle of the Hanjiang-to-Weihe River Valley Water Diversion Project is long, it will be built up until 2027.

Before the commissioning of the Hanjiang-to-Weihe River Valley Water Diversion Project, there is an urgent need to consider the appropriate addition of new water supply projects to safeguard the water demand for urban development, reduce the degree of water shortages, and provide emergency protection in the event of overhaul and emergencies of the existing water supply system.

17.2 Status of Urban Water Supply

In 2021, There are 256 water wells, 3 surface water sources, 11 water works, 4,218 km of water pipeline network, a water supply population of 6.95×10^6 people, and a public water supply popularization rate of 99% in Xi'an urban. The total water supply quantity was 6.49×10^8 m^3, and the water sales quantity was 5.57×10^8 m^3, with a loss rate of 9.76%. Among the components of the water sales quantity, the water consumption of residents accounted for 63% of the total water sales quantity. The maximum daily water supply quantity is 2.07×10^6 m^3/d. The following problems mainly exist in the urban water supply: high degree of water sources development, limited water supply capacity of water source projects; water transmission projects close to full capacity, making it difficult to make use of the water diverted from the Hanjiang River ahead of time; water works have been running at full capacity for a long time, and there is an urgent need to expand their scales; the water supply system is not developed in a balanced way, and there is insufficient water supply for the local highland districts and the new districts.

17.3 Water Demand Prediction

According to the Xi'an Urban Water Supply "14th Five-Year Plan", the interpolation method is used to get the water supply population in Xi'an City from 2022 to 2027 (the expected year of the water supply of Hanjiang-to-Weihe River Valley Water Diversion Project) (Table 17.1).

The average daily water demand and the maximum daily water demand in Xi'an urban area from 2022 to 2027 are predicted by the method of comprehensive urban

Table 17.1 Forecast result of water supply population in Xi'an urban area

Years	2022	2023	2024	2025	2026	2027
Population ($\times 10^4$ persons)	700.04	706.36	712.83	719.43	726.19	733.09

Table 17.2 Forecast result of Water Demand in Xi'an urban area

Water demand ($\times 10^4$ m^3/d)	Years	Comprehensive urban water consumption index method	Correlation of the proportion of comprehensive domestic water consumption method	Annual growth rate method
Average daily	2022	193.98	196.97	188.00
	2023	209.84	214.24	198.87
	2024	226.00	231.84	210.36
	2025	242.46	248.61	222.52
	2026	244.72	250.92	235.38
	2027	247.02	253.27	248.99
Maximum daily	2022	224.50	222.61	218.35
	2023	242.87	245.28	230.56
	2024	261.59	263.16	243.44
	2025	280.67	280.08	257.05
	2026	283.29	282.68	271.42
	2027	285.98	285.34	286.59

water consumption index, the method of correlation of the proportion of comprehensive domestic water consumption, and the method of annual growth rate (Table 17.2). The water demand prediction results of the 3 methods are relatively close, and the differences between the results are mainly due to the differences in the principles of prediction methods and the selection of prediction indicators. According to the review results and the relevant water supply planning, the comprehensive urban water consumption index method is chosen for the water demand prediction results.

17.4 Water Supply Capacity

Xi'an has formed a water supply pattern in which the surface water supply system of the Heihe River is the main source of water supply, supplemented by groundwater sources in the suburbs. There are 3 surface water sources: the Heihe River, the Lijiahe River, and the Chanhe River, and 5 groundwater sources along the Fenghe River, the Wehe River, the Chanhe River, the Bahe River, the Zaohe River, and the Laohe River, with a total water supply capacity of 2.26×10^6 m^3/d, and an actual

Table 17.3 Statistics of current water sources and water works in Xi'an urban area

Name of water sources	Name of water works	Designed water supply capacity ($\times 10^4$ m^3/d)	Actual water supply capacity ($\times 10^4$ m^3/d)
Surface water of the Heihe River system	Qujiang water works	60	60
	Leyouyuan water works	50	50
	Caotang water works	15	15
	Southwest suburb water works (Phase I, Stage 1)	20	20
Surface water of the Lijiahe River system	Bailuyuan water works	25	25
Surface water of the Chanhe River	Chanhe surface water works	6	2
Groundwater in Duan Village along the Bahe River	The second water works	50	50
Riverside water along the Fenghe River and the Zaohe River	The third water works		
Groundwater along the Weihe River	The fourth water works		
Groundwater of the Weihe River in the northwestern suburbs	The fifth water works		
Groundwater along the Laohe River and the Weihe River	Laowei water works		
Total		226	222

total water supply capacity of 2.22×10^6 m^3/d under extreme drought conditions. Statistics of the current urban water supply projects are shown in Table 17.3.

17.5 Water Supply and Demand Balance Analysis

If new construction and expansion projects are not considered, the daily water supply capacity of Xi'an urban area will not be able to meet the average daily water demand from 2024, and the daily water supply capacity will no longer be able to meet the maximum daily water demand from 2022. There is still a large gap between the water supply capacity of the existing water supply projects and the water demand in the planning years. The results of the primary supply and demand balance analysis for each year before the commissioning of the Hanjiang-to-Weihe River Valley Water Diversion Project are shown in Table 17.4.

Table 17.4 Primary supply and demand balance analysis in Xi'an urban area

Years	Water supply capacity ($\times 10^4$ m^3/d)			Average daily ($\times 10^4$ m^3/d)		Maximum daily ($\times 10^4$ m^3/d)	
	Surface water	Groundwater	Total	Water demand	Remaining	Water demand	Remaining
2022	172.00	50.00	222.00	193.98	28.02	224.50	−2.50
2023				209.84	12.16	242.87	−20.87
2024				226.00	−4.00	261.59	−39.59
2025				242.46	−20.46	280.67	−58.67
2026				244.72	−22.72	283.29	−61.29
2027				247.02	−25.02	285.98	−63.98

17.6 Water Supply Security Schemes

17.6.1 Programming Principles

(1) Fully integrated and moderately advanced;
(2) Prioritize the use of local water and fair use of external water;
(3) Make full use of the existing works and construct related works as appropriate;
(4) Put saving in the first place, efficient and economical (Zheng et al. 2006).

17.6.2 Water Supply Security Scheme in 2022

In 2022, the water supply capacity of all projects in Xi'an City will be 2.22×10^6 m^3/d, which can basically meet the maximum daily needs. Meanwhile, with the completion and operation of the Southwest suburb water works (Phase I, Stage 2) and the connecting project between the water works and Xi'an City at the end of June, the water supply capacity of Xi'an City during the peak period will reach 2.42×10^6 m^3/d, which can meet the emergency water demand in the year. The analysis of the secondary supply and demand balance of Xi'an City in 2022 is shown in Table 17.5.

Table 17.5 Secondary supply and demand balance analysis in 2022

Water supply capacity ($\times 10^4$ m^3/d)				Average daily ($\times 10^4$ m^3/d)		Maximum daily ($\times 10^4$ m^3/d)	
Surface water	Increased surface water	Groundwater	Total	Water demand	Remaining	Water demand	Remaining
172.00	20.00	50.00	242.00	193.98	48.02	224.50	17.50

17.6.3 Water Supply Security Scheme in 2023

In 2022, the water source of the Heihe River system has reached the designed water supply capacity, and although the water diverted from the Hanjiang River entered the Huangchi gully hub and connected with the Heihe River system, the lack of a new water pipeline will make it impossible to use the water diverted from the Hanjiang River. Considering the discrepancy between the forecasted water demand and the actual water demand, to ensure the safety of urban water utilization in 2023 and to give consideration to the increase of water utilization after 2023, groundwater sources in the northeastern suburb of the Bahe River and riverside water along the Laohe River and the Weihe River should be explored in 2023.

At present, the Northeast suburb groundwater source project (designed water supply capacity of 1×10^5 m^3/d) has the possibility of supplying water to Xi'an City in 2023. The riverside water along the Laohe River and the Weihe River are the existing water source (design water supply capacity of 1×10^5 m^3/d), affected by the limited capacity of the water pipeline, the water supply capacity is 5×10^4 m^3/d currently, there are remained capacity of 5×10^4 m^3/d. Therefore, from the point of view of the water source, the increased water supply capacity in 2023 is 1.5×10^5 m^3/d.

Currently, Xi'an has started the construction of Northeast suburb water works and the expansion of the supporting projects of Laowei water works, to make full use of these two plants before the peak water consumption period, the construction and expansion should be completed before the end of May 2023. At that time, the scale of the new water plants will be 1.5×10^5 m^3/d. The Northeast suburb water works is connected to the city's water supply pipeline of DN 1000, and the Laowei water works is connected to the Xi'an urban water supply network.

After the implementation of the water supply scheme in 2023, the water supply capacity in Xi'an urban area will reach 2.57×10^6 m^3/d, which can meet the emergency water demand in that year. A comparison of the increased water demand and supply capacity in 2023 is shown in Table 17.6. The analysis of the secondary supply and demand balance is shown in Table 17.7.

Table 17.6 Comparison of increased water demand and water supply capacity in 2023

Increased average daily water demand ($\times 10^4$ m^3/d)	Increased maximum daily water demand ($\times 10^4$ m^3/d)	Increased water supply capacity of water sources ($\times 10^4$ m^3/d)	Increased water supply capacity of water works ($\times 10^4$ m^3/d)	Increased water supply capacity ($\times 10^4$ m^3/d)
15.86	18.37	15.00	15.00	15.00

Table 17.7 Secondary supply and demand balance analysis in 2023

Water supply capacity ($\times 10^4$ m^3/d)			Average daily ($\times 10^4$ m^3/d)		Maximum daily ($\times 10^4$ m^3/d)	
Surface water	Groundwater	Total	Water demand	Remaining	Water demand	Remaining
202.00	55.00	257.00	209.84	47.16	242.87	14.13

17.6.4 Water Supply Security Scheme in 2024

The Bahe River surface water has an available quantity of 3.03×10^8 m^3. Before the completion of the Hanjiang-to-Weihe River Valley Water Diversion Project, the water source of Bahe water works is the surface water of the Bahe River, with a water supply capacity of 1×10^5 m^3/d. At present, the Bahe water works has the possibility of supplying water before the peak water consumption period in 2024.

The reservoirs of the Dayu water supply system have a capacity of 1.52×10^7 m^3. According to Xi'an City Water Supply's "14th Five-Year Plan", the Dayu water supply system can supply water quantity of 5×10^4 m^3/d to Xi'an City in 2024. At present, the Dayu water supply system is connected to the water supply pipeline of Leyouyuan water works, it can supply water to Leyouyuan water works. By replacing the water from the Heihe system with the water from the Dayu system, which will supply 5×10^4 m^3/d of water to Southwest suburb water works (Phase II, Stage 1).

Southwest suburb water works (Phase II, Stage 1) is scheduled to be completed in June 2023, which will add 3×10^5 m^3/d of water supply capacity. 1.5×10^5 m^3/d of water supply capacity will be increased in 2024 by utilizing the newly constructed Bahe surface water works and the expanded Southwest suburb water works (use 5×10^4 m^3/d in this year). There are two water supply trunk pipelines of DN 2000 planned for the Bahe water works site, and Southwest suburb water works has been connected to the Xi'an urban water supply pipeline (Xi'an Water (Group) Planning, Design, and Research Institute Co., Xi'an Water Supply Project Special Planning (Draft for Review) 2021).

After the implementation of the water supply scheme in 2024, the water supply capacity in Xi'an urban area will reach 2.72×10^6 m^3/d, which can meet the emergency water demand in that year. A Comparison of the increased water demand and supply capacity in 2024 is shown in Table 17.8. The analysis of the secondary supply and demand balance is shown in Table 17.9.

Table 17.8 Comparison of increased water demand and water supply capacity in 2024

Increased average daily water demand ($\times 10^4$ m^3/d)	Increased maximum daily water demand ($\times 10^4$ m^3/d)	Increased water supply capacity of water sources ($\times 10^4$ m^3/d)	Increased water supply capacity of water works ($\times 10^4$ m^3/d)	Increased water supply capacity ($\times 10^4$ m^3/d)
16.16	18.72	15.00	40.00	15.00

Table 17.9 Secondary supply and demand balance analysis in 2024

Water supply capacity ($\times 10^4$ m^3/d)			Average daily ($\times 10^4$ m^3/d)		Maximum daily ($\times 10^4$ m^3/d)	
Surface water	Groundwater	Total	Water demand	Remaining	Water demand	Remaining
217.00	55.00	272.00	226.00	46.00	261.59	10.41

17.6.5 *Water Supply Security Scheme in 2025*

By 2025, local water sources will no longer meet the increased water demand and will have to utilize the water diverted from the Hanjiang River. However, because the north trunk line and south trunk line have not yet been completed, the limited transmission capacity of the Heihe water supply pipeline and the diversion project of the Southwest suburb water works make it difficult to transport the water from Hanjiang River to the urban area of Xi'an, so a new transmission pipeline must be built.

At present, Xi'an has constructed a diversion channel for the Heihe River irrigation area, with a design diversion capacity of 3×10^5 m^3/d, and a length of 29.5 km of channel has been constructed. The end of the channel is only 12 km away from Southwest suburb water works in a straight line, and it has not been put into use. Considering the height of the end of the channel is lower than the height of Southwest suburb water works, the water delivery requires pressure, and bidirectional connecting pipeline between Southwest suburb water works and Doumen Reservoir has been built, so it is recommended that by the end of May 2025, a 15 km long water transmission line from the end of the the irrigation channel to the connecting pipeline of Southwest suburb water works and the Doumen Reservoir, and a 1 km long water transmission line from the Linjiawan confluence pond of the Hanjiang-to-Weihe River Valley Water Diversion Project to the beginning of the irrigation channel should be built. The water transmission line (Power China Northwest Engineering Corporation Limited 2021) will convey the water through the irrigation channel to Doumen Reservoir for storage before the peak of water consumption period, and supply water to the neighboring Xixian's fourth water works and Xixian's first water works during the peak of water consumption period, and then supply water to the urban area of Xi'an through the water supply pipeline network connecting Xixian New Area and Xi'an city.

The water supply scale of Xixian's first water works (Phase I) is 8×10^4 m^3/d, which is planned to be completed before 2025; the water supply scale of Phase II is 8×10^4 m^3/d, which is planned to be completed before 2035. The water supply scale of Xixian's fourth water works (Phase I) is 8×10^4 m^3/d, which is planned to be completed before 2025. Phase I of the two water works are both constructed with raw water transmission pipelines, water purification plants, and other ancillary facilities, the water supply sources are both the Hanjiang River water, which is introduced into Doumen Reservoir (Xi'an Water (Group) Planning and Design, and

Table 17.10 Comparison of increased water demand and water supply capacity in 2025

Increased average daily water demand ($\times 10^4$ m^3/d)	Increased maximum daily water demand ($\times 10^4$ m^3/d)	Increased water supply capacity of water sources ($\times 10^4$ m^3/d)	Increased water supply capacity of water works ($\times 10^4$ m^3/d)	Increased water supply capacity ($\times 10^4$ m^3/d)
16.46	19.07	137.00	30.00	24.00

Table 17.11 Secondary supply and demand balance analysis in 2025

Water supply capacity ($\times 10^4$ m^3/d)			Average daily ($\times 10^4$ m^3/d)		Maximum daily ($\times 10^4$ m^3/d)	
Surface water	Groundwater	Total	Water demand	Remaining	Water demand	Remaining
241.00	55.00	296.00	242.46	53.54	280.67	15.33

Research Institute Co., Special Planning for Urban Water Supply in Xixian New Area (Revision) 2021) through the south Line of the Hanjiang-to-Weihe River Valley Water Diversion Project. To meet the increased maximum daily water demand of 1.91×10^5 m^3/d in 2025, it is necessary to build Xixian's first water works (Phase II) of 8×10^4 m^3/d before the end of May 2025.

After the implementation of the water supply scheme in 2025, the water supply capacity of Xi'an urban area can reach 2.96×10^6 m^3/d, which can meet the emergency water demand in that year. A Comparison of the increased water demand and water supply capacity in 2025 is shown in Table 17.10. The analysis of secondary supply and demand balance is shown in Table 17.11.

17.6.6 Water Supply Security Scheme in 2026

In 2025, after the completion of the Heihe irrigation channel and the Southwest suburb water works—Doumen Reservoir connection project, the water supply capacity of the Xi'an urban area will have a large increase, and the average and maximum daily remaining water supply capacity at the end of 2025 will be larger and will be able to satisfy the increased average and maximum daily water consumption in 2026, so there will be no more new water source project, water transmission and distribution project, and water works in 2026. The analysis of secondary supply and demand balance in 2026 is shown in Table 17.12.

Table 17.12 Secondary supply and demand balance analysis in 2025

Water supply capacity ($\times 10^4$ m^3/d)			Average daily ($\times 10^4$ m^3/d)		Maximum daily ($\times 10^4$ m^3/d)	
Surface water	Groundwater	Total	Water demand	Remaining	Water demand	Remaining
241.00	55.00	296.00	244.72	51.28	283.29	12.71

Table 17.13 Secondary supply and demand balance analysis in 2025

Water supply capacity ($\times 10^4$ m^3/d)			Average daily ($\times 10^4$ m^3/d)		Maximum daily ($\times 10^4$ m^3/d)	
Surface water	Groundwater	Total	Water demand	Remaining	Water demand	Remaining
241.00	55.00	296.00	247.02	48.98	285.98	10.02

17.6.7 Water Supply Security Scheme in 2027

After the completion of the Heihe irrigation channel and the Southwest suburb water works—Doumen Reservoir connection project in 2025, the water supply capacity of Xi'an urban area will still be able to meet the increased average and maximum daily water consumption in 2027, so there will be no more new water source projects, water transmission and distribution projects, and water works in 2027. The analysis of the balance of secondary supply and demand of Xi'an urban area in 2027 is shown in Table 17.13.

17.6.8 Summary of Schemes

A summary of water supply guarantee schemes before the opening of the Hanjiang-to-Weihe River Valley Water Diversion Project is shown in Table 17.14 and Fig. 17.1.

17.7 Security Measures for Water Supply Safety

17.7.1 Constructing and Managing the Emergency Water Sources

It is still necessary to strengthen the management of the Shibianyu emergency water source before the completion of the Hanjiang-to-Weihe River Valley Water Diversion Project and to do a good job in the dispatch of Shibianyu Reservoir. The Lijiahe River system should use Luyuan Reservoir as its emergency water source. The management

Table 17.14 Summary of water supply guarantee schemes in Xi'an urban area

Years	Water source projects	Water transmission and distribution projects	Water works projects	Increased water supply capacity ($\times 10^4$ m^3/d)	Note
2022	Heihe River system	Hehei River diversion channel	Southwest suburb water works (Phase I, Stage II) (2×10^5 m^3/d)	20	
2023	Groundwater in the north-eastern suburbs	–	Northeast suburb water works (1×10^5 m^3/d), Laowei water works (5×10^4 m^3/d)	15	
2024	Bahe River surface water, Dayu water supply system, Heihe River system	Hehei River diversion channel, Dayu water system—Leyou yuan water works connection pipeline	Bahe water works (1×10^5 m^3/d), Southwest suburb water works (Phase I, Stage II + Phase II, Phase I) (5×10^4 m^3/d utilized this year)	15	Replacing the water from the Heihe system with the water from the Dayu system, supply water to the Southwest suburb water works (5×10^4 m^3/d)
2025	Heihe River system, Hanjiang-to-Weihe River Valley Water Diversion Project	Heihe River irrigation channel	Xixian's first water works (Phase I) (8×10^4 m^3/d) + (Phase II) (8×10^4 m3/d), Xixian's fourth water works (Phase I) (8×10^4 m^3/d)	24	Connecting the Heihe system and the Hanjiang-to-Weihe River Valley Water Diversion Project, building irrigation channel connection projects, and advancing the construction of Xixian's first water works (Phase II)

(continued)

Table 17.14 (continued)

Years	Water source projects	Water transmission and distribution projects	Water works projects	Increased water supply capacity ($\times 10^4$ m^3/d)	Note
2026	Heihe River system, Hanjiang-to-Weihe River Valley Water Diversion Project	Heihe River irrigation channel	–	0	There is no need to add new water supply projects, continue to utilize the remaining water supply capacity
2027	Heihe River system, Hanjiang-to-Weihe River Valley Water Diversion Project	Heihe River irrigation channel	–	0	There is no need to add new water supply projects, continue to utilize the remaining water supply capacity
Total				74	

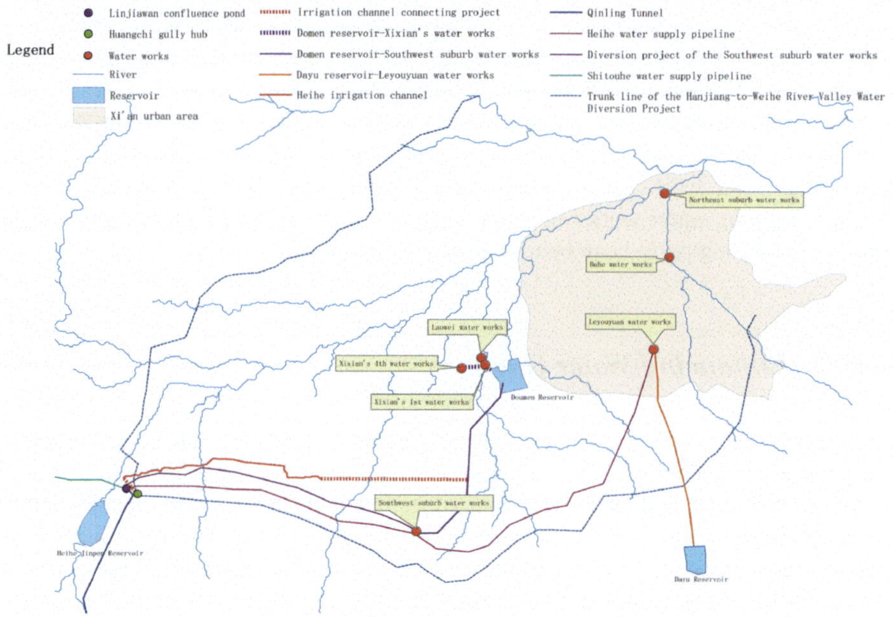

Fig. 17.1 Distribution of water supply guarantee projects in Xi'an urban area

of Dayu Reservoir, Xujiagou Reservoir, and Donggou Reservoir as the city's emergency water sources should be strengthened. The Taipingyu, Gaoguanyu, Tianyu, Juyu, and Ganyu water sources projects should be constructed, to improve the guaranteeing rate of the water supply of the Heihe system. Self-supplied wells with conditions can also be considered as emergency water sources, but the management of self-supplied wells needs to be improved (Chen et al. 2009).

17.7.2 Formating a Pattern of Mutual Security Between the Hanjiang-to-Weihe River Valley Water Diversion Project and the Heihe System

When the Heihe system needs to be repaired, the amount of water supplied from the Hanjiang-to-Weihe system to Southwest suburb water works, from Shibianyu Reservoir to Xi'an Ziwu water works, and from Jingyugou Reservoir to Hanjiang-to-Weihe Bahe water works, Qiujiang water works, and the South suburb water works should be increased. In case of failure in the section of the Hanjiang-to-Weihe system from Huangchi gully hub to Southwestern suburb water works, the Heihe system, Fenghe River storage reservoir, and the Laohe River storage reservoir can be utilized to supply water to Southwestern suburb water works, Shibianyu Reservoir can be utilized to supply water to the Xi'an Ziwu water works, and Jingyugou Reservoir can be utilized to supply water to Hanjiang-to-Weihe Bahe water works, Qiujiang water works, and South suburb water works. In case of maintenance of the Qinling Tunnel of the Hanjiang-to-Weihe River Valley Water Diversion Project, water can be transferred from the Heihe Confluence Pool and Jingyugou Reservoir into the Hanjiang-to-Weihe system for short-term emergencies. After the completion of the supporting water works of the Hanjiang-to-Weihe River Valley Water Diversion Project, Qiujiang water works and South suburb water works which running at full capacity for a long period can be updated in due course.

17.7.3 Enhancing Water Quality

Firstly, the water source monitoring and early warning system should be constructed and improved. Secondly, it is necessary to accelerate the transformation of the groundwater works, add the biological activated carbon filtration and other in-depth treatment processes to adapt to the water supply requirements. Thirdly, it is recommended that water works such as Southwest suburb water works and Bailuyuan water works, construct emergency treatment facilities with the main body of the powdered activated carbon, potassium permanganate, and the acid and alkali emergency dosing system, to meet the needs of emergency treatment under different pollution conditions (Guoguang et al. 2011). Heihe River irrigation channel is an open channel now, as the city emergency water diversion project in the planning years, it needs to strengthen the protection of water sources, which can be taken in advance to cover, fencing, close local segments, and other engineering measures, take water pollution prevention and control measures, and design emergency response to environmental emergencies, to ensure that the water quality of the water conveyance is stable and meets the water quality standards. In addition, the water quality of Shitouhe Reservoir can be improved in time after the Hanjiang-to-Weihe River Valley Water Diversion Project is put into operation (Xi'an Municipal Water Bureau 2022).

17.7.4 Improving the Water Transmission and Distribution Networks

Accelerating the construction and renovation of the water distribution network from each water works to Xi'an City, improving the coverage of the water distribution network, expanding the network with insufficient water transmission capacity as soon as possible, and reducing the leakage rate of pipeline (Ruibao et al. 2010). Aiming at the Aerospace base, Chang'an high area, and the second phase of Qujiang area in southeast Xi'an with high terrain and rapid urbanization, it is suggested that expanding the pressurized station and pumping station, pressurized water pipeline and distribution pipeline as soon as possible, and in the far future, the water shortage can be supplied by Hanjiang-to-Weihe Ziwu water works and other projects (Lijun et al. 2010).

17.8 Conclusions and Recommendations

(1) By the completion and joint operation of planned water sources projects, water transmission and distribution projects, and water works, the water supply capacity can meet the water demand of Xi'an City and the increased annual water demand under extreme drought conditions, and the remaining water can be supplied to adjacent districts and counties.
(2) Due to the rapid growth of the urban population in Xi'an, but the construction of water supply projects requires a period, there may be a lag in the water supply amount compared to the growth of water demand. Therefore, groundwater is still an important water source to make up for the water shortage in the next few years. Within the scope of the groundwater utilization control index, the emergency water supply capacity of the groundwater source can be appropriately explored in the peak time, then the groundwater can be replaced by the surface water after the commissioning of the Hanjiang-to-Weihe River Valley Water Diversion Project, to conserve the groundwater.
(3) Promoting water conservation in all sectors of society by raising awareness of water conservation, strengthening constraints, making up for shortcomings in facilities, and improving market mechanisms, so that the efficiency and effectiveness of water resources utilization can be greatly improved.

References

Chen X, Xu Y, Li S, et al (2009) Enhancing construction of water supply safety guarantee system in China. China Water Supply Drain 25(14):25–27

Guoguang C, Mingqun M, Jingru Q (2011) Strengthening water quality management to ensure water supply quality. China Water Wastewater 37(10):45–48

Lijun GUO, Caiyun SUN, Shuqin CHEN et al (2010) Study on the safety guarantee system of water supply in urban area of Xinmi City. J Water Resour Water Eng 21(3):5–9

PowerChina Northwest Engineering Corporation Limited (2021) Preliminary design report (finalized draft) for the Hanjiang-to-Weihe River Valley Water Diversion Project-Phase II

Ruibao J, Shaohua S, Wuchang S et al (2010) Study of state and guarantee countermeasures of Yinhuang water supply system. Water Wastewater Eng 36(z1):26–29

Wei T (2016) Analysis of the conditions for prior water supply of the Hanjiang-to-Weihe River Valley Water Diversion Project. Shaanxi Water Conserv 16(3):13–14

Xi'an Municipal Water Bureau (2022) Xi'an municipal development and reform commission. Xi'an urban water supply "14th five-year plan"

Xi'an Water (Group) Planning and Design, and Research Institute Co. Special Planning for Urban Water Supply in Xixian New Area (Revision) (2021)

Xi'an Water (Group) Planning, Design, and Research Institute Co. Xi'an Water Supply Project Special Planning (Draft for Review) (2021)
Zheng W, Zhulai W, Yong Z et al (2006) Study on emergency response and guarantee system of water-supply safety in city. J Catastrophology 21(2):106–109

Chapter 18
The Impact of Wind on Rainfall Distribution in Urban High-Rise Residential Areas

Xichao Gao, Xinyi Song, Zhiyong Yang, Pengfei Wang, and Kai Gao

Abstract The spatial distribution of rainfall in urban areas has a significant impact on urban rainfall-runoff processes. Wind can affect raindrop trajectories and thus affect the spatial distribution of rainfall. Due to the obstructive effect of building facades, the influence of wind on the distribution of rainfall is more pronounced in urban areas. In this study, we proposed a method to detect the spatial distribution of rainfall in urban high-rise residential areas based on computational fluid dynamics modeling. The variation of rainfall distribution and the ratio of rainfall on builds to that on the ground with changes in wind speed was analyzed using the proposed method. The results show that: (1) the spatial distribution of rainfall intensity on the ground in the high-rise residential area becomes increasingly uneven with the increase in the wind speed. (2) wind can lead to an increase in the total rainfall in the high-rise residential area by more than twice. (3) the ratio of rainfall on buildings to that on the ground increases significantly with the increasing wind speed.

Keywords Rainfall distribution · Wind impact · Urban areas

X. Gao · Z. Yang (✉) · K. Gao
State Key Laboratory of Simulation and Regulation of Water Cycle in River Basin, Institute of Water Resources and Hydropower Research, Beijing 100038, China
e-mail: yangzy@iwhr.com

X. Song
School of Hydraulic and Environmental Engineering, Changsha University of Science and Technology, Changsha 410114, China

P. Wang
Ministry of Emergency Management Big Data Center, Beijing 100013, China

H. Xu (ed.), *The 6th International Symposium on Water Resource and Environmental Management*, Environmental Science and Engineering,
https://doi.org/10.1007/978-3-031-55989-1_18

18.1 Introduction

Flooding has become one of the most serious natural disasters in the world. As a manifestation of flooding, urban flooding has garnered increasing attention in recent years (Luo et al. 2022; Schreider et al. 2000; Sohn et al. 2020). It's frequency and intensity have been confirmed to be increasing significantly with the development of urbanization and the warmer climate. Therefore, the flooding mitigation strategy in urban areas is currently a research hotspot (Gao et al. 2023). Numerical modelling is a crucial tool for guiding urban flood control and disaster reduction plan. Many models have been developed based on hydrological or hydraulic methods and been used in urban rainfall runoff simulations in recent decades (Rong et al. 2020; Guo et al. 2021). The studies show that the temporal and spatial distribution of rainfall have substantial impacts on urban flood simulation with numerical models (Kabir et al. 2022). In addition, some studies show that the proportion of rainfall on buildings and ground surface can affect the rainfall-runoff process significantly (Gao et al. 2021). Wind can alter the trajectory of raindrops. In areas with complex terrain, such as valley and urban areas, local wind fields are disturbed and distributed unevenly, altering the spatial distribution and the proportion on buildings and ground surface of the rainfall intensity. However, most studies about the influence of wind on rainfall focused on the rainfall distribution on building facades for a specific building from the perspective of building environment (Blocken and Carmeliet 2006). In this study, we analyzed the impact of spatial distribution on rainfall distribution in urban areas, especially for the proportion of rainfall on buildings and the ground surface, through computational fluid dynamics (CFD) methods.

The paper is organized as follows. Section 18.2 describes the study process and the CFD method. Section 18.3 presents the results of the impact of wind on the spatial distribution of rainfall based on the CFD simulation and a conclusion is given in Sect. 18.4.

18.2 Materials and Methods

18.2.1 Simulation Method of the Raindrop Trajectory

We used the CFD model to simulate raindrop trajectories in the wind, and then calculated rainfall intensities on building facets and the ground surface by counting the number of raindrops landing on them. The simulation of raindrop trajectories in the wind includes two parts: the simulation of the wind field in urban areas and the simulation of the movement of raindrops in the wind field. It should be noted that in order to streamline the simulation process, we assume that the raindrop moves after the wind field becomes stable.

The continuity equation of fluid, the Reynolds-averaged Navier–Stokes equations, and the standard $k-\varepsilon$ turbulence model were combined to describe the wind flow around buildings. The equations are as follows (Lakehal et al. 1995):

$$\frac{\partial U_i}{\partial x_i} = 0 \tag{18.1}$$

$$\frac{\partial U_i}{\partial t} + U_j \frac{\partial U_i}{\partial x_j} = -\frac{1}{\rho}\frac{\partial P}{\partial x_i} + \frac{\partial}{\partial x_i}\left(\nu \frac{\partial U_i}{\partial x_j}\right) + \frac{\partial}{\partial x_j}\left(-\overline{u_i u_j}\right) \tag{18.2}$$

$$\frac{\partial k}{\partial t} + U_j \frac{\partial k}{\partial x_j} = \frac{\partial}{\partial x_j}\left[\left(\nu + \frac{\nu_t}{\sigma_k}\right)\frac{\partial k}{\partial x_j}\right] - \overline{u_i u_j}\frac{\partial U_i}{\partial x_j} - \varepsilon \tag{18.3}$$

$$\frac{\partial \varepsilon}{\partial t} + U_j \frac{\partial \varepsilon}{\partial x_j} = \frac{\partial}{\partial x_j}\left[\left(\nu + \frac{\nu_t}{\sigma_\varepsilon}\right)\frac{\partial \varepsilon}{\partial x_j}\right] - C_{\varepsilon 1}\frac{\varepsilon}{k}\overline{u_i u_j}\frac{\partial U_i}{\partial x_j} - C_{\varepsilon 2}\frac{\varepsilon^2}{k} \tag{18.4}$$

$$-\overline{u_i u_j} = \nu_t\left(\frac{\partial U_i}{\partial x_j} + \frac{\partial U_j}{\partial x_i}\right) - \frac{2}{3}k\delta_{ij} \tag{18.5}$$

$$\upsilon_t = C_\mu \frac{k^2}{\varepsilon} \tag{18.6}$$

where U_i represents the average wind speed of direction $i\,(i = 1, 2, 3)$; u_i represents the turbulent speed of direction $i\,(i = 1, 2, 3)$; P represents the pressure; $-\overline{u_i u_j}$ represents the Reynolds stress; k and ε represent the turbulent kinetic energy and the dissipation rate of the turbulent kinetic energy, respectively; υ_t represents the turbulent eddy viscosity, and ν represents the kinematic viscosity. Coefficient values of the standard k−ε turbulence model indicated as follows:

$$C_\mu = 0.09, C_{\varepsilon 1} = 1.44, C_{\varepsilon 2} = 1.92, \sigma_k = 1.0, \sigma_\varepsilon = 1.3 \tag{18.7}$$

Upon obtaining the wind field distribution based on the above formulas, we can simulate the trajectory of raindrop motion through the application of the following kinematic theory. Assuming the raindrop is a rigid spherical particle, its movement in the wind field can be described by the following equations:

$$m\frac{d^2x}{dt^2} = 6\pi\mu r\left(U - \frac{dx}{dt}\right)\frac{C_D R}{24} \tag{18.8}$$

$$m\frac{d^2y}{dt^2} = 6\pi\mu r\left(V - \frac{dy}{dt}\right)\frac{C_D R}{24} - mg\left(1 - \frac{\rho_a}{\rho_w}\right) \tag{18.9}$$

$$m\frac{d^2z}{dt^2} = 6\pi\mu r\left(W - \frac{dz}{dt}\right)\frac{C_D R}{24} \tag{18.10}$$

$$m = \frac{4}{3}\pi r^3 \rho_w \tag{18.11}$$

$$R = \frac{2\rho_a r}{\mu}\sqrt{\left(\left(U - \frac{dx}{dt}\right)^2 + \left(V - \frac{dy}{dt}\right)^2 + \left(W - \frac{dz}{dt}\right)^2\right)} \tag{18.12}$$

where r represents the raindrop radius; ρ_a represents the density of air and ρ_w represents the density of water; μ represents the viscosity of air; U, V, W represent the velocity of wind of the x, y, z direction, respectively; m represents the raindrop mass; R represents the Reynolds number; and C_D is the drag coefficient representing the drag impact of the wind on the raindrop. In this study, the value for C_D is referenced from the study of Gunn and Kinzer (Gunn and Kinzer 1949).

18.2.2 Study Area

A typical urban high-rise residential area was selected as the study site. The distribution of buildings in this high-rise residential area is shown in Fig. 18.1. The study area has a length of 375 m and a width of 300 m. 6 buildings are evenly distributed in the study area with horizontal and vertical intervals of 50 m. All buildings have the same height of 100 m.

The wind field and the raindrop trajectories in the study area were simulated using the methods presented above. A three-dimensional CFD model was built. To mitigate the interference caused by factors such as recirculation of ambient air at the boundaries of the computational domain, the outlet of the computational domain is positioned at a distance 15H downstream from the building in the wind direction, where H represents the height of the buildings. A fixed total pressure boundary was applied at the outlet. The left and right boundaries of the computational domain

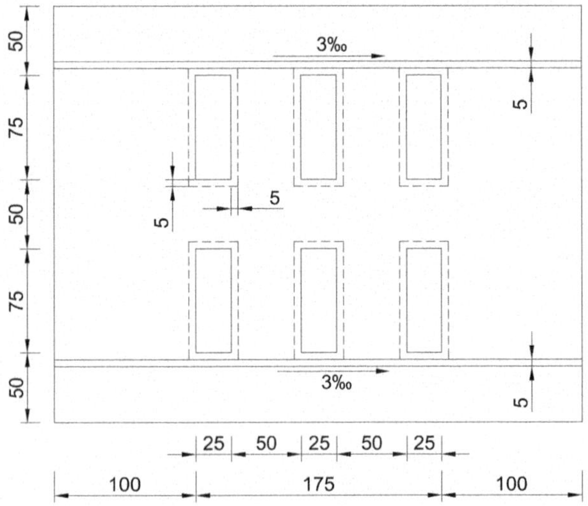

Fig. 18.1 Plan of the study area. Rectangles represent buildings

are set at a distance 3.5H beyond the left and right edges of the buildings in the building. Symmetry boundary conditions were employed on these two sides. The top boundary is set at a distance 2H above the top of the buildings and symmetry boundary conditions were used. The ground surface serves as the bottom boundary of the computational domain, and the no-slip wall boundary condition is applied. The surfaces of buildings also utilized the no-slip wall boundary condition. The near-wall flow was calculated using the standard wall function. The inlet of the computational domain is positioned at a distance 10H beyond the windward edge of buildings and the velocity inlet boundary condition was used. The velocity imposed on the computational inlet was calculated using the following equation:

$$v(z) = v_b(z/z_b)^{\alpha} \tag{18.13}$$

In which, z_b represents the standard reference height. v_b represents the wind speed at the standard reference height (reference wind speed). z and v represent the height and velocity, respectively. α represents the ground roughness index.

In this study, eight velocity scenarios were simulated to detect the impact on rainfall distribution of different wind speeds. The reference wind speed of the scenarios is 0, 3, 5, 7, 9, 10, 11, 13 m/s.

Raindrop particles were released from the top of the computational domain. All raindrops have the same diameter of 2 mm. Evaporation and motion dispersion of raindrops are ignored in the calculations.

18.3 Results

The distribution of rainfall intensities on the ground surface and roof surfaces under different wind speed scenarios is shown in Fig. 18.2. The figure shows that with the increase in the inlet wind speed, the spatial distribution of rainfall intensity on the ground in the high-rise residential area becomes increasingly uneven. The distribution of rain-free zones and areas of concentrated rainfall becomes more pronounced due to the obstruction by buildings and their interference with the wind field. The rainfall intensity on the roof surfaces of buildings exhibits an increasing trend with the increase in the inlet wind speed. As the value of the inlet reference wind speed increases from 0 to 13 m/s, the rainfall intensity on specific regions of the roof surfaces in the residential area can increase by more than tenfold.

The total rainfall received by the studied high-rise residential area is illustrated in Fig. 18.3. The figure shows that with the increase in the inlet wind speed, the total rainfall in the study area shows a trend of initial increase followed by leveling off. In the scenarios set for this study, when the wind speed is less than 7 m/s, the total rainfall received in the residential area significantly increases with the increase in inlet wind speed. However, when the wind speed exceeds 7 m/s, the change in total rainfall received by the high-rise residential area becomes less pronounced with increasing wind speed. In windy conditions, the rainfall received by the high-rise

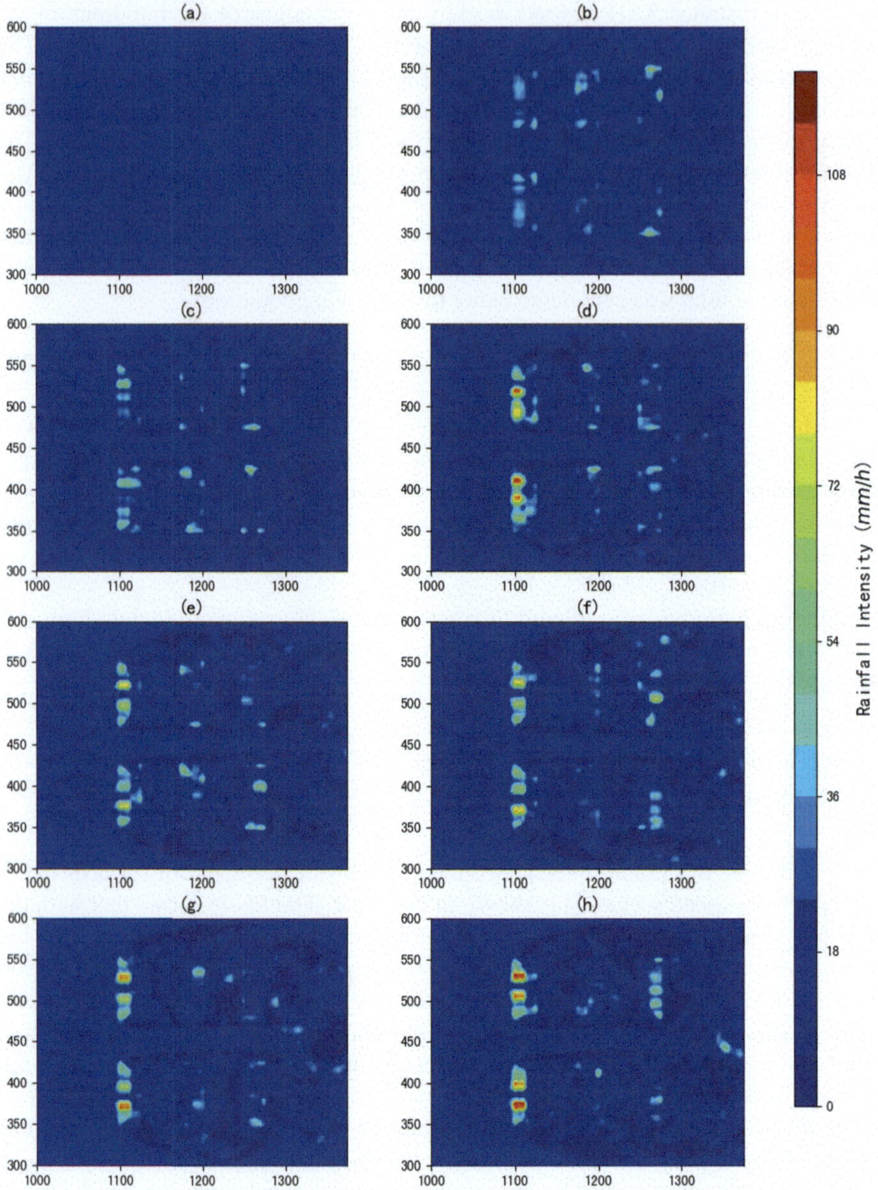

Fig. 18.2 The distribution of rainfall intensities on the surface ground and roof surfaces

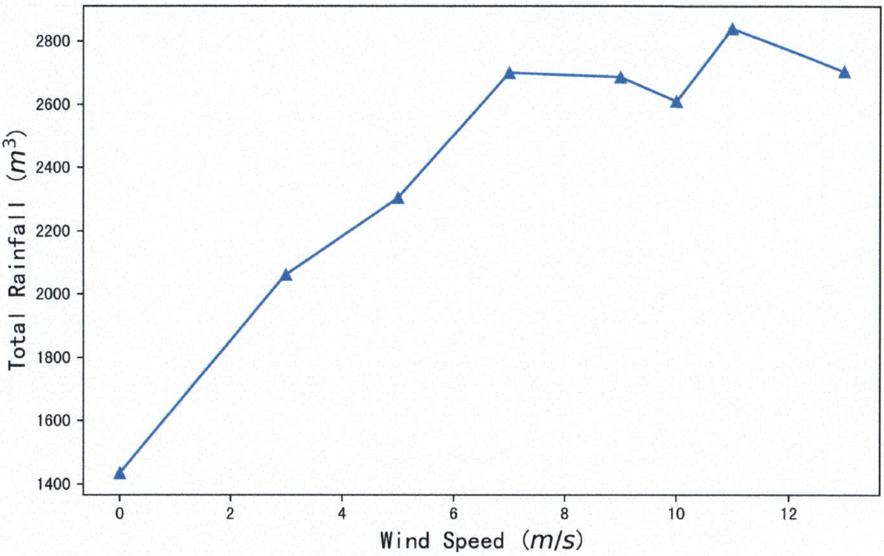

Fig. 18.3 The trend of the total rainfall with the increasing inlet wind speed

residential area can be up to twice as much as in calm conditions. The distribution of rainfall received by the ground and buildings is illustrated in Fig. 18.4. Similar to the variation in total rainfall in the high-rise residential area, the ratio of the rainfall received by buildings to that received by the ground shows a trend of initial increase followed by a gradual leveling off with increasing wind speed. As the wind speed increases from 0 to 12 m/s, the ratio of precipitation on buildings to that on the ground increases from 0.2 to 1.4. It implies that wind causes the transfer of rainfall from the ground to the building surfaces.

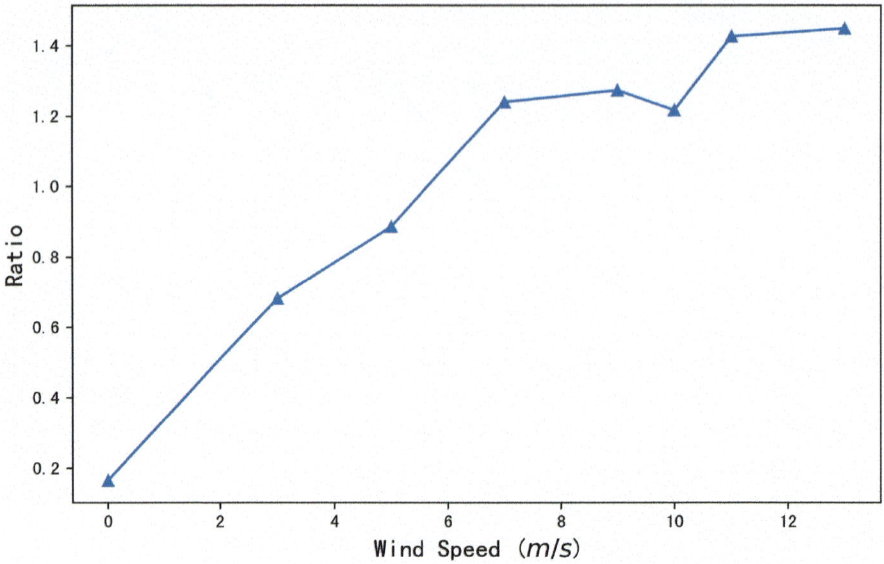

Fig. 18.4 The trend of the rate of rainfall on the ground to that on buildings with the increasing inlet wind speed

18.4 Conclusion

In this study, the impact of wind on rainfall distribution in high-rise residential areas was explored based on CFD modeling. The main results are: (1) wind can lead to an increase in the total rainfall in the high-rise residential area by more than twice. (2) the ratio of rainfall on buildings to that on the ground increases significantly with the increasing wind speed.

Competing Interests The authors declare that they have no conflict of interest.

Financial Support This study was financially supported by the National Key Research and Development Project (No. 2022YFC3002705-5), the National Natural Science Foundation of China (52,209,044), the Natural Science Foundation of Hunan Province (No. 2022JJ40492).

References

Blocken B, Carmeliet J (2006) On the accuracy of wind-driven rain measurements on buildings. Build Environ 41(12):1798–1810

Gao X, Yang Z, Han D et al (2021) The impact of wind on the rainfall–runoff relationship in urban high-rise building areas. Hydrol Earth Syst Sci 25(11):6023–6039

Gao X, Gao K, Yang Z et al (2023) A conceptual model to estimate flood damages for large-scale cities. Urban Clim 50:101574

Gunn R, Kinzer GD (1949) The terminal velocity of fall for water droplets in stagnant air. J Meteorol 6(4):243–248
Guo Z, Leitao JP, Simões NE et al (2021) Data-driven flood emulation: speeding up urban flood predictions by deep convolutional neural networks. J Flood Risk Manag 14(1):e12684
Kabir T, Pokhrel Y, Felfelani F: On the precipitation-induced uncertainties in process-based hydrological modeling in the Mekong River Basin. Water Resour Res 58(2):e2021WR030828
Lakehal D, Mestayer PG, Edson JB et al (1995) Eulero-Lagrangian simulation of raindrop trajectories and impacts within the urban canopy. Atmos Environ 29(23):3501–3517
Luo P, Luo M, Li F et al (2022) Urban flood numerical simulation: research, methods and future perspectives. Environ Model Softw 156:105478
Rong Y, Zhang T, Zheng Y et al (2020) Three-dimensional urban flood inundation simulation based on digital aerial photogrammetry. J Hydrol 584:124308
Schreider SY, Smith DI et al (2000) Climate change impacts on urban flooding. Clim Change 47:91–115
Sohn W, Kim JH, Li MH et al (2020) How does increasing impervious surfaces affect urban flooding in response to climate variability? Ecol Ind 118:106774

Chapter 19
From the Perspective of High-Risk Leisure Environment: A Study of the Impact of Green Education Literacy on River Water, Ecology, Surrounding Rural Economic Development, and Youth Well-Being

Hsiao-Hsien Lin, Chih-Cheng Lo, Yuan-Shing Perng, Jianguang Zhao, Zhongxuan Jin, Heyong Wei, Changqing Guo, Po-Hsuan Wu, Su-Fang Zhang, Penghui Liu, and Chin-Hsien Hsu

Abstract This study aims to discuss the impact of green education literacy on lake water, rural economic development, and youth well-being in a high-risk leisure environment. A mixed methods approach was adopted. First, a questionnaire was developed and 650 samples were tested using SPSS 26.0 statistical software for basic statistical testing and Pearson's correlation analysis. Semi-structured interviews

---

H.-H. Lin · Z. Jin · H. Wei · C. Guo · P. Liu (✉)
School of Physical Education, Guangdong Province, Jiaying University, Meizhou City 514015, China
e-mail: lrlph@hotmail.com

H.-H. Lin
e-mail: chrishome12001@yahoo.com.tw

Z. Jin
e-mail: jinzhongxuan1209@163.com

H. Wei
e-mail: kybjwh@163.com

C. Guo
e-mail: 554014641@qq.com

C.-C. Lo
Department of Industrial Education and Technology, National Changhua University of Education, Changhua City 500207, Taiwan
e-mail: charleslo@cc.ncue.edu.tw

Y.-S. Perng
Department of Forestry, National Chung Hsing University, No. 145, Xingda Rd., South Dist., Taichung City 402, Taiwan
e-mail: ysperng@nchu.edu.tw

H. Xu (ed.), *The 6th International Symposium on Water Resource and Environmental Management*, Environmental Science and Engineering,
https://doi.org/10.1007/978-3-031-55989-1_19

were then conducted to collect opinions from experts, tourists, residents and others. Finally, multiple verification analysis was used for discussion. The results found that although the existing economic, construction and watersports industry development scales around rural rivers can promote green leisure education literacy to drive rural development effectiveness. However, river water, ecological and green space resources can temporarily satisfy youth leisure needs. However, if river watersports leisure industries can be further increased, and related watersports leisure activities or tourism developed, it can provide youth with more diverse leisure choices, increase leisure benefits, and obtain feelings of happiness and well-being.

Keywords Leisure environment · Green literacy · Water sustainable economy · Youth · Well-being

19.1 Introduction

Tourism refers to people leaving their place of residence and engaging in leisure sports or tourism activities that satisfy the public's demand for recreational activities based on the natural ecology, rural economy, and society of a specific space or regional environment (Yang et al. 2010). Lakes, with their extensive catchment areas, abundant forests, and rich ecology and water resources (Yanda and Madulu 2005), possess ample water resources. The surrounding areas also boast abundant forest vegetation, a mild climate, and diverse ecology (Ma et al. 2023), meeting the essential conditions for human development. With the advancement of science and technology and the pursuit of a high quality of life, there has been an increased emphasis on leisure and tourism awareness. People aim to secure stable and safe lake water resources through water conservancy construction to preserve ecological diversity. The objective is to establish lake water parks with functions such as flood control, hydropower generation, and water storage (Zhuang 2016), catering to human needs for fishing, hunting, irrigation, and water usage. This initiative propels

J. Zhao
School of Economics and Management, Jiaying University, Meizhou City 514015, Guangdong Province, China
e-mail: 18643277717@163.com

P.-H. Wu
Apex Circuit (Thailand) Co., Ltd, Samut Sakhon 74000, Thailand

S.-F. Zhang
Tourism Management, Athena Institute of Holidtic Wellness, WuYi University, No. 26, WuYi Avenue, Wuyishan 354300, Fujian Province, China

C.-H. Hsu (✉)
Department of Healthcare Industry Technology Development and Management, National Chin-Yi University of Technology, Taichung 41170, Taiwan
e-mail: hsu6292000@yahoo.com.tw

the development of rural leisure sports or tourism industries, enhancing people's recreational benefits and well-being (Frantál and Urbánková 2017). This leads to a human-centric approach centered around lakes and surrounding rural areas to drive tourism development, shaping regional lake tourism economies and establishing models for leisure sports consumption (Rubin 1993).

The Meijiang River belongs to the Wujiang River System in the Yangtze River Basin, with a length of 69 km and a catchment area of 2143 km^2. It has an average annual flow rate of 25 cubic meters per second, harboring substantial water resources. As a tributary of the Hanjiang River, located in the western part of Guangdong Province, the Meijiang River traverses a total of 343 km within Meizhou City (Wang et al. 1997). With its abundant water resources and diverse ecology, it provides benefits for irrigation, domestic water use, and industrial development. The government has actively invested funds in the renovation of existing public facilities, the improvement of lake water quality, the promotion of environmental conservation awareness, and the utilization of nearby tourism resources to drive rural development (Ma et al. 2023). This has ultimately resulted in the acquisition of rich leisure and tourism resources, fostering improvements in local development (Wang et al. 1997; Lin et al. 2018). Even during the pandemic, it continues to attract an annual average of 1.363 million visitors, generating a tourism output value exceeding USD 11.316 billion, with an annual growth rate of 6.2% (Xue 2023; Meizhou Municipal People's Government, Guangdong Province, China 2021). This indicates that current development not only promotes industry and local economic growth but also enhances the quality of life for residents (Ma et al. 2023; Xue 2023; Meizhou Municipal People's Government, Guangdong Province, China 2021). However, while the promotion of the leisure and tourism industry in rural areas can stimulate leisure trends, enhance local reputation, facilitate cultural exchange and economic development, and improve living environments (Ma et al. 2023; Xue 2023; Meizhou Municipal People's Government, Guangdong Province, China 2021; Huang 2021), it also increases the benefits people derive from tourism or leisure activities. This alleviates stress, satisfies the psychological need for learning and exploration, and achieves the physiological effect of promoting health (Yu et al. 2021a; Ivanova et al. 2021; Wang et al. 2021). Nevertheless, development also raises local consumption costs, leads to cultural and security conflicts, and harms the quality of water, ecology, air, and rural living environments (Xue 2023; Meizhou Municipal People's Government, Guangdong Province, China 2021; Huang 2021). This results in youth developing a distrust of the current living environment and experiencing negative emotions (Scott et al. 1998), thereby increasing leisure risks. Local areas urgently need to find safe and stable improvement strategies to reconstruct tourism and leisure consumption models with the aim of providing people with relief from pressure, a sense of pleasure, and improvements in physical and mental health (Yu et al. 2021a; Ivanova et al. 2021; Wang et al. 2021). Therefore, we believe that identifying friendly tourism behaviors, building safe and diverse living environments, and creating pleasant and comfortable leisure models for young people, while also fostering a sense of happiness, will be crucial issues.

Green education refers to educational courses or activities conducted in an energy-saving, carbon-reducing manner or using green energy sources (Agarwal et al. 2013). Green literacy is an environmental literacy model encompassing cognition, attitudes, and skills under the concept of carbon reduction (Maurer and Bogner 2022; UNWTO. Sustainable Development Goals 2015). Meanwhile, green leisure behavior refers to leisure decision-making behaviors established under a low-carbon environmental model (Hu et al. 2013). Scholars believe that education promotion activities, founded on correct low-carbon leisure cognition, attitudes, and skills based on the concept of low-carbon literacy (Agarwal et al. 2013; Maurer and Bogner 2022; UNWTO. Sustainable Development Goals 2015; Hu et al. 2013; Chen 2014), can lead to a proper understanding of green tourism, sound judgment, and the planning of eco-friendly tourism behaviors (Can and Hongbing 2011). These actions can make people willing to continuously pay attention to carbon reduction and the effectiveness of ecological conservation (Zhang et al. 2021), and take practical actions to achieve green leisure or tourism consumption behaviors for low-carbon and conservation ecological sustainability goals (Yu et al. 2021b) during leisure activities. Scholars believe this process will help youth establish correct principles of action or beliefs, obtain judgment and execution capabilities that can be fully utilized in leisure or tourism activities to implement low-carbon leisure goals (Can and Hongbing 2011; Zhang et al. 2021). Moreover, these behaviors can bring greater economic, social, and environmental benefits to rural areas and provide people with higher quality tourism experiences and pleasure (Zhang et al. 2021; Yu et al. 2021b). This is because establishing low-carbon behaviors can reduce energy losses and promote economic development (Vasiliu et al. 2020). It can also establish proper social operation mechanisms for life, enterprise development, and society (Chen and Li 2019). Furthermore, it assists the government and the public in jointly shaping clean and sanitary living environments while maintaining conditions for sustainable human and ecological development (Chen and Li 2019; Xiao et al. 2021; Kaffashi and Shamsudin 2019). Therefore, we believe that promoting green education literacy can improve local economic, social, environmental, and ecological development as well as build safe and suitable leisure environments to enhance youth well-being. Thus, we hypothesize that green literacy education has a consistent and positive impact on village economic (H1), social (H2), environmental (H3), and ecological (H4) development, while youth well-being will be positively influenced by the status quo of economic, social, environmental, and ecological development.

However, scholars believe that the actual effectiveness of implementation can easily deviate from the initial predictions based on literature, theories, or decisions (Lin et al. 2018; Yu et al. 2021a). These issues often arise due to human or other external factors, leading to different positive or negative outcomes (Lin et al. 2018; Yu et al. 2021a). Typically, these phenomena or results need to be confirmed over time (Yu et al. 2021a; Yang et al. 2023), and the real answers are often discovered through the feelings and feedback of actual participants. This process involves adopting multi-perspective research methods (Lin et al. 2018, 2023; Yu et al. 2021a; Yang et al. 2023). In addition, current research primarily discusses topics such as life attitudes and well-being (Deb et al. 2020), tourism and ecological conservation (Yuxi and Linsheng

2020), tourism and renewable energy (Calderón-Vargas et al. 2019), policies and public welfare (Samuel and Hadjar 2016), individual cognition and leisure decisions (Yang et al. 2018), low-carbon policy assessment (Meyer and Truccone-Borgogno 2022), low-carbon consumption awareness (Yu et al. 2021c), low-carbon tourism facilities (Alba and Jencka 2020; Han and Li 2021), and low-carbon services and consumer behaviors (Chang et al. 2020). There is limited discussion using Meizhou City as a case to examine the impact of low-carbon literacy on economic, social, environmental, and ecological development, as well as youth well-being. Therefore, we believe that this issue will contribute to the government and youth paying more attention to green literacy education policies and effects, aiming for sustainable human and ecological development. This discussion holds significance and social contribution value. This is the main purpose and significance of this study.

19.2 Methodology

19.2.1 Research Process

This study focuses on local youth near the Meijiang River in Meizhou City, investigating the impact of low-carbon leisure education on river and rural economic development, the leisure industry, environmental factors, and youth well-being in a high-risk environment. After reviewing relevant literature, a questionnaire was designed based on literature related to the Meijiang River, Meizhou City, well-being, low-carbon literacy, rural economics, social aspects, environmental factors, and ecological development, among others (Yang et al. 2010, 2023, 2018; Yanda and Madulu 2005; Ma et al. 2023; Zhuang 2016; Frantál and Urbánková 2017; Rubin 1993; Wang et al. 1997, 2021; Lin et al. 2018, 2023; Xue 2023; Meizhou Municipal People's Government, Guangdong Province, China 2021; Huang 2021; Daily 2023; Scott et al. 1998; Yu et al. 2021a, b, c; Ivanova et al. 2021; Agarwal et al. 2013; Maurer and Bogner 2022; UNWTO. Sustainable Development Goals 2015; Hu et al. 2013; Chen 2014; Can and Hongbing 2011; Zhang et al. 2021; Vasiliu et al. 2020; Chen and Li 2019; Xiao et al. 2021; Kaffashi and Shamsudin 2019; Deb et al. 2020; Yuxi and Linsheng 2020; Calderón-Vargas et al. 2019; Samuel and Hadjar 2016; Meyer and Truccone-Borgogno 2022; Alba and Jencka 2020; Han and Li 2021). From August to November 2022, 800 questionnaires were distributed, resulting in 650 valid responses (81.25%). Data analysis was conducted using SPSS 26.0, followed by interviews to gather additional perspectives from participants. Finally, all data were compiled, and a triangulation analysis method was employed for discussion (Gursoy et al. 2002; Janesick et al. 2000). The study framework is presented in Fig. 19.1.

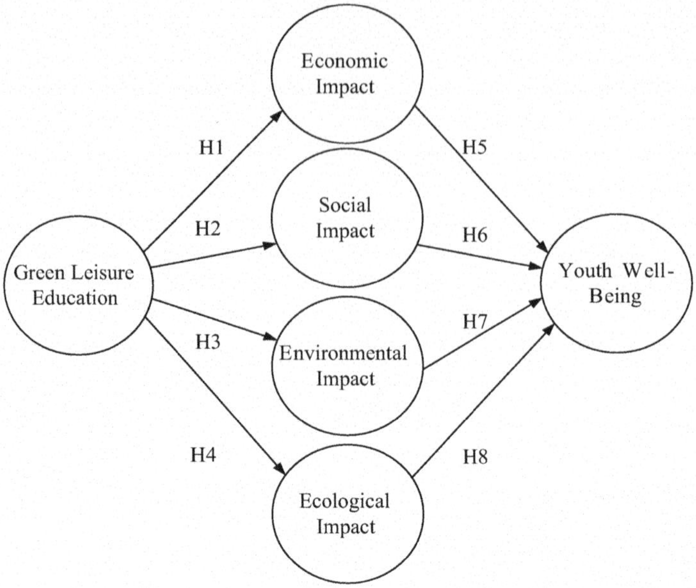

Fig. 19.1 Framework

19.2.2 Research Tools, Analysis and Methods

A mixed research method was adopted in this study. First, the current development status of Meizhou City was reviewed. Then, the questionnaire tool was compiled based on literature on green literacy, rural economic, social, environmental and ecological development, etc. (Yang et al. 2010, 2023, 2018; Yanda and Madulu 2005; Ma et al. 2023; Zhuang 2016; Frantál and Urbánková 2017; Rubin 1993; Wang et al. 1997, 2021; Lin et al. 2018, 2023; Xue 2023; Meizhou Municipal People's Government, Guangdong Province, China 2021; Huang 2021; Daily 2023; Scott et al. 1998; Yu et al. 2021a, b, c; Ivanova et al. 2021; Agarwal et al. 2013; Maurer and Bogner 2022; UNWTO. Sustainable Development Goals 2015; Hu et al. 2013; Chen 2014; Can and Hongbing 2011; Zhang et al. 2021; Vasiliu et al. 2020; Chen and Li 2019; Xiao et al. 2021; Kaffashi and Shamsudin 2019; Deb et al. 2020; Yuxi and Linsheng 2020; Calderón-Vargas et al. 2019; Samuel and Hadjar 2016; Meyer and Truccone-Borgogno 2022; Alba and Jencka 2020; Han and Li 2021). SPSS 26.0 statistical software was utilized to validate the questionnaire tool. Wang et al. believe that for issues with test results meeting the criteria of Kaiser-Meyer Olkin (KMO) > 0.06 and p value <0.01 ($p < 0.01$) in the Bartlett's test, it indicates that the scale is suitable for factor analysis (Ajzen 1991). Next, if α is found to be greater than 0.60, the issue is considered to be highly reliable (Bower et al. 2019) and can be further analyzed (Civil Code of the People's Republic of China 2021). According to the above criteria, the analysis data of all issues met the specifications. It can be determined that this questionnaire demonstrates high reliability and can be utilized for further analysis. The analysis results are shown in Table 19.1.

We followed up with interviews of 9 participants from diverse backgrounds, including scholars, local government officials, tourism operators, visitors, and residents. A semi-structured interview approach was employed, presenting individual perspectives based on the questionnaire analysis results. Finally, all gathered information was classified, organized, integrated, and compared. A triangulation analysis method was then utilized for discussion. Details of the interviewees and topics can be found in Table 19.2.

19.2.3 Ethical Considerations

This study was conducted in accordance with Articles 1004 and 1009 of the Civil Code of China (Civil Code of the People's Republic of China 2021) and was approved by the Academic Ethics Committee of Jiaying University Physical Education College (No. 131/322E1860). During the research process, investigators first explained the research topic and participants' rights and obligations with a research statement before the investigation (research statement placed on the first page of the questionnaire/online questionnaire). Then, volunteers were asked to sign an informed consent form before commencing the survey procedure under principles of fairness, impartiality, and transparency (Bower et al. 2019). Therefore, we believe the experimental procedure meets local government legal requirements (Civil Code of the People's Republic of China 2021) as well as the Declaration of Helsinki (Taiwan Medical Association 2014).

19.3 Analysis

19.3.1 Background Analysis

Firstly, basic statistical testing was conducted using SPSS 26.0 on the 650 questionnaires to analyze the sample background. It was found that there were 165 residents (25.5%) and 485 visitors (74.5%). There were 300 males (46.3%) and 350 females (53.7%). It shows that most participants in the questionnaire had backgrounds as young female and visitors.

Table 19.1 Questionnaire analysis

Construct		KMO (Bartlett/ DF)	Issues	α
Green leisure literacy	Green leisure cognition	0.448 (787.377/ 6)*	Not littering (GAw1); Choosing low-carbon leisure activities (GAw2); Taking public transportation (GAw3); Purchasing products with less packaging (GAw4)	0.600–0.702
	Green leisure attitudes	0.767 (991.34/ 10)*	Eco-friendly (GAt1); Engaging in low-carbon leisure is convenient (GAt2); Reasonable pricing (GAt3); Low-carbon leisure is enjoyable (GAt4); Low-carbon leisure brings happiness (GAt5)	0.796–0.853
	Green leisure behaviors	0.747 (755.36/ 3)*	Willingness to bring own sports equipment and tableware (GB1); Willingness to participate in low-carbon products (GB2); Willingness to choose low-carbon transportation (GB3)	0.842–0.898
	Economic impact	0.792 (1492.496/6)*	Increasing infrastructure construction (Ei1); Increasing industries (Ei2); Developing creative products (Ei3); Expanding leisure industry business opportunities (Ei4)	0.799–0.808
	Social impact	0.898 (929.44/ 28)*	Improving service quality (Si1); Improving activity quality (Si2); Increasing leisure opportunities (Si3); Enhancing public willingness to participate in community development (Si4); Increasing youth's willingness to return hometown (Si5); Preserving historical sites (Si6); Disappearance of local characteristic buildings (Si7); Friendly interactions (Si8)	0.790–0.802
	Environmental impact	0.600 (40.667/ 3)*	Impacting community environment (Eni1); Vehicle exhaust emissions pollution (Eni2); Sufficient number of public toilets (Eni3)	0.701–0.718

(continued)

Table 19.1 (continued)

Construct		KMO (Bartlett/DF)	Issues	α
	Ecological impact	0.690 (174.873/6)*	Increasing river waste (Eci1); Surface oil pollution (Eci2); Excessive vegetation exploitation (Eci3); Increasing vegetation waste (Eci4)	0.700–0.702
	Well-being	0.610 (408.503/3)*	Relaxing spiritually (Wb1); Liking my life (Wb2); Strengthening sense of life participation (Wb3); Improving social interactions (Wb4)	0.738–0.750

*p < 0.000

Table 19.2 Interviewee Background Introduction

Professional background	Identity	Professional background	Identity	Professional background	Identity
Sustainable development, education	Professor	Decision-making and administration	Village chief	Tourism industry	Industry staff
Leisure management, economics	Professor	University student majoring in sports	Visitor	University student majoring in finance	Visitor
Convenience store service staff	Resident	University student majoring in tourism	Resident	Volunteer	Resident

19.3.2 Correlation Analysis on Green Leisure Education Literacy, Riverside Rural Economic, Social, Environmental, Ecological Development and Youth Well-Being

Next, we analyzed the impact of low-carbon leisure awareness on the economic, social, environmental and ecological development of lakeside rural areas. As shown in Table 19.3, we found that although low-carbon leisure education literacy had no significant impact on rural development ($p > 0.05$), this result verifies that hypotheses 1–4 are not valid.

Finally, we analyzed the impact of riverside rural economic, social, environmental and ecological development on youth well-being. We discovered that rural environmental and ecological development had a significant negative impact on youth well-being ($p < 0.01$). This outcome confirms that hypotheses 5–6 are not valid while hypotheses 7–8 are valid.

Table 19.3 Correlation analysis of green leisure education literacy on riverside rural economic, social, environmental and ecological development

Facet	Economics	Society	Environment	Ecology
Green leisure education literacy	−0.030	0.032	−0.049	0.059
Youth well-being	−0.043	0.043	**−0.122***	**0.107***

* $p < 0.05$; ** $p < 0.01$

19.4 Discussion

19.4.1 Analysis of the Impact of Low-Carbon Leisure Education Literacy on Riverside Rural Economic, Social, Environmental and Ecological Development

An analysis was conducted on the impact of low-carbon leisure education literacy on the economic, social, environmental, and ecological development effectiveness of the rural areas around the river. The results revealed that low-carbon leisure education literacy did not exert a significant influence on the economic, social, environmental, and ecological development of the rural areas around the river, contrary to the findings in the literature (UNWTO. Sustainable Development Goals 2015; Can and Hongbing 2011).

The author concludes that while low-carbon leisure education primarily aims to cultivate public awareness of energy-saving and carbon reduction (UNWTO. Sustainable Development Goals 2015), establish concepts of carbon reduction consumption (Hu et al. 2013), and encourage green leisure behaviors (UNWTO. Sustainable Development Goals 2015; Can and Hongbing 2011). Therefore, individuals with low-carbon leisure awareness are expected to facilitate the development of principles or beliefs in low-carbon consumption, acquire judgment and execution capabilities, and apply them effectively in leisure or travel activities to implement low-carbon leisure goals (UNWTO. Sustainable Development Goals 2015; Can and Hongbing 2011). However, the Meijiang River flows through underdeveloped urban areas, with Meizhou City having limited industries related to riverside leisure or tourism. The opportunities for riverside leisure sports activities are scarce, and the river lacks distinctive features for watersports or tourism. Consequently, even with the promotion of low-carbon leisure education literacy, it fails to significantly influence the local economic, social, environmental, and ecological development. This leads young people to perceive the inability to directly impact rural economic, social, environmental, and ecological development as the main reason for the ineffectiveness of constructing low-carbon leisure awareness.

19.4.2 Analysis of the Impact of Riverside Rural Economic and Ecological Development on Youth Well-Being After Promoting Low-Carbon Leisure Education Literacy

The analysis revealed the influence of riverside rural economic and ecological development on youth well-being following the promotion of low-carbon leisure education literacy. The results indicate that, despite the enhancement of low-carbon leisure education literacy, riverside rural economic and social development showed no impact on youth happiness. Only rural environmental and ecological development demonstrated a significant influence.

The author concludes that while low-carbon leisure awareness can mitigate the environmental impact of tourism development, preserve local social and cultural uniqueness, maintain ecological balance, and stimulate local economic growth (Maurer and Bogner 2022; UNWTO. Sustainable Development Goals 2015; Can and Hongbing 2011), the local community maintains simplicity in its traditions, and the level of urban development is modest. Additionally, there is a scarcity of river-based leisure industries or activities, resulting in limited opportunities for youth to engage in watersports and cultivate leisure awareness. Consequently, despite ongoing improvements in local economic infrastructure, social order, and the reinforcement of Hakka cultural features, the leisure and living needs of the youth remain unmet, hindering their pursuit of happiness.

However, due to the abundant river water resources and rich ecology, along with well-planned green spaces along the riverbanks, an abundance of fisheries and recreational areas has been cultivated, providing a diverse range of leisure activity options. Only through the establishment of a safe living and leisure environment and the maintenance of diverse ecological development can youth obtain leisure benefits from riverside greenspaces, water, and ecological resources. This enables them to experience feelings of pleasure and well-being, even if they cannot directly acquire resources from local economic and social development outcomes to fulfill their personal living and recreational needs (Fig. 19.2).

19.5 Conclusion

Although green leisure education literacy has the potential to establish youth green consumption awareness, implement low-carbon consumer behaviors, and promote local economic and recreational industry growth, the overall scales of urban economic development and watersports activities industries can prevent riverside rural areas from leveraging green leisure education literacy to stimulate rural economic, social, environmental and ecological growth. However, even with small-scale rural economic development and few river watersports leisure activities and industries locally, youth leisure needs can still be met through water, ecological

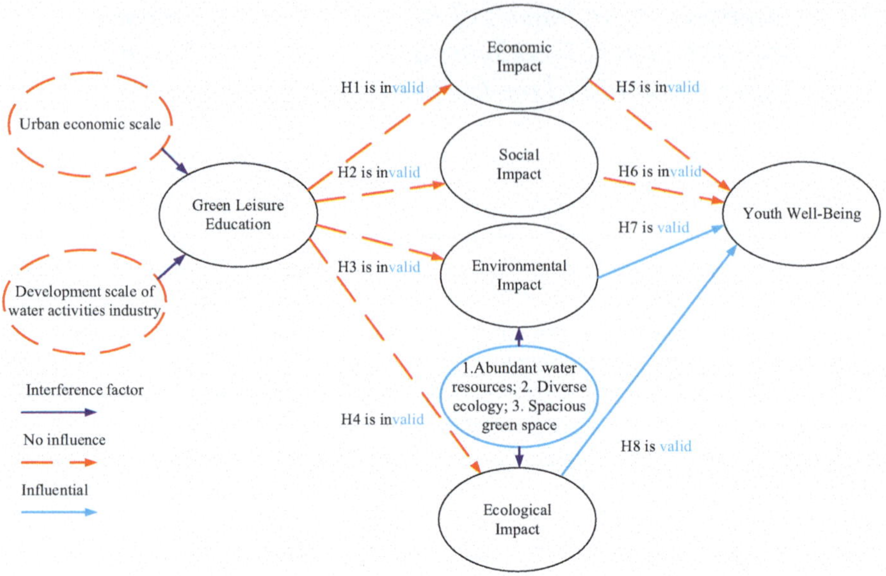

Fig. 19.2 Analysis chart of influencing factors related to low-carbon leisure education literacy, river ecology, rural economic development, and youth well-being

and greenspace resources to attain recreational sports benefits and happiness for well-being.

References

Agarwal S, Basu K, Nath A (2013) Green computing and green technology based teaching learning and administration in higher education institutions. Int J Adv Comput Res 3(11):295–303. https://citeseerx.ist.psu.edu/document?repid=rep1&type=pdf&doi=be7c6463f7c6bb3227ae91676df270ad8c3ad233

Ajzen I (1991) The theory of planned behavior. Organ Behav Hum Decis Process 50:179–211. https://doi.org/10.1016/0749-5978(91)90020-T

Alba de MMG, Jencka NJ (2020) A study on low-carbon aviation fuels in the United States: economic potential and challenges facing consumer awareness and favorability (Master Thesis). Norwegian School of Economics, Norwegian. https://openaccess.nhh.no/nhh-xmlui/bitstream/handle/11250/2678371/masterthesis.pdf?sequence=1

Bower I, Tucker R, Enticott PG (2019) Impact of built environment design on emotion measured via neurophysiological correlates and subjective indicators: a systematic review. J Environ Psychol 66:101344. https://doi.org/10.1016/j.jenvp.2019.101344

Calderón-Vargas F, Asmat-Campos D, Carretero-Gómez A (2019) Sustainable tourism and renewable energy: binomial for local development in Cocachimba, Amazonas, Peru. Sustainability 11(18):4891. https://doi.org/10.3390/su11184891

Can H, Hongbing D (2011) The model of developing low-carbon tourism in the context of leisure economy. Energy Procedia 5:1974–1978. https://doi.org/10.1016/j.egypro.2011.03.339

Chang SH, Hernández-Díaz RJ, Lo WS (2020) The impact of low-carbon service operations on responsible tourist behavior: the psychological processes of sustainable cultural tourism. Sustainability 12(12):4943. https://doi.org/10.3390/su12124943

Chen LJ (2014) Low carbon traveler's leisure benefit and willingness to pay (Master thesis). National Dong Hwa University, Hualien

Chen W, Li J (2019) Who are the low-carbon activists? Analysis of the influence mechanism and group characteristics of low-carbon behavior in Tianjin, China. Sci Total Environ 683(15):729–736. https://doi.org/10.1016/j.scitotenv.2019.05.307

Civil Code of the People's Republic of China (2021) Title IV: Personality rights: Chapter II. Right to life, body and health, Articles 1004, 1109. Standing Committee of the National People's Congress, People's Republic of China

Daily M. Fight the floods! Continuous heavy rainfall caused disasters in many places. Party members and cadres at all levels and rescue forces in Meizhou quickly moved to the front line of flood prevention and disaster relief. https://mp.weixin.qq.com/s?__biz=MzA4MzI5NzQwOA==&mid=2652940378&idx=2&sn=f4714ed730e6ecb0c213a4acf07b3640. Accessed 20 Nov 2023

Deb S, Thomas S, Bose A, Aswathi T (2020) Happiness, meaning, and satisfaction in life as perceived by Indian university students and their association with spirituality. J Relig Health 59:2469–2485. https://doi.org/10.1007/s10943-019-00806-w

Frantál B, Urbánková R (2017) Energy tourism: an emerging field of study. Curr Issue Tour 20(13):1395–1412. https://doi.org/10.1080/13683500.2014.987734

Gursoy D, Jurowski C, Uysal M (2002) Resident attitudes: a structural modeling approach. Ann Tour Res 20(1):79–105. https://doi.org/10.1016/S0160-7383(01)00028-7Get

Han L, Li L (2021) Sustainable development of tourism under the background of low-carbon and green economy. Adv Mater Sci Eng 1–8. https://doi.org/10.1155/2021/8587024

Hu MLM, Horng JS, Teng CCC, Yen CD (2013) Assessing students' low carbon literacy by Ridit IPA approach. J Hosp Leis Sport Tour Educ 3:202–212. https://doi.org/10.1016/j.jhlste.2013.09.006

Huang JD (2021) Flood prevention plan for Meijiang, Tingjiang, Chengjiang, and Shikuo Rivers in Meizhou City. Guangdong Academy of Water Resources and Hydropower Research, 2021

Ivanova M, Ivanov IK, Ivanov S (2021) Travel behaviour after the pandemic: the case of Bulgaria. Anatolia 32(1):1–11. https://doi.org/10.1080/13032917.2020.1818267

Janesick VJ (2000) The choreography of qualitative research design: minuets, improvisations, and crystallization. In: Denzin NK, Lincoln YS (eds) Handbook of qualitative research. Sage, Thousand Oak, CA, pp 379–399

Kaffashi S, Shamsudin MN (2019) Transforming to a low carbon society; an extended theory of planned behaviour of Malaysian citizens. J Clean Prod 235(20):1255–1264. https://doi.org/10.1016/j.jclepro.2019.07.047

Lin HH, Lee SS, Perng YS, Yu ST (2018) Investigation about the impact of tourism development on a water conservation area in Taiwan. Sustainability 10(7):2328. https://doi.org/10.3390/su10072328

Lin HH, Shen CC, Ling Y, Lin CY (2023) Effect of traditional Chinese medicine massage on physical and mental health of middle-aged and elderly women. Biotechnol Genet Eng Rev 1–26. https://doi.org/10.1080/02648725.2023.2186495

Mao S, Ye Q, Yan T. Meijiang irrigation district project is under full construction. http://jx.people.com.cn/BIG5/n2/2022/1109/c190267-40186863.html. Accessed 20 Nov 2023

Maurer M, Bogner FX (2022) Green awareness in action of saving energy in school life: modeling environmental literacy in theory and practice experience. In: Lackner M, Sajjadi B, Chen WY (eds) Handbook of climate change mitigation and adaptation. Springer, Cham. https://doi.org/10.1007/978-3-030-72579-2_157

Meizhou Municipal People's Government, Guangdong Province, China (2021) The 14th Five-Year Plan for National Economic and Social Development of Meijiang District, Meizhou City and the Outline of Long-term Goals for 2035. www.meijiang.gov.cn/attachment/0/106/106745/2201984.pdf

Meyer LH, Truccone-Borgogno S (2022) Legitimate expectations: assessing policies of transformation to a low-carbon society. Environ Values. https://doi.org/10.3197/096327122X16386102424001

Rubin BM (1993) Economic impacts of global climate change on the Great Lakes Region. In: Digital library of the commons conference paper, pp 6–8. https://hdl.handle.net/10535/8251

Samuel R, Hadjar A (2016) How welfare-state regimes shape subjective well-being across Europe. Soc Indic Res 129:565–587. https://doi.org/10.1007/s11205-015-1125-0

Scott S, Jackson S, Backett-Milburn K (1998) Swings and roundabouts: risk anxiety and the everyday worlds of children. Sociology 32(4):689–705. https://doi.org/10.1177/0038038598032004004

Taiwan Medical Association (2014) World Medical Association Declaration of Helsinki. Taiwan Med J 57(5):266–269. https://rec.chass.ncku.edu.tw/sites/default/files/2013%E8%B5%AB%E7%88%BE%E8%BE%9B%E5%9F%BA%E5%AE%A3%E8%A8%80_%E5%8F%B0%E7%81%A3%E9%86%AB%E7%95%8C.pdf

UNWTO. Sustainable Development Goals (2015). https://sustainabledevelopment.un.org/. Accessed 19 May 2023

Vasiliu D, Bucse A, Lupascu N et al (2020) Assessment of the metal pollution in surface sediments of coastal Tasaul Lake (Romania). Environ Monit Assess 192:749. https://doi.org/10.1007/s10661-020-08698-0

Wang Y, Di Y, Ye J, Wei W (2021) Study on the public psychological states and its related factors during the outbreak of coronavirus disease 2019 (COVID-19) in some regions of China. Psychol Health Med 20(1):12–33. https://doi.org/10.1080/13548506.2020.1746817

Wang J, Wang BS, Luo ZQ (1997) Yangtze River dictionary. Wuhan Publishing House. ISBN, 754301615X, 9787543016156

Xiao J, Zhen Z, Tian L, Su B, Chen H, Zhu AX (2021) Green behavior towards low-carbon society: theory, measurement and action. J Clean Prod 278(1):123765. https://doi.org/10.1016/j.jclepro.2020.123765

Xue M. Total investment is 4.379 billion yuan! Jiangxi Ganzhou Meijiang irrigation district project started. https://www.seetao.com/details/190497.html. Accessed 20 Nov 2023

Yanda PZ, Madulu NF (2005) Water resource management and biodiversity conservation in the Eastern Rift Valley Lakes, Northern Tanzania. Phys Chem Earth Parts A/B/C 30(11–16):717–725. https://doi.org/10.1016/j.pce.2005.08.013

Yang Z, Cai J, Sliuzas R (2010) Agro-tourism enterprises as a form of multi-functional urban agriculture for peri-urban development in China. Habitat Int 34(4):374–385. https://doi.org/10.1016/j.habitatint.2009.11.002

Yang J, Zhou Q, Liu X, Liu M, Qu S, Bi J (2018) Biased perception misguided by affect: How does emotional experience lead to incorrect judgments about environmental quality? Glob Environ Chang 53:104–113. https://doi.org/10.1016/j.gloenvcha.2018.09.007

Yang L, Zheng YZ, Lin HH, Chen IS, Chen KY, Li QY, Tsai IE (2023) Under the risk of COVID-19 epidemic: a study on the influence of life attitude, leisure sports values, and workplace risk perception on urban development and public well-being. Sustainability 15(10):7740. https://doi.org/10.3390/su15107740

Yu JH, Lin HH, Lo YC, Hsu CH, Liang Y, An Z (2021a) Is the travel bubble under COVID-19 a feasible idea or not? Int J Environ Res Pub Health 18(11):5717. https://doi.org/10.3390/ijerph18115717

Yu Z, Liu E, Lin Q, Zhang E, Yang F, Wei C, Shen J (2021c) Comprehensive assessment of heavy metal pollution and ecological risk in lake sediment by combining total concentration and chemical partitioning. Environ Pollut 269:116212. https://doi.org/10.1016/j.envpol.2020.116212

Yu Z, Liu E, Lin Q, Zhang E, Yang F, Wei C, Shen J (2021b) Comprehensive assessment of heavy metal pollution and ecological risk in lake sediment by combining total concentration and chemical partitioning. Environ Pollut 269:116212. https://doi.org/10.1016/j.envpol.2020.116212

Yuxi Z, Linsheng Z (2020) Identifying conflicts tendency between nature-based tourism development and ecological protection in China. Ecol Ind 109:105791. https://doi.org/10.1016/j.ecolind.2019.105791

Zhang J, Ma L, Li J (2021) Why low-carbon publicity effect limits? The role of heterogeneous intention in reducing household energy consumption. Energies 14(22):7634. https://doi.org/10.3390/en14227634

Zhuang W (2016) Eco-environmental impact of inter-basin water transfer projects: a review. Environ Sci Pollut Res 23:12867–12879. https://doi.org/10.1007/s11356-016-6854-3

Chapter 20
Saussure Semiotic Theory Analysis of Public Service Advertisements: A Case Study of CCTV Water Resource PSAs (2021–2023)

Bingbing Su, Jialu Lin, Zihao Huang, Mengyi Chen, Yijin Wang, and Guansu Wang

Abstract Public service advertisements are crucial in regulating general behavior, guiding positive public opinion, and promoting new social norms. This study aims to explore the role of symbolic expression in conveying information related to water resource protection in public service advertisements. The research focuses on water resource-related public service advertisements broadcasted by China Central Television (CCTV) from 2021 to 2023. Employing a qualitative analysis approach, the study utilizes Saussure's Semiotics theory to analyze and elucidate the verbal, colors, images, and auditory symbols used in three selected advertisements. The research findings indicate that public service advertisements employ a comprehensive range of symbols, contributing to the enrichment of cultural content and achieving the goal of high-quality communication. Appropriately using symbols can enhance the audience's receptivity to public service advertisements' favorable societal development concepts. This study is expected to inspire advertising creators to choose symbols and expressive methods more easily understood by the audience when crafting advertisements, thereby contributing to more effective communication.

Keywords Public service advertisement · Symbolic expression · Saussure's semiotics theory · Water resource protection · CCTV

B. Su · J. Lin · Z. Huang · M. Chen · Y. Wang · G. Wang (✉)
School of Humanities, Zhuhai College of Science and Technology, Zhuhai, Guangdong, China
e-mail: s2128778@siswa.um.edu.my

G. Wang
Asia-Europe Institute, Universiti Malaya, Kuala Lumpur, Malaysia

H. Xu (ed.), *The 6th International Symposium on Water Resource and Environmental Management*, Environmental Science and Engineering,
https://doi.org/10.1007/978-3-031-55989-1_20

20.1 Introduction

China experiences a highly uneven distribution of water resources, with per capita water availability at only 28% of the world average, posing a significant constraint on economic and social development (Bai et al. 2021). Accelerating a fundamental transformation in water resource utilization and enhancing efficiency constitute crucial elements of implementing a comprehensive conservation strategy. Since the era of reform and opening up, China's urban clusters have witnessed rapid economic and social development, accompanied by a surge in population and high concentration. Coupled with geographical factors, this has resulted in water scarcity in urban clusters far surpassing other regions. The inadequacy of water resources, the continuous degradation of water environments, and the escalating demand for water resources due to economic and social development exert immense pressure on rapidly growing urban clusters, further impeding overall socioeconomic progress (Jiao et al. 2020). Consequently, the urgency of conserving water resources cannot be overstated. In the 19th National Congress of the Communist Party of China report, there is a clear call for advancing comprehensive resource conservation and recycling alongside implementing a national water-saving initiative.

As a national-level media, China Central Television (CCTV) bears the historic mission of conveying the Chinese voice and telling China's story, possessing significant authority, great attention, and crucial influence in the hearts of the Chinese people. It serves as a vital platform for promoting traditional virtues, inheriting cultural legacies, fostering social harmony, and embodying the core socialist values (Chen 2023). In recent years, CCTV has launched a series of public service advertisements focused on water resources, aiming to enhance public awareness of cherishing water resources.

Symbols, more specifically, can take the form of written verbal, images, sounds, smells, tastes, behaviors, or physical objects (Juniartha et al. 2022). The Saussure's Semiotic Theory is one of the most important theories in semiotic theory, positing that symbols comprise two essential elements: the signifier and the signified. The signifier represents the form adopted by the symbol, while the signified is the concept or object it represents in verbal and non-verbal communication (Saussure 1983). The study suggested that verbal symbols connect not between things and names but between pictures and sound images (Holdcroft 1991). Consequently, it is proposed to retain the use of the term 'symbol' to represent the entirety of the idea, with 'signified' and 'signifier' replacing 'concept' and 'sound-image,' respectively, and emphasizing that the connection between the signifier and the signified is arbitrary. He highlighted that visual and auditory symbols encompass visible elements such as images and colors, and non-verbal symbols like actions and textual (dialogue) symbols. These symbols, laden with rich meanings, endow audiences with diverse interpretations, shaping values and perspectives by disseminating visual and auditory symbols. Providing information to the reader is a crucial strategy advertisers employ, so this viewpoint plays a significant role in advertising (Wahyuni et al. 2022).

Previously, scholars have applied Saussure's semiotic theory to research commodity packaging design and the symbolic analysis of the live broadcast text of the PyeongChang Winter Olympics opening ceremony. The findings indicate a close connection between classic domestic packaging design and socio-cultural contexts (Wang et al. 2022). Visual and auditory symbols, encompassing speech and image symbols, subtly convey national ideologies and values, playing a crucial role in cultural identity and intercultural connections (Ma 2019). Moreover, Scholars approached the analysis of 37 public service advertisements aired during the Spring Festival Gala from a narrative perspective. They summarized narrative characteristics, structures, the dominance of a zero view, and the diverse use of symbols. Additionally, they explored the construction pathways of cultural identity regarding home, hometown, and national culture through spatial, behavioral, and symbolic dimensions (Yuan and Hu 2013). Therefore, to enhance the efficacy of advertising communication, a deep understanding of semiotic theory, the audience needs, and the revelation of underlying meanings behind advertisements are crucial, offering the potential for new development spaces and foundations in advertising (Zhou and Zhang 2014).

This study employs Saussure's semiotic theory, specifically focusing on the signifier and signified, to analyze the artistic expressions of symbols in CCTV water resource public service advertisements from 2021 to 2023. The study aims to explore the role of these symbols in conveying information about water resource protection from the perspectives of verbal, color, image, and auditory symbols.

20.2 Methodology

This study selects the water resource-related public service advertisements broadcasted on CCTV from 2021 to 2023 as case studies. The data is sourced from water conservation campaigns jointly initiated by the Ministry of Water Resources of the People's Republic of China and the National Office of Water Conservation, collaborating with CCTV, during the National Water Week events in 2021, 2022, and 2023. CCTV is a prominent news and opinion platform in China, playing various roles in disseminating news, societal education, cultural entertainment, and information services (Chen 2023). Public service advertisements from CCTV are widely recognized for their high credibility, significant influence, and extensive reach, making them representative.

Data collection involves several steps. Firstly, public service advertisements were searched on the official websites. Secondly, the advertisements were downloaded. Thirdly, the relevant and essential information within the scope of the study was read and recorded. The data was categorized into verbal, color, image, and auditory symbols.

In analyzing the data, a qualitative approach was employed. The data analysis consisted of several steps. Initially, theoretical research was conducted on verbal, color, image, and speech symbols. Subsequently, the meanings of verbal, color, image, and auditory symbols were individually examined. Finally, the implications of verbal, color, image, and auditory symbols were analyzed based on Saussure's semiotic theory. This methodological approach aims to provide a comprehensive understanding of the symbolic elements present in CCTV water resource public service advertisements, utilizing a qualitative framework and drawing on the insights offered by semiotic theory.

20.3 Result and Discussion

The analysis of CCTV's water resources public service advertisements from 2021 to 2023, based on Saussure's semiotic theory, encompasses four perspectives: verbal symbols, color symbols, visual symbols, and auditory symbols. The following conclusions are drawn.

20.3.1 Verbal Symbols—Gradual Simplification for Enhanced Comprehensibility

In terms of verbal symbols, advertisements can incorporate text with excellent word choices designed to captivate the reader's interest (Wahyuni et al. 2022). This study extracted representative verbal symbols from three selected advertisements (see Table 20.1), and the findings indicate that CCTV water resources public service advertisements in 2021 start with the premise of a single droplet of water, emphasizing the concept of "gathering." The aim is to draw the audience's attention to the value derived from the simple act of "saving a droplet of water everyday" and to cultivate a habit of water conservation. In contrast, the 2022 advertisements focus on the younger demographic, specifically targeting children to instill the notion of conserving water resources. This strategic approach aims to popularize and universalize the concept of water conservation, thereby maximizing water resource utilization efficiency. Moving into 2023, the public service advertisements adopt straightforward and plain language, explicitly conveying that, in the context of the new historical backdrop, there is a collective need to enhance water conservation efficiency in a civilized manner, emphasizing this as a responsibility for each individual. It is evident that over the past three years, CCTV's water resources public service advertisements have gradually simplified and diminished verbal intricacies to make the content more accessible and acceptable to the general public.

Table 20.1 The signifier and signified of verbal symbols in public service advertisements

Advertisements	Verbal symbols	Signifier	Signified
CCTV Saving Water Public Service Advertisement (2021)	1. Save a drop of water every day 2. Each person saves a drop every day. With 1.4 billion people over 365 days, it accumulates to 26,000 tons, enough to provide a year's worth of drinking water for 47,000 people 3. Even a small water droplet carries profound significance	1. Should save a drop of water every day 2. If 1.4 billion people save just one drop of water daily, the total saved amounts to 26,000 tons, which can provide hope for life, sustaining 47,000 people for a year" 3. Saving even a small droplet of water is a noble quality	1. In our daily lives, it is essential to be mindful of water that can be conserved, to use water resources correctly, starting with small actions, and to promptly prevent the senseless loss of water resources 2. Many places on Earth are facing water scarcity, and people are living in challenging conditions. Saving water resources can, to some extent, help other forms of life 3. Although saving a tiny droplet of water may seem insignificant in daily life, conserving water resources is a commendable act. Regardless of the quantity of water saved, the action itself deserves appreciation
CCTV Saving Water Public Service Advertisement (2022)	1. Start water conservation from childhood 2. Every drop counts, contributing to rivers and oceans 3. Building a better homeland depends on each of us	1. Water conservation is not just for adults; it's crucial to guide children to use water responsibly from an early age, avoiding the unnecessary depletion of water resources 2. Every drop of water, no matter how small, contributes to the formation of rivers. Building a habit of water conservation is essential 3. Creating a favorable living environment requires the active participation of everyone	1. We need to instill the idea in the next generation that "water resources are precious and should not be wasted." Teaching children to use every drop of water wisely, engraving the concept of "water conservation" in their minds, and passing it on to the next generation is crucial for advocating civilized water use on a broad scale, maximizing the value of every drop of water 2. Saving a single drop of water may appear meaningless, but if this action is repeated millions of times, millions of drops of water can form a river. Conserving water resources can exceed our imagination 3. Losing water resources means we cannot survive, let alone build a home. Therefore, each of us is responsible for conserving clean water and contributing to our home's beautiful environment

(continued)

Table 20.1 (continued)

Advertisements	Verbal symbols	Signifier	Signified
CCTV Saving Water Public Service Advertisement (2023)	1. Water nourishes all things without contention; life is inseparable from water 2. However, water is not limitless. Conserving water is everyone's responsibility 3. Be a water-saving individual in this era of civilization. In the new generation, be even more water-conscious	1. Water has the ability to nourish all things, and the survival of everything is inseparable from water 2. Water resources won't last indefinitely. Each individual bears the responsibility to conserve and protect water resources 3. This implies inviting the audience to become individuals with quality and morality, reducing water consumption and losses. It emphasizes that, in the current developmental context, we need to intensify efforts in water conservation	1. Emphasizing the importance of water for the survival of all life on Earth underscores the critical role water plays in every aspect of existence 2. Water is one of the essential material resources for human survival and development, a vital component of the Earth. Everyone living on Earth must be responsible for conserving and protecting water resources to ensure long-term survival 3. This is an appeal to improve the efficiency of using clean water and protect it. The phenomenon of water scarcity has persistently existed and is gradually worsening, posing a threat to human production and life

20.3.2 Color Symbols—Emotional Connection Between Colors in Advertisements and the Audience

Color is more than just a visual stimulus. It is intricately linked to emotions and mental states (Valdez and Mehrabian 1994). Therefore, the signifier of a color symbol includes specific colors presented in an image, warm or cool tones, or the three attributes of color (hue, saturation, and brightness). The signified, on the other hand, encompasses the sensations, emotions, or cultural meanings associated with specific colors and the concepts that public service advertisements aim to convey to the audience.

Li (2010) pointed out that the emotional impact of color symbols primarily arises from the psychological associations generated by the subjective intentions of color users. These psychological associations encompass concrete, abstract, and psychological associations. Color symbols serve as emotional symbols, expressing the inner emotions of a population. The emotional aspect of color symbols is the expressive nature of color, determined by various attributes of color. These attributes link individuals to their physiological, psychological, and social utility properties, evoking specific emotions. In public service advertisements, color symbols convey concepts, produce emotional resonance, and capture the audience's attention.

The following analysis will focus on color symbols in three water resource public service advertisements from warm and cool tones, color hues, and brightness perspectives.

20.3.2.1 Analysis of Warm and Cool Tones in Color

Color symbols serve as emotional mediators, sharing similarities and commonalities between the experiences and emotional expressions associated with color (Li 2010). Public service advertisements often employ warm tones to connect with people's visual perceptions, allowing them to grasp the concepts the advertisements aim to convey. Warm tones imply cultural connotations such as warmth, enthusiasm, and affinity. Public service advertisements frequently use warm colors like beige and pink, combining them with friendly and warm scenes and linking them to cultural connotations of human warmth and emotion (Zhang 2017).

(See Fig. 20.1). The overall color tone of the scene leans towards warmth, including beige home decor and the female protagonist wearing a pink outfit. The overall low saturation tones give the audience a gentle and mild impression. (See Fig. 20.2). The main subject of the scene is a young girl in a pink outfit, her father's hand, and an open tap. The warm tones, including the yellowish skin, pink clothing, and light yellow background, are used in both scenes. The warm color symbols in these two public service advertisements effectively connect with people's visual perceptions. The warm tones naturally evoke associations with the beauty and happiness associated with abundant water resources, aligning with the water conservation

Fig. 20.1 Screen from CCTV's 2021 water resource public service advertisement (00:11)

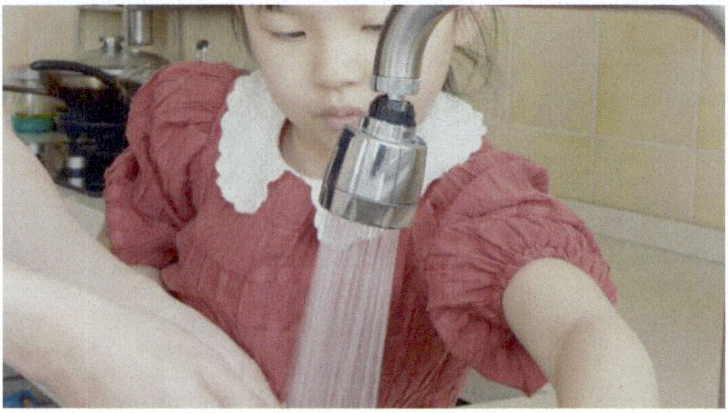

Fig. 20.2 Screenshot from CCTV's 2023 water resource public service advertisement (00:22)

concept the advertisements aim to convey. Coupled with other non-verbal and verbal symbols, these warm tones deepen the idea of water conservation in the audience's minds.

20.3.2.2 Analysis of Color Brightness

In color contrast, lacking brightness contrast can diminish the impact and make visuals appear pale and feeble (Shi 2014). Concerning brightness, colors with higher brightness are more noticeable than those with lower brightness (Huang 2022). Public service advertisements often employ high-brightness color symbols to enhance the visual impact on the audience, emphasizing the main subject the advertisements seek to portray. This, in turn, facilitates a stronger emotional connection between the intended message and the audience.

Fig. 20.3 Screenshot from CCTV's 2022 water resource public service advertisement (00:20)

As observed (see Fig. 20.3), the scene predominantly features the high-brightness color green, representing plants and clear water. Green is one of the most common colors found in nature, associated with natural elements such as plants, trees, and grass. Therefore, green is often utilized to express concepts of nature, vitality, and environmental protection. The high brightness of green enhances visual impact and prompts the audience to associate it with an environmentally friendly ecosystem, thereby fostering psychological connections with the conservation of "green water" and water conservation.

(See Fig. 20.4). The brightness level of colors is adjusted in the background's rocks and sky to highlight the main subject—water. The white water flow is the most prominent area of the scene, strongly capturing the audience's attention and making it easier for them to focus on water as a critical element. In the brightness scale of colors, white occupies the highest position, representing the maximum brightness of light without any mixture of other colors. Among the three elements of color, white has the highest brightness (Li 2019). This advertisement strategically uses the white water flow to evoke emotional associations, naturally prompting viewers to develop the concept of cherishing clean water resources.

20.3.2.3 Analysis of Color Hue Contrast

Hue contrast is a crucial visual element in film and television advertisements that can influence the audience's emotions and attention. Essentially, hue contrast involves substantial differentiation of different hues to create diverse visual impacts on the audience (Li 2019). In public service advertisements, hue contrast is often used to establish harmonious and comfortable color combinations, attracting viewers to watch the advertisements and enhancing the acceptability of the conveyed concepts, generating emotional resonance with these concepts.

(See Fig. 20.5). The main subject of the scene consists of green hills and clear blue water and sky. The overall presentation involves the two primary colors—green and blue. In this public service advertisement, the two primary colors, blue and green,

Fig. 20.4 Screenshot from CCTV's 2023 water resource public service advertisement (00:02)

Fig. 20.5 Screenshot from CCTV's 2022 water resource public service advertisement (00:44)

are complementary hues, ensuring a balanced overall color scheme in the ad and avoiding an overly chaotic or monotonous appearance. The blue clear water and green hills represent natural scenes in the image. In other words, blue and green can also be seen as representing the protected clean water and green hills, respectively. This coordinated composition forms a harmonious color combination, enhancing the audience's acceptance of the viewpoint conveyed in this water resource public service advertisement—cherishing water sources and conserving water. Consequently, the visual stimulation from the color symbols connects with the emotional resonance of valuing and conserving water resources.

20.3.3 Image Symbols—Comprehensive Application from an Objective Perspective

CCTV's water resource advertisements utilizing objective perspectives in image presentation. The objective perspective involves shooting and depicting scenes objectively, providing a lens that reflects the objectivity and rationality of characters. For instance (see Figs. 20.6 and 20.7), to convey the importance of "life, inseparable from water," the photographer selected scenes of a snail drinking water on a plant and the growth of trees. They opted for natural, intuitive, and straightforward angles. In this process, they did not influence the development of events or express their opinions. By presenting the current state, causes, and consequences of events, the audience is immersed in the experience, allowing them to interpret and reflect based on what they see.

An objective perspective can be used not only for capturing stationary objects but also for filming moving objects. This involves using a static frame and background as a reference, allowing static elements to complement the moving scenes. This approach reflects authenticity and provides an objective record of the speed and rhythm of the object's movement. Simultaneously, it enhances the dynamics of the photographed object, as seen in Fig. 20.8, where a fixed lens with an objective perspective captures the dynamic movement of water.

In addition to capturing movement, CCTV public service advertisements are adept at using an objective perspective to shoot panoramic views, such as distant mountains or expansive forests. After viewing these scenes, it feels like observing a landscape painting, effectively conveying information about the tranquility, lack of pollution, and environmental friendliness of water sources in CCTV's water resource public service advertisements (see Fig. 20.9).

Furthermore, the advertisements utilize numerous fixed shots from an objective perspective, focusing on facial expressions and actions to portray the characters better. From subtle changes in facial expressions to movements, the audience can

Fig. 20.6 Screenshot from CCTV's 2023 water resource public service advertisement (00:06)

Fig. 20.7 Screenshot from CCTV's 2023 water resource public service advertisement (00:09)

Fig. 20.8 Screenshot from CCTV's 2023 water resource public service advertisement (00:01)

perceive the changing psychological states of the characters in the advertisement (Fig. 20.10).

In conclusion, objectivity is a crucial means of gaining public trust, and the objective perspective is widely applied in CCTV public service advertisements. The objective perspective excels in objectivity and authenticity, serving an educational role from the perspective of image expression.

Fig. 20.9 Screenshot from CCTV's 2022 water resource public service advertisement (00:43)

Fig. 20.10 Screenshot from CCTV's 2021 water resource public service advertisement (00:24)

20.3.4 Auditory Symbols—Evoking Resonance Through Auditory Effects

Auditory symbols in advertisements refer to the use of human voices, including narration, singing, rhythm, and more. In this study, we explored the use of lively and smooth instrumental music in these public service advertisements to complement the visual effects and set an appropriate emotional tone. The soft ambient music evokes a sense of tranquility in harmony with nature, aligning with the themes of cherishing and protecting water resources. For example, in the 2023 advertisement, sounds of splashing water, droplets, or flowing water are incorporated, metaphorically symbolizing the value of precious water resources and the passage of time. Combining visual and auditory symbols conveys a persuasive message: the importance of cherishing and protecting our water resources for future generations. Symbolic representations, metaphors, and music create an emotional experience beyond the limitations of language. CCTV's continuous use of vocal symbols in water resource public service advertisements over three years serves to guide audience emotions and convey ideologies.

Examining the tone of the voiceover narration, for instance, in the 2021 water resource public service advertisement, the narrator states, "Pooling the strength of every droplet, converging into a magnificent force." The tone of this narration is serious and powerful, addressing the scarcity of water resources. This form of presentation emphasizes the active involvement of every individual in water resource protection. It encourages the audience to understand that only through collective effort can a tremendous force be formed, allowing each droplet of water to fulfill its inherent value. This narration style undoubtedly enhances the persuasiveness and emotional impact of the advertisement, making it easier for the audience to comprehend and accept the water conservation ideology it conveys.

From the perspective of vocal music, in the 2022 CCTV public service advertisement, children's singing is bright, pure, soft, and complete, conveying warmth and healing. The timbre of these child voices, interwoven with other musical elements, forms a unique symbolic system. Through the expression of children's singing, the advertisement conveys the message of water resource protection, deeply embedding this critical issue into the hearts of families and children. It raises awareness among people about protecting water resources for future generations, stimulating environmental consciousness within society. Through the use of auditory symbols, the water resource public service advertisement successfully conveys the concept of protecting water resources, prompting thoughtful consideration and action from a broad audience. This initiative fully demonstrates the profound impact of public service advertisements, effectively touching the emotions deep within individuals and eliciting a positive response to water resource protection.

Analyzing the musical accompaniment in the 2023 CCTV public service advertisement, piano music beats form the main composition of the background music. The rhythmic and lively beats create a sense of flowing like a clear stream in people's ears. The crisp sound of the piano, combined with other soothing musical elements, enhances the rhythm of the music. Simultaneously, this music complements the gentle pace of the visuals, forming a highly rhythmic public service advertisement. Through the combination of music and visuals, the advertisement successfully captures the audience's attention and sparks resonance with the public service ideology.

20.4 Conclusion

In conclusion, this research, conducted through the lens of Saussure's Semiotics theory, delved into the realm of public service advertisements, focusing specifically on the water resource campaigns aired on China Central Television (CCTV) over the past three years. The study analyzed linguistic symbols, visual elements, and the overall semiotic strategies employed in these advertisements. The research findings indicate that the gradual simplification of linguistic symbols contributes to better understanding. Color symbols excel in using warm tones and high color contrast to evoke emotional resonance. Objective perspectives in visual symbols effectively stimulate subjective water-saving awareness. Audio symbols play a crucial role in complementing visual symbols and enhancing resonance. These results suggest that a comprehensive use of multiple symbols in advertisements can convey public service messages holistically and effectively, guiding the audience to deeply resonate with water resource conservation.

This study contributes to understanding how semiotic elements, rooted in Saussure's Semiotics theory, can effectively communicate complex societal issues such as water conservation. However, it is essential to acknowledge certain limitations that may impact the generalizability of the findings. Firstly, the focus on CCTV water resource advertisements within a specific time frame may limit the broader applicability of the results to different cultural contexts or media platforms. The exclusive reliance on Saussure's Semiotics theory might overlook the nuances introduced by other semiotic theories, such as Peircean semiotics or cultural-specific interpretations.

In terms of future prospects, researchers might delve deeper into the intersectionality of semiotic elements, considering the dynamic interplay between linguistic, visual, and auditory symbols. Comparative studies across different cultures and media platforms could provide a more comprehensive understanding of the universal and culture-specific aspects of effective public service communication.

References

Bai Y, Dai J, Huang W, Tan T, Zhang Y (2021) Water conservation policy and agricultural economic growth: evidence of grain to green project in China. Urban Clim 40:100994

Chen X (2023) Research applying traditional cultural elements in CCTV public service advertising planning. Editorial 04:172–174

Holdcroft D (1991) Saussure: signs, system and arbitrariness. Cambridge University Press

Huang B (2022) How to use suitable colors in advertising design. Footwear Technol Des 2(23):30–32

Jiao S, Chen L, Wang A, Zhang C, Yin Y, Zhang X, Zhao R (2020) Research on changes and driving forces of water resources ecological footprint in Central Plains urban agglomeration. Yellow River 42(5)

Juniartha IW, Sulatra IK, Andriyani AD (2022) Discovering the meaning of signs in public service advertisements. Transl Linguist (Transl) 2(2):45–53

Li Q (2010) On the color symbols conveyed in commercial display space Master's thesis, Nanjing Forestry University

Li X (2019) Analysis of the application of color semantics in modern graphic advertising design. Collection 8

Ma J (2019) Semiotic analysis of the live broadcast text of the opening ceremony of the PyeongChang Winter Olympics. Master's thesis, Chengdu Institute of Physical Education

Saussure FD (1983) Theory of signified and signifier. McGraw-Hill Book Company, Geneva

Shi J (2014) Talking about the contrast and harmony of advertising design colors. Art Technol 6:248–248

Valdez P, Mehrabian A (1994) Effects of color on emotions. J Exp Psychol Gen 123(4):394

Wahyuni NMS, Sulatra IK, Pratiwi DPE (2022) A semiotic analysis of public service advertisements. J Humanit Soc Sci Public Adm Manag (HUSOCPUMENT) 2(2):113–120

Wang L, Chen F, Ma K (2022) Research on the packaging design of White Rabbit toffee from the perspective of Saussure's semiotics. Packag Eng Art Ed 43(14):238–245

Yuan A, Hu X (2023) The construction of cultural identity of public service advertisements from the perspective of narrative theory—taking CCTV Spring Festival Gala public service advertisements from 2013 to 2023 as an example. Media Forum 6(15):12–14+18

Zhang L (2017) Research on the color expression of public service advertisement. In: 2017 international conference on social sciences, arts and humanities

Zhou M, Zhang Y (2014) Semiotic interpretation of the TV commercial "Chopsticks." Engl Sq Acad Res (8):20–21

Chapter 21
Morphological Change and Quantification of Uncertainty in Remotely-Sensed Measurement of Indus River

Muhammad Waseem Boota, Shan-e-hyder Soomro, Haoming Xia, Yaochen Qin, Chaode Yan, Syed Shahid Azeem, and Jikun Xu

Abstract The increasing accessibility and resolution of remote sensing (RS) images of the Earth's images has significantly improved the accuracy and spatio-temporal range with which we can analyse the channel movement and width variations. We summarized and evaluated current best practices, tested novel techniques to deal with the error and uncertainty, and present recommendations for futuristic study utilizing RS observations of Indus River migration and width variations in this paper. This study also deals with the morphological changes in the Indus River. It was observed that mean root mean square error - RMSE (Landsat 8) value for difference between actual centerline and remote sensing delineated centerline across all years was found to be 0.127 km in the study reach. The error found in Landsat 7 delineated centerline was more as compared to Landsat 8 and Landsat 5, due to the scan line error. In addition, Remote sensing delineated river widths were compared with the actual cross-section surveyed on or near to image acquisition date. The coefficient of

Muhammad Waseem Boota, Chaode Yan are equally contributed.

M. W. Boota · H. Xia (✉) · Y. Qin (✉)
College of Geography and Environmental Science, Henan University, Kaifeng 475004, China
e-mail: xiahm@vip.henu.edu.cn

Y. Qin
e-mail: qinyc@henu.edu.cn

M. W. Boota · C. Yan · J. Xu
College of Water Conservancy and Transportation, Zhengzhou University, Zhengzhou 450001, China

S. Soomro
College of Hydraulic and Environmental Engineering, China Three Gorges University, Yichang 443002, China

S. S. Azeem
Center of Excellence in Water Resources Engineering, University of Engineering and Technology, Lahore, Pakistan

H. Xu (ed.), *The 6th International Symposium on Water Resource and Environmental Management*, Environmental Science and Engineering,
https://doi.org/10.1007/978-3-031-55989-1_21

correlation between actual and remote sensing delineated river widths comes out ($R^2 = 0.8665$), and RMSE was found to be 115.4 m. Lastly, we advocate concentration on enhancing the simplicity, generalization, and open-source prospects of programs and tools used to calculate the channel planform variations and spatially varying error and uncertainty, resulting in single platform for measuring and comparing outcomes.

Keywords Uncertainty in RS images · Morphological change · Conceptual model · Landloss · Indus river

21.1 Introduction

Channels for rivers are one of the more changing features on the planet's surface, moving horizantally over valley bottoms in convert and occasionally disastrous manners over episodes, decade, and periodic periods. Remote sensing (RS) images are frequently used to define channel bars in order to compute variations in channel morphological study (Donovan et al. 2019; Donovan and Belmont 2019), which includes lateral variations and width changes (Cheng et al. 2006). These measurements serve as a foundation for comprehending effective erosion mitigation throughout the vegetation corridor (Piégay et al. 2005), as well as input for sediment load (Belmont et al. 2011) and automated classification of single-threaded versus multiple-threaded channel planforms (Rowland et al. 2016). Such measurements also help us to understand the crucial concerns involving erosional risks caused by moving streams, influence of anthropogenic changes to fluvial basins, and riparian habitat management (Rhoades et al. 2009). To ensure efficient channel corridor management requires the proper quantification and report of error and uncertainty in channel migration measurements. In this way, we can describe the inappropriate advising costly channel and vegetation management techniques, such as banks stability and intrusive remedations standards. The wider accessibility and resolution of aerial photography, satellite imagery, unmanned aerial vehicle (UAV) imagery, and LiDAR or digital elevation/ terrain models (DEM/ DTMs) of the Ground surface have significantly improved the accuracy and spatio-temporal extent with which we can evaluate the river movement (Harpold et al. 2015). Notwithstanding a wealth of RS data and unique possibilities by constantly improving programmings, the investigation of fluvial regime based on RS lacks a solid and reliable framework for assessing and dealing with error and uncertainty (Schook et al. 2017). Numerous research has recently advanced our knowledge of error and uncertainty in measurements using remote sensing imaging (Piégay et al. 2005; Shumway et al. 2002), but no complete methodology has been established for river planform change. For morphological variations, temporal uncertainty is usually negligible because to image acquisition time is generally known (Donovan et al. 2019). The precise computing error is critical not only for screening away unverified measurements, as well as for maintaining accurate assessments (Rowland et al. 2016). While uncertainty cannot be fully described, findings should be restricted based on both (upper, lower bounds) of error or emphasizing regions where the measurements are higher or lesser reflective of unpredictible variations to provide end users with a judgement of reliability. For any application, proper analysis and revealing uncertainty can effectively enhance the reliability of outputs and subsequent solutions. Research seeking to cover huge spatiotemporal scales may thus choose for automatic delineations, thus sacrificing precision to enhance the scope of analysis and manage time. One limitation is that this kind of research must take place along adequate large and active rivers, because manual delineations might increase the spatial rangelower-order channels delinations to automatic techniques (Rowland et al. 2016). Manual techniques increase overall delineations when performed by an expert user. Thus, this is the first time research which deals with the error and uncertainty in the remotely-sensed images and morphological changes in the Indus River.

21.2 Case of Indus River and Acquisition of Data

Indus River is the largest river of Pakistan with a length of 2900 km, and computed sediment load to the Arabian Sea prior to human-induced interferences was about 250 × 106 mt/yr (Mahar et al. 2022), as shown in Fig. 21.1. Landsat data (2007 and 2010) flood events were used to determine the schannel numbers in the LIR (Carling et al. 2018). It was observed that five channels were active during the dry season and six channels during the monsoon periods. Mean yearly monsoon values (1961–2020) occurred in Guddu Barrage with a runoff value of 318 to 17,082 $m^3.sec^{-1}$, with mean upstream water level of 74.62 m to 78.33 m. Thus, mean head difference (Guddu Barrage) between upstream and downstream is 1.65 m to 0.67 m. In addition, 2-yr return period value at Sukkur wa observed 17,630 $m^3.sec^{-1}$. The exceedance probability Pi (X > = xt) and non-exceedance probability Pn (X = < xt) at 2-year RP is observed to be 0.485 and 0.515. The peak flow values at 50-year return period is 35,373 $m^3.sec^{-1}$ with Pi (0.019) and Pn (0.981). It was also observed thatpeak flow at 100-yr return period wa 37,865 $m^3.sec^{-1}$ with Pi (0.010) and Pn (0.990). It is evident from the collected data (1963–2020) that percentage of occurrence is about 67% for collectively high flood, very high flood and exceptionally high flood events at Guddu Barrage, depicted in Fig. 21.2 (Table 21.1).

21.3 Materials and Methods

21.3.1 Flood Frequency Analysis in Study Area

The flood frequency analysis by using the Gumble distribution was selected for the Sukkur and Kotri stations. It is pertinent to mention here that Gumble distribution is the best-fit distribution based on Chi-square test (not illustrated) but explained in

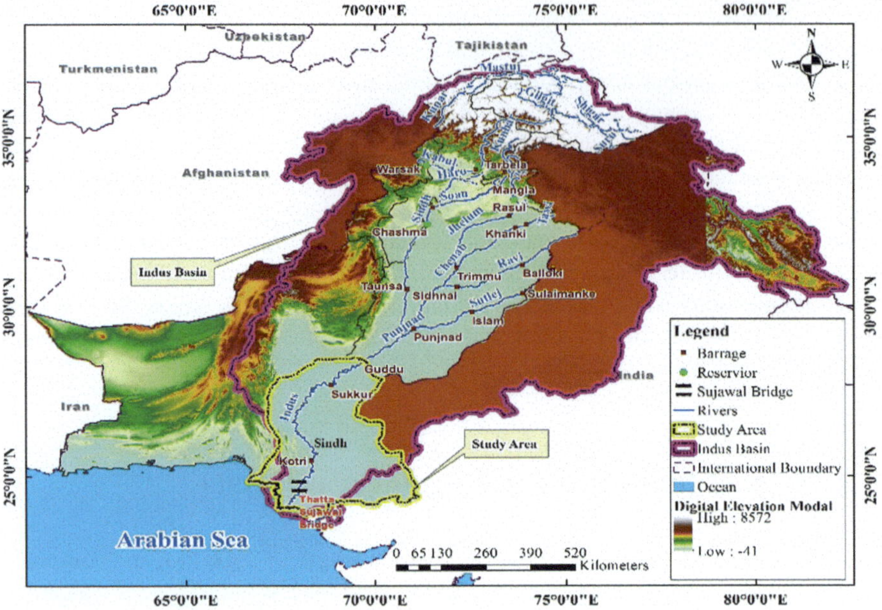

Fig. 21.1 Map representing Indus river basin location used in the current study

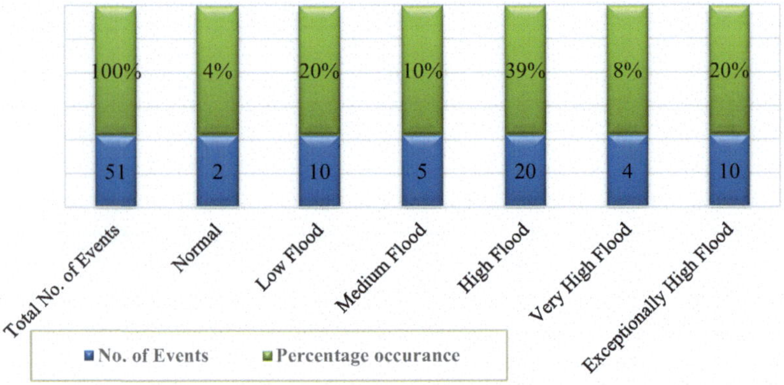

Fig. 21.2 Flood limits of Lower Indus River at Guddu Reach

Naz et al. (2019). The probable runoff values at different return periods by using Gumble distribution is represented in Fig. 21.3a, b. In contrast, relationship between runoff and reduced variate is represented in Fig. 21.4a, b).

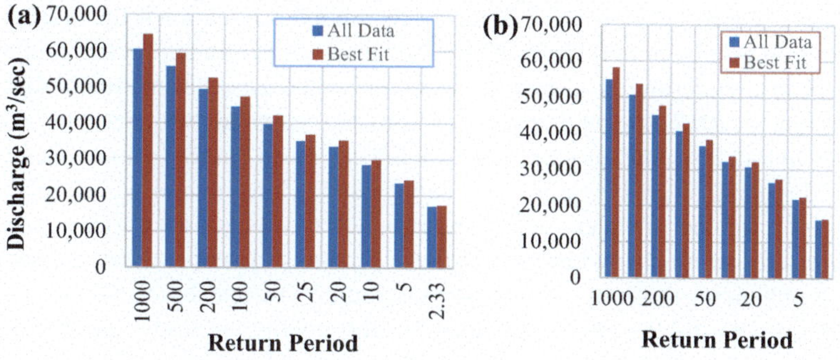

Fig. 21.3 Runoff frequency analysis at **a** Sukkur **b** Kotri

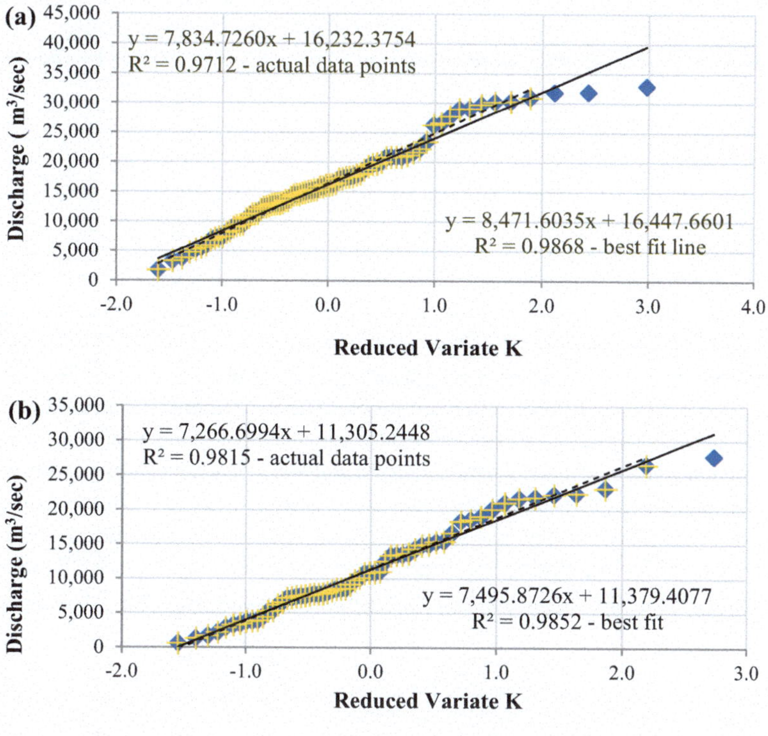

Fig. 21.4 Relationship between runoff and reduced variate at **a** Sukkur **b** Kotri

21.3.2 Quantification of Error and Uncertainty Using Remote Sensing Images

The criteria to assess uncertainty for river migration from manual delineations based on high-resolution Google imagery and Landsat imagery were carried out for study reach. It is crystal clear from the literature that manual delineation is the most precise and common (Blundell and Opitz 2006), but semi- and fully automated techniques are useful in saving time by removing the complicated aspect of manual delineation (Schwenk et al. 2017). A small amount of study has assessed the uncertainty components for manualluy and automatic river delineation and categrozitions. Automated channel delineations are becoming commonly used, while manual-based delineation is much precise and prevalent and, therefore, act as validation for evaluating error of automated findings (Rowland et al. 2016). Manual delineation error can be minimized by enhancing the digitization scale (i.e. zoomed-in) and lesser extent—Landsat scale. Herein, we did the manual delineations throughout the study in order to attain precision based on the literature mentioned above, and error and uncertainty using remote sensing measurement of channel planform change was observed with respect to different scenarios. First, channel centerlines were manually delineated for Landsat and high-resolution Google imagery. Valley line was drawn, covering the Landsat and Google imagery centerlines. The offsets from centerlines to valley line were drawn in ArcGIS, and RMSE were computed from centerline offsets by the formula given as.

$$RMSE = \sqrt{\frac{1}{N}\sum_{i=1}^{N}(P_i - O_i)^2} \tag{21.1}$$

where Pi is the predicted value for the ith observation in dataset, Oi is the observed value for the ith observation in dataset, and N is the sample size. Handling values below Level of error detection (LeD) threshold is called non-detects (NoD). The NoD are commonly handled in morphology by reducing NoD or substitute the NoD with 0, 0.5, 0.7, $1/\sqrt{(2)}$ of LeD limitis, and keeping NoD measurement to bracket leads with variety of bounds (upper & lower) based on error degree. These approaches bring error in the estimation of total variation, error and skewed statistically parameters by incorporating a disproportionatly numbers of arbitrarily variables. Thus, we proposed three different methods in the study area such as, Maximum Likelihood Estimation (MLEs)—Eq. 21.1, imputation via Regression on Order Statistics (ROS) (T. C. E. versus Time 2009)–Eq. 21.2, and finally Kaplan–Meier (KM)—Eq. 21.3, given as.

$$L(\vartheta) = \prod_{i=1}^{n} \frac{1}{\sqrt{2\pi}} e^{-\frac{(X_i - \vartheta_0)^2}{2\vartheta_1}} \tag{21.2}$$

where L (ϑ) is the likelihood for a continuous variable, ϑ is a vector with two values, with detail description is provided in T. C. E. versus Time (2009).

$$\begin{aligned} pe_i &= pe_i + \frac{A_i}{A_i + B_i}(1 - pe_{i+1}) \\ pd_y &= (1 - pe_i) + \left(\frac{j}{A_i + 1}\right).(pe_i - pe_{i+1}) \end{aligned} \tag{21.3}$$

where, pej denotes the proportion of the sample exceeding the ith RL. Once the exceedance probabilities are computed, plotting positions for detects i.e., cumulative probabilities on a probability plot, can be calculated with equation pdy.

21.4 Results and Discussion

21.4.1 Morphological Change of Indus River

The Indus at upstream of Guddu Barrage is complicated sand bed river network with stochastic aspect, although river network can be predicted to some extent. As observed in numerous other rivers, the channel capacity increased (Taunsa) following the severe flood episodes, and it is evident that extension of channel area owing to riverbank erosion and out-weighing bankline deposition. The landuse map of selected study reach is provided (Fig. 21.5), and river network transfer the imposed flows and sedimentation at downstream during HF episodes and hence the river network is sensitive to anomalous monsoonal exceptionally HF episodes. As the river channel network capacity increases during LF and HF periods (as it also accounts for peak annual runoff), but the increase in capacity is substantial when compared to the range of gauged flood peak runoff values. Belmont et al. (2011) has noted that an alluvial river widened significantly as a result of large runoff; channel narrowing is not justifiable to accommodate smaller runoff. However, the fact that HF episodes are significantly accommodate the floodplains implies that the system is not able to convey HF flows in-channel, as observed for other peak floods, and in our study HF change the channel morphology. In below Fig. 21.6, the blue symbols represent disproportionate land loss during a selected flood event in 2020. The purple symbols illustrate the right and left banks of monsoonal flood conditions (1986–2020), and y-error bars represent the variations across the study reach cross-sections in LIR.

Based on the fact that land loss is the pressing concern for the survival of the Indus River delta, we define "land" (and its attributes values such as altitude, magnitude of land loss, land use variation) as the target attributes for integrating the natural and manmade components throughout the vegetation. The resilience structure was modified based on the previous study on USA river study, and this strategy is ideal, theoretical and contains on hypothesis. In the Indus River, land loss is connected with the flood events and sediment load (SSL, BL) as well as human-induced factors which affect the river shape changes. Human-induced activities include the construction of levees, canals, upstream dams and in coastal regions; the sea level rise also affect the river morphology. Indus River delta coupled with natural-manmade dynamics structure is given in Fig. 21.7. The Indus River delta and coupled natural manmade structure has eight components linked with the central response factor (land loss or water change). Flood episodes are major factors that affect the sediment load in fluvial and deltaic regions (Fig. 21.6).

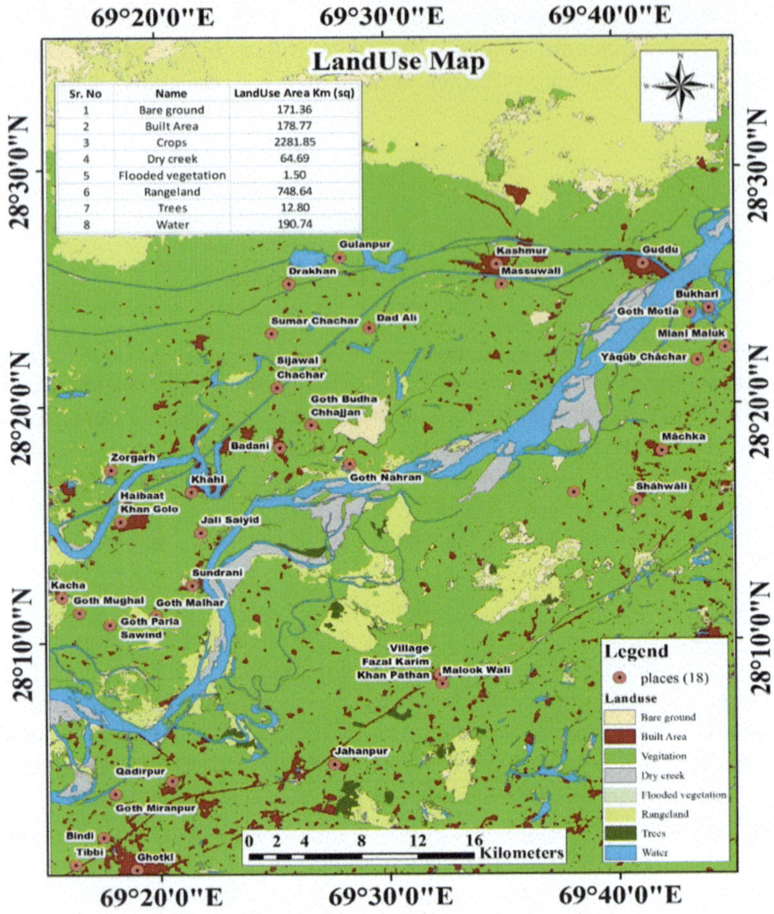

Fig. 21.5 Example of processed data in the start reach during 2018

21.4.2 Morphological Change of Indus River

The error and uncertainty in RS measurement for morphology change of LIR were analyzed based on literature, quantification the factors that influencing the magnitude, location, and types of offset from digitization. Average *RMSE* (Landsat 8) value for difference between actual centerline and remote sensing delineated centerline across all years was found to be 0.127 km (Fig. 21.8). The error found in Landsat 7 delineated centerline was more as compared to Landsat 8 and Landsat 5, due to the scan line error. Remote sensing delineated river widths were compared with the actual cross-section surveyed on or near to image acquisition date. For brevity, graphical representation of Landsat 5 is not illustrated here but concisely here, and statistical analysis was also completed to strengthen the error and uncertainty section.

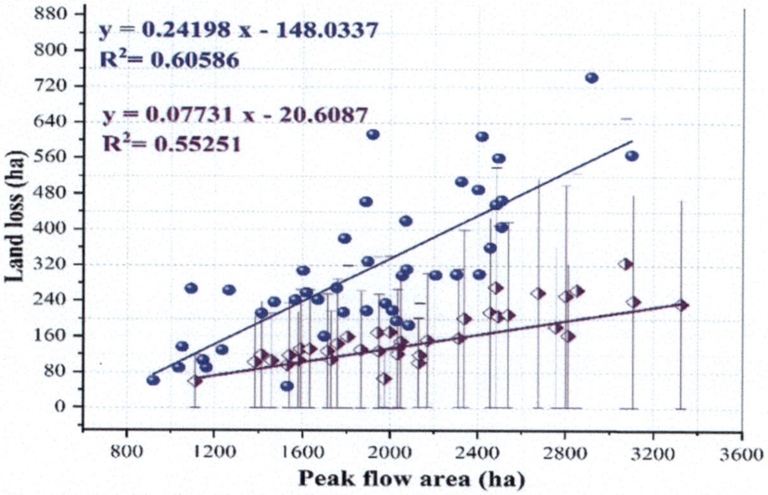

Fig. 21.6 Area flooded during annual peak flow and vegetated land loss each year

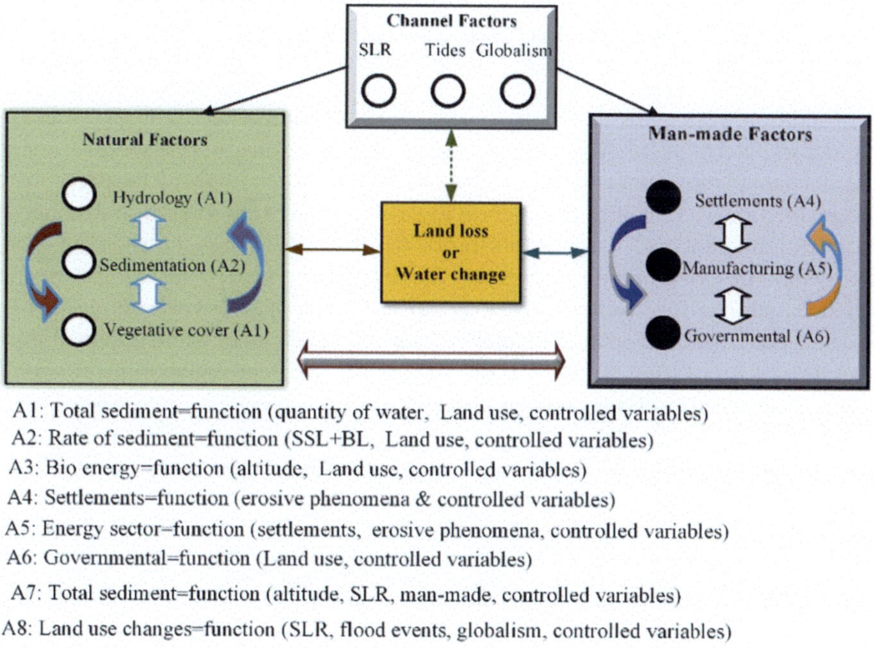

Fig. 21.7 Conceptual model of deltaic coupled natural-manmade strategy in Indus River

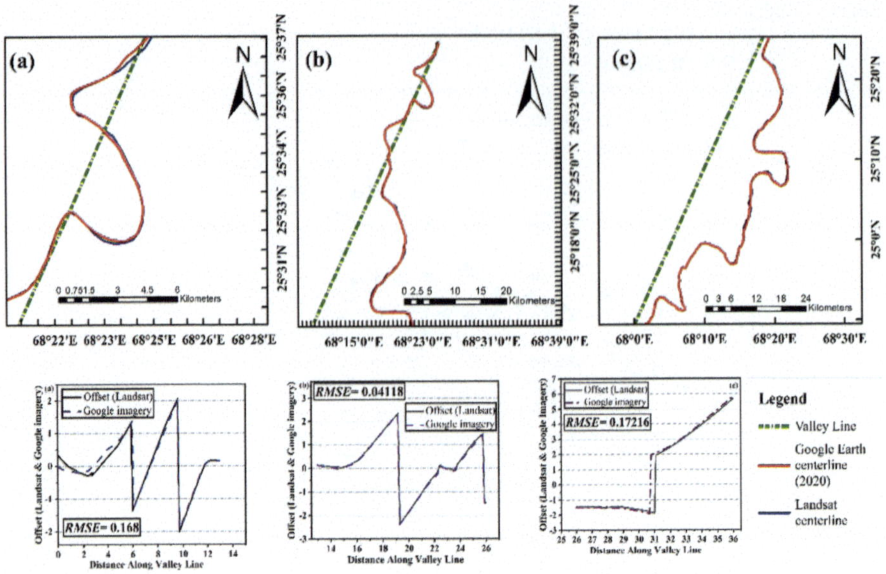

Fig. 21.8 Statistical (*RMSE*) analysis of Landsat 8 in different sections

In addition, systematical pattrens in percentage of retain observations were tested, in order to determine whether characteristics such as acquisition period, RS image size, or channel movement distances have substantial association with the percentage of remained measurements. The natural logarithm of average migration distance is the most significant in all these factors ($p < 0.001$, $R^2 = 0.8901$), as shown in Fig. 21.8. The method described here for assessing the digitizating uncertainty can be used to various situations for investigating potential differences by Werbylo et al. (2017), who observed observation at particular cross-sectional channel width obtained from several digitizers differ up to 19% of river width, whereas channel-mean widths indicated no significant discrepancies (Fig. 21.9).

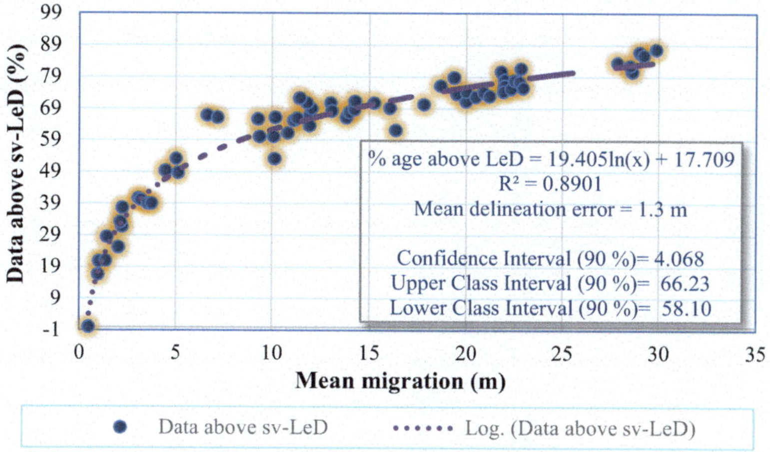

Fig. 21.9 Relationship between mean migration and percentage of retained measurement

21.5 Conclusion

This research shows that successive flood episodes during the LF period do not affect the channel erosion of the selected study reach. Contrarily, monsoonal flood episodes cause significant channel erosion and channel system adjustment. The study also shows that Landsat analysis shows that the formation and abandonment of channels are not linked with the flood episodes during the LF period, and significant morphological variations occurred during the HF period (Monsoonal flood). The error and uncertainty in the remotely-sensed images show that Mean *RMSE* (Landsat 8) value for difference between actual centerline and remote sensing delineated centerline across all years was found to be 0.127 km. The error found in Landsat 7 delineated centerline was more as compared to Landsat 8 and Landsat 5, due to the scan line error. Remote sensing delineated river widths were compared with the actual cross-section surveyed on or near to image acquisition date. The coefficient of correlation between actual and remote sensing delineated river widths comes out ($R^2 = 0.8665$), and RMSE was found to be 115.4 m. Future studies should strive to build on the recommendations offered above in order to enhance the error and uncertainty in the fluvial change of channel measurements. There has been limited agreement on how to utilize information gained from more than two decades of study on determining and evaluating the error and uncertainty in river channel migration.

Table 21.1 Data sets and acquisition time of satellite images

Data set	Spatial & temporal resolution	Source	HF period	LF period	HF runoff (m3/s)	LF runoff (m3/s)
Hydro-meterological data	1975–2021	PIDs				
Sediment data	1941–2021	PIDs				
Geomorphological data of LIR	1986 (60 m)	Landsat-TM	13/08/1986	11/02/1986	33224	677
	1992 (30 m)	Landsat-TM	18/09/1992	03/01/1992	30778	737
	2010 (30 m)	Landsat-TM	08/08/2010	09/02/2010	32529	738
	2015 (30 m)	Landsat-OLI/TIRS	10/08/2015	19/12/2015	30881	682
	2020 (30 m)	Landsat-OLI/TIRS	10/08/2020	18/11/2020	31800	708

Note HF: high flood, LF: low flows, LIR: Lower Indus River

Acknowledgements This research was supported by "National Natural Science Foundation of China (42271118)", "Science Foundation of Zhengzhou University, with grant (i) 32340370/22 Academician Team Research Launch Project (No. 13432340370), and (ii) First-class Project Special Funding of Yellow River Laboratory (No. YRL22IR13)".

References

Belmont P, Gran KB, Schottler SP, Wilcock PR, Day SS, Jennings C, Lauer JW, Viparelli E, Willenbring JK, Engstrom DR, Parker G (2011) Large shift in source of fine sediment in the Upper Mississippi River. Environ Sci Technol 45(20):8804–8810

Blundell JS, Opitz DW (2006) Object recognition and feature extraction from imagery: The Feature Analyst approach. Int. Arch. Photogramm. Remote Sensing and Spatial Information Sciences. 36(4), 42

Carling PA, Trieu H, Hornby DD, Huang HQ, Darby SE, Sear DA, Hutton C, Hill C, Ali Z, Ahmed A, Iqbal I, Hussain Z (2018) Are equilibrium multichannel networks predictable? The case of the regulated Indus River. Pakistan. Geomorphol. 302:20–34

Cheng Q, Ko C, Yuan Y, Ge Y, Zhang S (2006) GIS modeling for predicting river runoff volume in ungauged drainages in the Greater Toronto Area. Canada Comput Geosci 32(8):1108–1119. https://doi.org/10.1016/j.cageo.2006.02.005

Donovan M, Belmont P, Notebaert B, Coombs T, Larson P, Souffront M (2019) "Accounting for uncertainty in remotely-sensed measurements of river planform change. Earth Sci Rev 193:220–236

Donovan M, Belmont P (2019) Timescale dependence in river channel migration measurements.: Earth Surf. Process. Landforms. 44(8), 1530–1541

Harpold AA, Marshall JA, Lyon SW, Barnhart TB, Fisher BA, Donovan M, Brubaker KM, Crosby CJ, Glenn NF, Glennie CL, Kirchner PB, Lam N, Mankoff KD, McCreight JL, Molotch NP, Musselman KN, Pelletier J, Russo T, Sangireddy H, Sjöberg Y, Swetnam T, West N (2015) Laser vision: lidar as a transformative tool to advance critical zone science. Hydrol Earth Syst Sci 19(6):2881–2897

Mahar GA, Zaigham NA, Azam M, Hussain SI (2022) Morphological evolution of Indus shelf region under the influence of hydrodynamic conditions in 20th century. J Sea Res 184:102216

Naz S, Baig MJ, Inayatullah S, Siddiqi TA, Ahsanuddin M (2019) Flood risk assessment of guddu barrage using gumbel's distribution. Int J Sci 8(04):33–38

Piégay H, Darby SA, Mosselmann E, Surian N (2005) The erodible corridor concept: applicability and limitations for river management. River Res Appl 21:773–789

Rhoades EL, O'Neal MA, Pizzuto JE (2009) Quantifying bank erosion on the South River from 1937 to 2005, and its importance in assessing Hg contamination. Appl Geogr 29(1):125–134

Rowland JC, Shelef E, Pope PA, Muss J, Gangodagamage C, Brumby SP, Wilson CJ (2016) A morphology independent methodology for quantifying planview river change and characteristics from remotely sensed imagery". Remote Sens Environ 184:212–228

Schook DM, Rathburn SL, Friedman JM, Wolf JM (2017) A 184-year record of river meander migration from tree rings, aerial imagery, and cross sections. Geomorphology 293:227–239

Schwenk J, Khandelwal V, Fratkin M, Kumar V, Foufoula-Georgiou E (2017) High spatiotemporal resolution of river planform dynamics from Landsat: The RivMAP toolbox and results from the Ucayali River. Earth Space Sci. 4(2):46–75

Shumway RH, Azari RS, Kayhanian M (2002) Statistical approaches to estimating mean water quality concentrations with detection limits. Environ Sci Technol 36(15):3345–3353

T. C. E. versus Time (2009) Statistical analysis of groundwater monitoring data at RCRA facilities. Unified Guidance of Environmental Protection Agency EPA 530 /R-09-007

Werbylo KL, Farnsworth JM, Baasch DM, Farrell PD (2017) Investigating the accuracy of photointerpreted unvegetated channel widths in a braided river system: a Platte River case study. Geomorphol 278:163–170

Chapter 22
Comparison Between Manual Method and Flow Injection Assay for Determination of Total Phosphorus in Water

Hui-li Gao and Liu Yang

Abstract The methods for total phosphorus measurement in water primarily involve manual method and flow injection assay. This study compared and analyzed the application characteristics of these two methods by examining their principles, reagent preparation, sample analysis, detection limits, precision, accuracy, and real sample testing. The results indicated that both methods exhibited a linear correlation coefficient of $\geq$0.999. The manual method had a detection limit of 0.007 mg/L, while the flow injection method had a detection limit of 0.002 mg/L. For the two methods, precision and accuracy analyses using three standard samples with different concentrations showed relative errors (REs) and relative standard deviations (RSDs) consistently below 4 and 5%, respectively. The adding standard recovery for samples for the two methods both ranged from 96.91 to 105.04%, meeting the quality requirements for water sample analysis outlined in the DZ/T0130-2006 "Quality Management Specification for Geological and Mineral Laboratories". Overall, both methods demonstrate similar precision and accuracy, but the flow injection method exhibited higher sensitivity and simplicity, significantly enhancing the analytical efficiency in practical work.

Keywords Manual method · Flow injection assay · Total phosphorus

22.1 Introduction

Phosphorus is an essential element for life and constitutes a crucial component of basic biochemical substances. As a major limiting factor in ecosystems, phosphorus directly or indirectly controls the biogeochemical cycles of many other elements

H. Gao · L. Yang (✉)
Shenyang Center, China Geological Survey, Shenyang 110034, Liaoning, China
e-mail: 282412082@qq.com

H. Xu (ed.), *The 6th International Symposium on Water Resource and Environmental Management*, Environmental Science and Engineering,
https://doi.org/10.1007/978-3-031-55989-1_22

(Wang et al. 2006; Xu et al. 2008). In environmental water bodies, phosphorus influences primary productivity, determining the biomass and nutrient structure within aquatic systems. However, excessive phosphorus input can lead to environmental issues such as eutrophication (Wu et al. 2010). According to the data from the National Surface Water Environmental Quality Monitoring Network (cnemc.cn), the proportion of eutrophic lakes and reservoirs among 58 key water bodies monitored from 2012 to 2020 increased from 27.1% in 2012 to 34.5% in 2020. Research also indicates that the occurrence and severity of water eutrophication are related to factors such as runoff volume, water depth, turbidity, temperature, light conditions, and the content and ratio of elements such as carbon, nitrogen, silicon, and phosphorus, with total phosphorus concentration being the most critical factor (Xu et al. 2021). Current environmental standards in China, including the "Surface Water Environmental Quality Standards" (Environmental Quality Standards for Surface Water (GB3838-2002)) and the "Comprehensive Emission Standards for Pollutants" (Comprehensive Emission Standards for Pollutants (GB8978-1996)), list total phosphorus as a mandatory parameter for water quality assessment. Therefore, accurately analyzing and monitoring total phosphorus levels in environmental water bodies is of great significance for water environmental protection and management in China.

Total phosphorus in water includes various forms of phosphorus dissolved in water and attached to solid particles such as bacteria, colloids, and phytoplankton (Wu et al. 2022). The analysis and detection of total phosphorus typically involve converting different forms of phosphorus into orthophosphate by oxidative digestion, followed by measuring the concentration of orthophosphate in the digested sample (Jilin University, Wuhan University, Nankai University, et al. Inorganic Chemistry, Higher Education Press 2004). Currently, the prevailing standard methods for determining total phosphorus in water are manual methods (Water quality—Determination of total phosphorus—Ammonium molybdate spectrophotometric method (GB11893-89)) and flow injection assay (Water quality—Determination of phosphate and total phosphorus—Continuous flow-ammonium molybdate spectrophotometric method (HJ670-2013)). Based on the application of these two methods in practical testing work, this study selected three standard samples with high, medium, and low concentrations, as well as five actual surface water samples, and determined the total phosphorus content in the samples using both manual and flow injection methods. Furthermore, the application characteristics of these two methods were compared and analyzed in terms of reagent preparation, sample analysis, detection limits, precision, accuracy, and real sample testing.

22.2 Experimental Section

22.2.1 Manual Method

Measurement Principle. Under neutral conditions, the sample was digested with potassium persulfate to oxidize all phosphorus into orthophosphate. In an acidic medium, orthophosphate reacted with ammonium molybdate, and in the presence of antimonic salt, formed phosphomolybdate heteropoly acid, which was immediately reduced by ascorbic acid, producing a blue complex. The absorbance of the blue complex was measured at 880 nm by comparing the color of a known concentration standard solution with the color of the sample/reagent of unknown concentration. The theoretical principle followed was that when light passes through a colored solution, a portion of the light is absorbed; the amount of light intensity absorbed depends on the concentration of the colored substance, the type of solution, and the length of the optical path. For dilute solutions, this conforms to Lambert–Beer's law:

$$A = \mathrm{Ig}(I_0/I) = c \cdot \varepsilon \cdot d \tag{22.1}$$

$$\text{or}\ \ C = A/\varepsilon \cdot d \tag{22.2}$$

where:

A = absorbance

I_0 = original light intensity (measurement value)

I = transmitted light intensity (measurement value)

C = concentration of the measured substance

ε = molar absorption coefficient (constant)

d = optical path length of the flow cell (constant).

Instruments and Reagents

UV–Visible spectrophotometer (Specord210plus);

High-pressure steam sterilizer (MLS3750);

Potassium persulfate: 5 g potassium persulfate (K2S2O8) was dissolved in water and diluted to 100 mL;

Ascorbic acid: 10 g ascorbic acid (C6H8O6) was dissolved in water and diluted to 100 mL;

Molybdate: 13 g ammonium molybdate was dissolved in 100 mL water, and 0.35 g antimony potassium tartrate was dissolved in 100 mL water. The ammonium molybdate solution was gradually added to 300 mL 1:1 sulfuric acid with continuous stirring, and then the antimony potassium tartrate solution was added and mixed well.

Sample Testing. Initially, 25 mL of water sample was taken put into a 50 mL colorimetric tube, and 4 mL of potassium persulfate solution was added. Next, the content in the colorimetric tube was digested in a high-pressure sterilizer (pressure reached 1.1 kg/cm^2) for 30 min and cooled, and then the tube was removed. After digestion, the sample was diluted with deionized water to 50 mL, and 1 mL of ascorbic acid was introduced. After 30 s, 2 mL of molybdate solution was added, mixed well, and let stand for 15 min. The light absorption value was determined using a spectrophotometer at 700 nm wavelength with water as reference, and the phosphorus content in the sample was obtained by referring the working curve.

22.2.2 Flow Injection Assay

Measurement Principle. The sample and reagent were pushed into the chemical reaction module by a peristaltic pump, flowing continuously in a closed pipeline. The flow was separated by air bubbles at regular intervals and mixed in a specific sequence and proportion. After complete reaction and coloration, the mixture entered the flow detection cell for photometric detection. The specific chemical reaction involved the sample undergoing ultraviolet digestion, oxidizing all forms of phosphorus into orthophosphate. The orthophosphate concentration in the digested sample was then measured, and the absorbance was measured at 880 nm. The flow injection assay uses a photometer to measure not the light intensity directly, but the current generated by the light beam in the photodetector, which is proportional to the light intensity. The content of the measured substance in the sample is automatically determined by data conversion.

Instruments and Reagents. A FUMI Flow Injection Analyzer (Shanghai Fumi Technology Co., Ltd.), an automatic sampler, an ultraviolet digester, an analysis module, a spectrophotometer, a data software processing unit, and a 50 mm flow cell were employed in this setup.

Potassium persulfate: 15 g of potassium persulfate was dissolved in deionized water and diluted to 500 mL.

FFD6 (lubricating agent, ensuring bubble shape is round): 2 mL of FFD6 was dissolved in 1000 mL of deionized water.

Ammonium molybdate: First, 40 mL of sulfuric acid was slowly added to 800 mL of deionized water with continuous stirring and cooled to room temperature. Then, 4.8 g of ammonium molybdate was dissolved in this solution, and 2 mL of FFD6 was added. Next, the solution was diluted to 1000 mL with deionized water, mixed well, and stored in a brown bottle.

Potassium antimony tartrate: Potassium antimony tartrate was dissolved in 80 mL of deionized water and made up to 100 mL.

Ascorbic acid: 18 g of ascorbic acid was dissolved in approximately 600 mL of deionized water, and 20 mL of potassium antimony tartrate solution was added. The solution was diluted to 1000 mL, mixed well, and stored at low temperature in a brown bottle.

Sample Testing. The various solution lines were connected according to the instrument requirements and placed in the corresponding reagent bottles. After the baseline stabilized, blank and sample determinations were performed. The main procedural

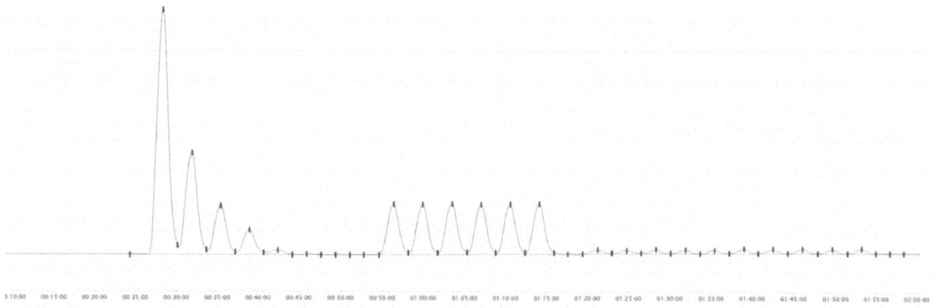

Fig. 22.1 Analytical atlas of flow injection assay

parameters for this method were as follows: sample uptake time of 90 s, sampler needle cleaning time of 90 s, data acquisition frequency of 50 ms, heating temperature of 145 °C. The data graph is illustrated in Fig. 22.1.

22.3 Results and Analysis

22.3.1 Determination of Standard Curve

In a 200 mL volumetric flask, 2 mL of total phosphorus solution standard substance GBW(E)083310 (1000 mg/L) was added and diluted to the mark with deionized water, preparing a standard intermediate solution of 10 mg/L concentration. In a 100 mL volumetric flask, 20 mL of the total phosphorus standard intermediate solution was added and diluted to the mark with deionized water, preparing a total phosphorus standard working solution of 2 mg/L concentration.

Working Curve of Manual Method. Seven 25 mL graduated tubes with stopper were taken, filled with 0.00, 0.50, 1.00, 3.00, 5.00, 10.00, and 15.00 mL of the total phosphorus standard working solution, respectively, and topped up to 25 mL with deionized water. Following the manual procedure, using water as a reference, absorbance was measured. After subtracting the blank test absorbance, a working curve was plotted against corresponding phosphorus content (Fig. 22.2). The test results indicate a good linearity of manual method for detecting total phosphorus in the range of 0.00 to 1.2 mg/L. The working curve is expressed as $y = 0.705*x-0.0017$, with $R^2 = 0.9999$ (Table 22.1).

Working Curve of Flow Injection Assay. Five 100 mL volumetric flasks were added with 0.00, 0.10, 0.50, 1.00, 2.00, and 5.00 mL of the standard intermediate solution, respectively, and the solutions were diluted to the graduation mark with deionized water. Following the analytical procedure, for the convenience and accuracy of peak detection, the determination was conducted from high concentration

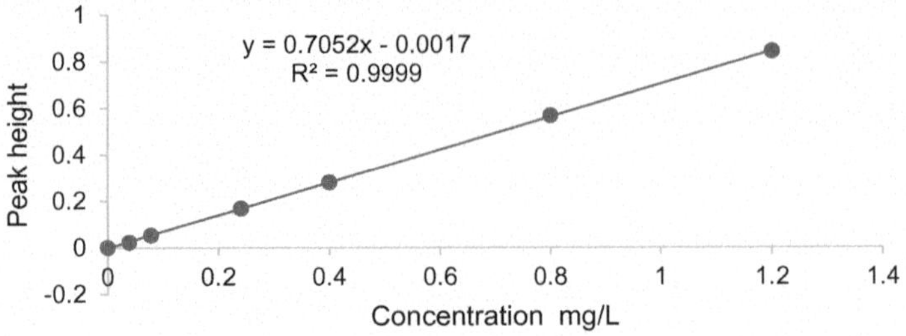

Fig. 22.2 Standard curve for manual total phosphorus measurement

Table 22.1 Standard curve for manual total phosphorus measurement

Standard	Sample volume	Phosphorus content (ug)	Concentration (mg/L)	Absorbance	Original absorbance
Standard blank-0	0.00	0	0.00	0.0000	0.0034
Standard solution-1	0.50	1	0.04	0.0226	0.0259
Standard solution-2	1.00	2	0.08	0.0535	0.0568
Standard solution-3	3.00	6	0.24	0.1678	0.1711
Standard solution-4	5.00	10	0.40	0.2799	0.2832
Standard solution-5	10.00	20	0.80	0.5667	0.5700
Standard solution-6	15.00	30	1.20	0.8420	0.8453

to low concentration. The working curve was plotted with the peak height of the signal value as the ordinate and the corresponding total phosphorus concentration as the abscissa (Fig. 22.3). From the test results, the flow injection assay exhibited good linearity within the range of 0.00 to 1.0 mg/L in measuring total phosphorus in water. The working curve is expressed as $y = 0.3262 * x + 0.0005$, with $R^2 = 0.9999$ (Table 22.2).

22.3.2 Detection Limits of Two Methods

The validation of detection limits for total phosphorus in water by manual and flow injection methods follows the methods specified in the "Technical Guidelines for Standard Formulation of Environmental Monitoring Analysis Methods" HJ-2020. Following all steps of sample analysis, seven sample blank tests were performed.

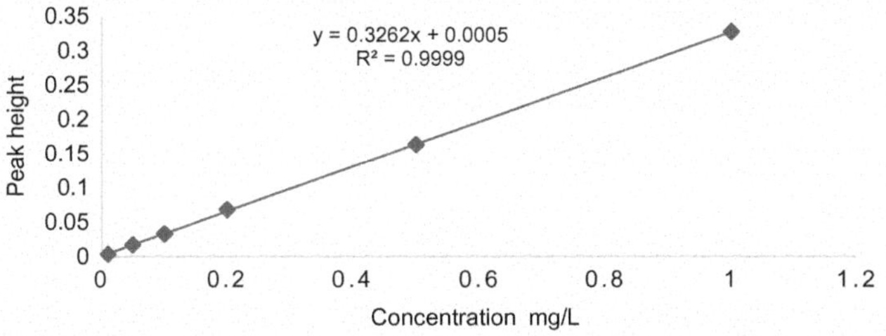

Fig. 22.3 Standard Curve of Total Phosphorus Measured by Flow Injection Assay

Table 22.2 Standard curve of total phosphorus measured by flow injection assay

No	Tube	Peak height	Concentration (mg/L)
Standard-0	1	0.0005	0
Standard-1	2	0.1629	0.5
Standard-2	3	0.0678	0.2
Standard-3	4	0.0326	0.1
Standard-4	5	0.0166	0.05
Standard-5	6	0.0031	0.01

The measurement results were converted into concentration levels in the samples to calculate the standard deviation (SD) of the seven parallel determinations, and the method detection limit was calculated according to MDL = t(n−1, 0.99) * S. MDL refers to the method detection limit; n is the number of parallel determinations of the sample; t represents the t-distribution value (unilateral) when the degree of freedom is n−1 and the confidence level is 99%, can be queried using the t-value table in the standard; S is the SD of n times of parallel determinations. The detection limits of the two methods in this experiment were calculated with n = 7 and the value of t(n−1, 0.99) as 3.143. Taking four times the detection limit as the lower limit of determination, specific calculation results are shown in Table 22.3. The detection limit of total phosphorus measured by the traditional method was 0.007 mg/L, with a lower limit of determination of 0.028 mg/L. The detection limit for total phosphorus measured by the flow injection assay was 0.002 mg/L, with a lower limit of determination of 0.008 mg/L. The detection limit of these two methods was lower than the detection limit of 0.01 mg/L in the traditional method standard GB11893-89 and the detection limit of 0.005 mg/L in the flow injection assay standard HJ671-2013, respectively.

Table 22.3 Results of method detection limit determination

Sample no.	Blank concentration of manual method (mg/L)	Blank concentration of flow injection assay (mg/L)
1	0.0150	0.0020
2	0.0197	0.0011
3	0.0150	0.0010
4	0.0200	0.0017
5	0.0170	0.0014
6	0.0160	0.0017
7	0.0150	0.0000
Average value	0.0168	0.0013
SD	0.0022	0.0007
Method detection limit	0.0069	0.0021
Lower limit of determination	0.0277	0.0083

22.3.3 Precision and Accuracy of Two Methods

Following the determination method outlined in HJ168-2020 regarding precision and accuracy, three standard phosphorus samples with different concentrations (0.218, 0.722, 0.101 mg/L) from the Standard Sample Research Institute of the Environmental Development Center of the Ministry of Ecology and Environment were selected. Six parallel determinations were conducted for each sample throughout the entire process. The average value, SD, and relative standard deviation (RSD) of each concentration sample were calculated (Table 22.4). For the traditional method, the determination of low-concentration standard phosphorus samples yielded a relative error (RE) of 2.01% and an RSD of 2.17%. The mid-concentration standard samples showed an RE of 0.46% and an RSD of 1.94%. The high-concentration standard samples resulted in an RE of 3.26% and an RSD of 4.05%. In the case of the flow injection assay, the determination of low-concentration standard phosphorus samples resulted in an RE of 1.09% and an RSD of 1.73%. The mid-concentration standard samples exhibited an RE of 1.22% and an RSD of 2.30%. The high-concentration standard samples showed an RE of 2.34% and an RSD of 4.23%. It can be observed that the determination results of total phosphorus standard samples with different concentrations (low, mid, and high) using both methods fall within their respective standard value ranges. Moreover, they demonstrated good precision and accuracy.

22.3.4 Analysis of Test on Real Samples

Five surface water samples from Shenyang were collected, and manual and flow injection methods were employed to conduct recovery determinations on the samples.

Table 22.4 Results of precision and accuracy determination

Determination method		Manual method			Flow injection assay		
		Low	Mid	High	Low	Mid	High
Measured value (mg/L)	1	0.2176	0.7305	0.0992	0.2174	0.7250	0.0973
	2	0.2268	0.7388	0.1099	0.2246	0.7384	0.1095
	3	0.2232	0.7018	0.0995	0.2217	0.6995	0.0993
	4	0.2293	0.7104	0.1044	0.2247	0.7016	0.1042
	5	0.2189	0.7103	0.1071	0.2160	0.6979	0.1052
	6	0.2186	0.7203	0.1057	0.2179	0.7169	0.1048
Mean value (mg/L)		0.2224	0.2224	0.7187	0.1043	0.2204	0.7132
Standard value (mg/L)		0.2180	0.2180	0.7220	0.1010	0.2180	0.7220
RE (%)		2.0153	2.0153	0.4576	3.2636	1.0917	1.2175
SD (mg/L)		0.0048	0.0048	0.0139	0.0042	0.0038	0.0164
RSD (%)		2.1734	2.1734	1.9393	4.0535	1.7265	2.2956

The adding standard recovery rates were calculated based on the measurement results of the water samples before and after adding standard substance. The adding standard recovery rates of the water samples obtained from manual method ranged from 96.91 to 105.04%, while for flow injection assay, the rates ranged from 97.11 to 102.79% (Table 22.5). Both methods met the requirements of the "Methods for the Monitoring and Analysis of Water and Wastewater", where the adding standard recovery rates should fall within the range of 90–110% when the total phosphorus concentration exceeds 0.025 mg/L.

Table 22.5 Test on real samples

Sample no.	Manual method			Flow injection assay		
	Sample concentration (mg/L)	Adding standard concentration (mg/L)	Adding standard recovery (%)	Sample concentration (mg/L)	Adding standard concentration (mg/L)	Adding standard recovery (%)
1	0.0546	0.1	102.05	0.0455	0.1	102.79
2	0.0545	0.1	96.91	0.0454	0.1	97.11
3	0.0547	0.1	100.49	0.0548	0.1	98.30
4	0.0552	0.1	99.23	0.0551	0.1	99.11
5	0.0545	0.1	105.04	0.0458	0.1	99.76

22.4 Conclusions

This study conducted a comprehensive comparison between the manual method and the flow injection assay method for testing total phosphorus in water. In the determination of total phosphorus in surface water, the manual method exhibited good linearity within the range of 0.00–1.2 mg/L, with R2 = 1. On the other hand, the flow injection method also showed good linearity within the range of 0.00 to 1.0 mg/L, with an R2 = 0.9999. The detection limit for total phosphorus using the manual method was 0.007 mg/L, while the flow injection method achieved a detection limit of 0.002 mg/L. Both methods have detection limits below the national standards, but the flow injection method, with its lower detection limit, demonstrated a relatively higher sensitivity. Precision and accuracy of both traditional and flow injection methods were evaluated by analyzing standard samples of three different concentrations. The results indicated that both methods yielded values within the specified range of standard values, with RE <4%, RSD <5%, and sample recoveries ranging from 96.91 to 105.04%. These outcomes met the quality requirements for water sample analysis outlined in the "Quality Management Specification for Geological and Mineral Laboratories" (DZ/T0130-2006). From the test results, it is evident that both methods exhibited good precision and accuracy, meeting the requirements of daily analytical testing. While traditional analysis methods are mature, they involve intricate steps and complex operations, demanding a high level of proficiency from the testing personnel. In comparison, the flow injection method offers simplicity in operation, enabling automated testing analysis with good repeatability, low consumption of reagents and samples, and minimal pollution as reactions take place within sealed analytical pipelines. Considering the need to enhance the efficiency of environmental monitoring analysis and the modernization requirements of ecological environmental monitoring, flow injection assay is expected to find broader applications in the detection of water total phosphorus.

References

Comprehensive Emission Standards for Pollutants (GB8978-1996)

Environmental Quality Standards for Surface Water (GB3838-2002)

Jilin University, Wuhan University, Nankai University et al (2004) Inorganic chemistry. Higher Education Press, pp 492–499

Real time data—China environmental monitoring station. cnemc.cn

Wang Q, Jiang X, Jin XC et al (2006) Phosphorus forms and distribution and correlation of bioavailable phosphorus in sediments of different trophic levels in the Taihu Lake. J Lake Sci 18(2):120–126

Water quality—Determination of phosphate and total phosphorus—Continuous flow-ammonium molybdate spectrophotometric method (HJ670-2013)

Water quality—Determination of total phosphorus—Ammonium molybdate spectrophotometric method (GB11893-89)

Wu Y, Deng TL, Xu Q et al (2010) Research progress on the occurrence forms and analytical methods of phosphorus in water environments. Rock Mineral Anal 29(5):557–564

Wu D, Wang K, Liu Y et al (2022) Research progress on detection methods of total phosphorus in water. Indus Water Treat 42(7):52–58

Xu Q, Wu Y, Liao MX et al (2008) Research progress on nitrogen and phosphorus speciation analysis methods in water environment. Rock Mineral Anal 27(2):103–107

Xu J, Yuan DX, Wang T et al (2021) Research progress on total methods and instruments for phosphorus analysis in environmental water samples. J Instrum Anal 40(6):861–868

Chapter 23
Review of Drought Monitoring and Forecast in Inner Mongolia and Construction of an Intelligent Drought Platform

Shuixia Zhao, Yingjie Wu, Shengjie Cui, Wenjun Wang, and Wei Li

Abstract Research on monitoring and early warning of drought in Inner Mongolia is important for ecosystem conservation. The current study and the existing deficiencies in drought monitoring and forecasting, especially in pastoral areas of Inner Mongolia have been reviewed. The structure and function of the Self-developed Inner Mongolia Drought Intelligent Platform were also been presented. Results show that drought monitoring is mainly based on the drought indices, which include the meteorological indices and remote sensing monitoring indices. Most of the meteorological indices focus on precipitation and evapotranspiration, which have not considered the conditions of effects of soil moisture and vegetation status on drought stress. The composite drought index for pastoral areas has been studied, but its generalizability needs further exploration. Remote sensing drought monitoring makes up for the problems on the scale of meteorological monitoring but is affected by the large-scale accuracy, and it is necessary to research the coupling of multi-remote sensing data or the coupling of ground and remote sensing. Numerical and statistical forecasting methods are chiefly used for drought forecasting, however, there are still some discrepancies in realizing accurate dynamic prediction of drought at present. The Self-developed Inner Mongolia Drought Intelligent Platform includes drought monitoring, prediction, drought risk, and resilience assessment, which strongly support the operational application of drought prevention and disaster mitigation in Inner Mongolia.

S. Zhao (✉) · Y. Wu · S. Cui (✉) · W. Wang · W. Li
Yinshanbeilu Grassland Eco-Hydrology National Observation and Research Station, China Institute of Water Resources and Hydropower Research, Beijing 100038, China
e-mail: zhaosx@iwhr.com

S. Cui
e-mail: nmndcsj@163.com

S. Cui
College of Water Conservancy and Civil Engineering, Inner Mongolia Agricultural University, Inner Mongolia, Hohhot 010018, China

H. Xu (ed.), *The 6th International Symposium on Water Resource and Environmental Management*, Environmental Science and Engineering,
https://doi.org/10.1007/978-3-031-55989-1_23

Keywords Inner Mongolia pastoral area · Drought monitoring · Drought prediction · Drought intelligent platform

23.1 Introduction

Drought and its disastrous consequences are recognized and studied as global problems (Sheffield and Wood 2012). Engineering and non-engineering technologies are used to reduce, alleviate, or partially control the hazards of drought and water shortage to humans. Basic drought science research has analyzed the factors underlying the occurrence and development of drought and water shortage, established evaluation indices, and clarified drought-related economic loss (Belal et al. 2014; Liu et al. 2019). In addition, considering the delay and randomness of the occurrence and development of drought-related disasters, probability and grey system theories have been applied to further study the trend of climate change in arid areas and the correlation of this change with drought disasters, as well as countermeasures to reduce these disasters (Luo et al. 2023).

Studies addressing these issues have begun; some are intensive and others have been applied to production practice. These studies have contributed to improving disaster prevention, resistance, and reduction. However, previous studies on drought tended to focus on agricultural systems, with scant systematic research data on drought in pastoral areas. A literature search indicates that relevant studies on drought in pastoral areas in China have mainly involved drought monitoring, drought prediction and early warning, and the construction of drought intelligent platforms.

This review summarizes the progress in drought research. The Inner Mongolia drought intelligence platform established by the Institute of Pastoral Hydraulic Research, MWR is used as an example to evaluate the drought research in pastoral areas of Inner Mongolia.

23.2 Overview of Drought Monitoring Developments

23.2.1 Drought Monitoring

Drought is a slow-changing natural phenomenon. Real-time monitoring involves the calculation of relevant drought indicators based on real-time hydrological and meteorological measurements that include precipitation, evaporation, soil moisture content, and river runoff, and the analysis of drought indicators to assess the current drought grade. Recently, several countries have undertaken operational research on real-time drought monitoring according to the national conditions and needs of different industries and a three-dimensional drought monitoring pattern for ground, aviation, and space have been developed (Rad et al. 2021). Examples include the United States

Drought Monitor (https://droughtmonitor.unl.edu) (Svoboda et al. 2002) and systems in China that include a drought and flood monitoring system (Changhan et al. 1998) and the Tianjian hydrometeorological integrated business application system for flood control and drought relief (Liu et al. 2016).

Drought monitoring is mainly based on data sources, including ground and remote sensing monitoring. The following subsections provide more detail.

23.2.1.1 Ground Monitoring of Drought

Drought monitoring indices are mainly established based on the monitoring data of ground meteorological stations and soil moisture stations. These indices include the Integrated Drought Condition Index, Standardized Precipitation Index (SPI), Standardized Precipitation Evapotranspiration Index (SPEI), Palmer Drought Severity Index (PDSI), Precipitation Anomaly Index (PA), Composite Meteorological Drought Index (MCI), and Potential Evapotranspiration (PET). The applicability and limitations of each drought monitoring index have been studied (Table 23.1).

Based on the analysis of existing meteorological monitoring indicators, some scholars have established a drought index model of Inner Mongolian grasslands

Table 23.1 Applicability and limitations of drought monitoring indices

Drought Monitoring Index	Applicability	Limitations
Potential Evapotranspiration (PET)	The calculation method is simple and the data are easy to obtain	The single parameter cannot comprehensively reflect the drought (Zhao et al. 2021)
Standardized Precipitation Index (SPI)	The calculation method is simple, data are easily accessible, and differences in precipitation use can be eliminated	SPI does not consider the effects of other factors, such as temperature and evaporation on drought (Naresh et al. 2009)
Standardized Precipitation Evapotranspiration Index (SPEI)	The index is a multi-time scale comprehensive evaluation of evapotranspiration and other factors that influence drought	Many parameters are not suitable for calculation (Beguería et al. 2014)
Palmer Drought Index (PDSI)	The index considers the effects of evapotranspiration, precipitation, and runoff on drought	There are limitations in time and space (Alley 1984)
Precipitation Anomaly Index (PA)	Calculation of the percentage of precipitation anomalies is simple and the required data are easy to obtain	The index does not reflect the internal mechanism of drought (Shen et al. 2019)
Composite Meteorological Drought Index (MCI)	Multiple time scales of the index take into account the impact of drought factors	Many parameters are present, hence, not suitable for calculation (Feng et al. 2023)

from the perspective of soil, hydrology, and plants according to the characteristics of pastoral areas. Two examples are a drought grade index based on the interaction of three vulnerability dimensions (Liu et al. 2013) and a case study of the variability and characteristics of soil water dynamics at the field scale using a physics-based MIKE SHE model combined with field monitoring and computer simulation (Hao et al. 2016).

Differences in response to rainfall and sensitivity to drought exist in different seasons and under different ground conditions. Ground monitoring can accurately and scientifically monitor drought in the study area by considering meteorological data and influencing factors of the underlying surface (Liu et al. 2016). However, this ground monitoring cannot analyze the degree of feedback of vegetation to drought and cannot adequately describe the drought (Zargar et al. 2011).

23.2.1.2 Drought Monitoring by Remote Sensing

Monitoring of drought remote sensing mainly comprises visible, near-infrared, thermal infrared, and microwave remote sensing (Liu et al. 2017). China's drought remote sensing monitoring technology has gradually shifted from experimental research to practical applications (Jingxuan et al. 2009). However, drought is complex and involves hydrology, meteorology, biophysiology, water resource management, and economic and social factors. Various indicators must be considered comprehensively. Therefore, establishing an effective remote sensing monitoring and forecasting system for droughts remains challenging.

In remote sensing drought monitoring of the Inner Mongolia Autonomous Region, Landsat, and Moderate Resolution Imaging Spectroradiometer (MODIS) products have been mainly used to conduct a variety of index analyses and evaluations. As an example, Dong and Hu (2022) calculated the fraction of vegetation cover based on Landsat image data and developed a monitoring model for vegetation cover change. Another example, Wei et al. (2021) used MODIS data from Inner Mongolia to perform a comparative analysis of the microwave comprehensive drought index, optimized vegetation drought index, optimized meteorological drought index, scale drought condition index, and comprehensive drought index (SDI).

Currently, comprehensive drought monitoring methods based on ground monitoring and remote sensing are scarce, especially comprehensive drought monitoring indices for grasslands in pastoral areas (Zhao et al. 2021). Systematic research on comprehensive monitoring and evaluation index systems for grassland drought monitoring based on ground and remote sensing is urgently required.

23.3 Overview of the Development of Drought Prediction Methods

The frequent occurrence of drought-related disasters highlights the importance of accurately predicting the dynamic processes of drought occurrence and development and the consequences that include the decline and even extinction of life in the affected regions (Sheffield and Wood 2008). Drought prediction methods can be classified into numerical and statistical (Weare 2006). The principles of meteorology have been used to establish a numerical prediction model (Krishnamurti and Bounoua 2018). This model is suitable for predicting large-scale systems and the extension time should not be excessively long (Pu and Kalnay 2019). The statistical prediction method uses mathematical statistics to establish mathematical models for predicting the degree of future drought.

Numerical and statistical forecasting methods for meteorological and agricultural droughts are relatively mature. Among them, common nonlinear forecasting methods for meteorological drought forecasting mainly include the grey system forecasting method, which has been adopted in several studies (Jiang and Chen 2009; Darvishi et al. 2017; Luo et al. 2022). A time series analysis method has also been used (Davis and Rappoport 1974; Murthy et al. 2016). A third nonlinear forecasting method is the fractal theory, which has been adopted in other studies (Li et al. 2015; Kang et al. 2021; Adnan et al. 2021; Başakın et al. 2021).

Agricultural drought is the result of the comprehensive effects of meteorological conditions, hydrological environment, soil matrix, water conservancy facilities, crop varieties and growth conditions, crop layout, and farming methods (Daryanto et al. 2015). The mechanism of agricultural drought is more complex and comprehensive than that of meteorological drought. Currently, a few relevant studies are available on grassland drought prediction in pastoral areas. Such drought prediction is mainly based on meteorological, soil, and remote sensing data. Selection of appropriate drought indicators and the applications of linear regression, Gauss process, M–K trend test, moving average, and least square methods have been used to forecast drought in the study areas. Meteorological and remote sensing data were used to forecast drought in Iran, Ethiopia, and other regions based on one or more of these methods (Pulwarty and Sivakumar 2014; Barrett et al. 2020; Zarei et al. 2020; Demisse et al. 2018). Accurate forecasts for the pastoral area of Inner Mongolia have been made based on historical meteorological data (Hao et al. 2009; Ma et al. 2016).

All the above studies on drought prediction in pastoral areas were based on a single meteorological drought index and did not consider the drought characteristics of different underlying conditions. Thus, they were one-sided to a certain extent. In particular, systematic research is required for drought prediction and early warning of pastoral grasslands. China's drought prediction mainly focuses on meteorological drought levels and lacks prediction of drought events after the interaction of different types of drought. Accurate and timely drought prediction remains a crucial challenge for scientists owing to the slow process of drought evolution and difficulty in judging the occurrence time.

23.4 Construction of the Inner Mongolia Drought Intelligent Platform

Based on the principles of unity, the combination of advanced nature and practicality, safety and reliability, high scalability, manageability, openness, compatibility, and inter-connectivity, the Grassland Drought Monitoring, Forecasting and Evaluation System in Inner Mongolia platform was developed by the Institute of Parstoral Hydraulic Research, MWR in China. The goal is to effectively provide drought relief in pastoral areas.

23.4.1 Platform Structure and Process

The platform is divided into the login and system parts. Six components are present in the system: a database, data processing, drought index calculation, drought prediction, drought risk assessment, and drought resistance evaluation. Each part has its own display area and calculation process and is finally displayed through the visualization system. The platform flowchart is presented in Fig. 23.1.

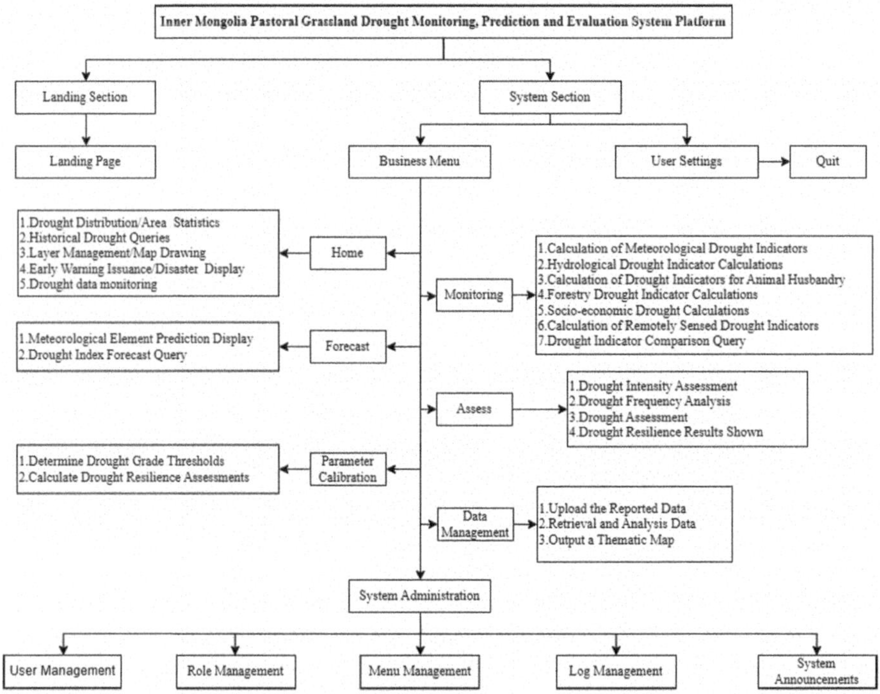

Fig. 23.1 Flowchart of the platform

Database. Based on the business requirements of drought early warning prediction and evaluation applications and the functional requirements of system construction in the Inner Mongolia pastoral (forest) areas, the relevant data required for system construction were sorted and analyzed. The five categories of data include basic background, drought monitoring, drought monitoring, drought prediction, drought assessment, and other data related to drought resistance.

The basic background data includes the administrative boundaries of Inner Mongolia, land use, grassland distribution, river system distribution, meteorological stations, hydrological (reservoir) monitoring stations, and basic location data from self-built meteorological stations. Drought monitoring data includes that needed to calculate the meteorological drought, hydrological drought, pastoral drought, forestry drought, socioeconomic drought, and remote sensing monitoring drought indices. Drought forecast data mainly includes meteorological forecast data. Drought assessment data mainly refers to the two types of drought classification and area statistical data after classification.

Drought index calculation. Based on the regional characteristics of grassland and forest areas in Inner Mongolia, eighteen drought indicators were selected for the platform based on six aspects of drought: meteorological, hydrological, pastoral, forestry, socioeconomic, and remote sensing. Quantitative monitoring of drought events involved the entire region, pastoral areas, and forest areas of Inner Mongolia.

Data processing. According to the meteorological, hydrological, CLDAS, and remote sensing data in the database, the platform can calculate each drought index through configuration. On this basis, spatial interpolation (supporting inverse distance, spline function, trend surface, and Kriging interpolation method), grade evaluation, storage, and output of the calculated results are performed at the site scale. Results of the drought monitoring at the provincial, city, and county levels can be generated.

Drought prediction. Relevant data are obtained from the database and the drought monitoring and analysis products are output in the form of charts and statistical data according to different periods, such as 24 h and 3, 5, and 10 d, including precipitation and temperature forecasts and drought monitoring index prediction. The information reflects the actual situation of drought and the future short-term trends of drought development and enables drought monitoring and prediction analysis.

Drought Assessment. This part of the platform is based on the drought monitoring and threshold calibration results. Drought intensity and frequency analyses were performed by calculating the drought frequency and station ratio. Through connection with the flood control and drought relief data, reported information on the actual drought situation in various places is obtained by manual interaction, and a list is displayed. The display and query of the drought distribution diagram are realized with the support of a geographic information system.

Drought Resistance Evaluation. This part of the platform includes regional background conditions, water conservancy project conditions, economic development levels, scientific and technological development levels, drought resistance management service levels, and comprehensive drought resistance evaluations. The system uses the analytic hierarchy process to construct an evaluation system for regional drought resistance. On the system page, the weight of each evaluation layer index is obtained by inputting the comparison score of the corresponding index, and the fuzzy evaluation method is used. Combined with the monitoring data of each index, the evaluation results of regional drought resistance are obtained.

23.4.2 Platform Function Display

Home Page. After logging in, the home page of the system opens by default. The page displays the latest drought distribution map and statistics for the area of each level of drought.

Monitoring. The user switches to the "Monitoring" module through the menu bar in the upper part of the system. This module performs the calculation, results display, and index comparison of various drought indicators. The platform calculates and displays the results of six meteorological drought indicators, four hydrological drought indicators, two pastoral drought indicators, two forestry drought indicators, one socioeconomic drought indicator, and three remote sensing drought indicators at different timescales.

The platform can use the inverse distance weighting method and the Kriging interpolation method to interpolate the calculated results to obtain the drought distribution results in the pastoral area of Inner Mongolia.

The platform can also monitor and calculate different drought indicators in the Inner Mongolian pastoral areas based on the database and display them. Examples are shown in Figs. 23.2, 23.3 and 23.4.

The platform can also compare the same region, different drought indicators, different scales, and different times (Fig. 23.5a); the same region, same index, same time, and different scales (Fig. 23.5b); and the same region, same time, same scale, and different indicators (Fig. 23.5c).

Prediction. By collecting real-time rainfall and temperature forecast data, the forecast module of the platform can realize the spatial display of the forecast results of meteorological elements and the drought index, as shown in Figs. 23.6 and 23.7.

Evaluation. This part of the module includes the query and spatial display of the results of the drought assessment and resistance evaluation (Figs. 23.8, 23.9, 23.10 and 23.11). The evaluation of drought includes the evaluation of drought intensity and analysis of drought frequency (Fig. 23.8a, b). The analysis module of drought frequency includes "frequency analysis" and "station ratio analysis" (Fig. 23.9a, b).

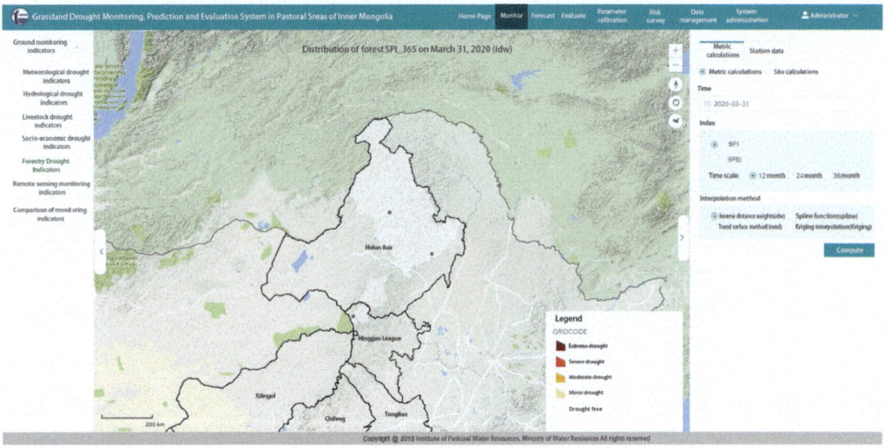

Fig. 23.2 Display of forestry drought index

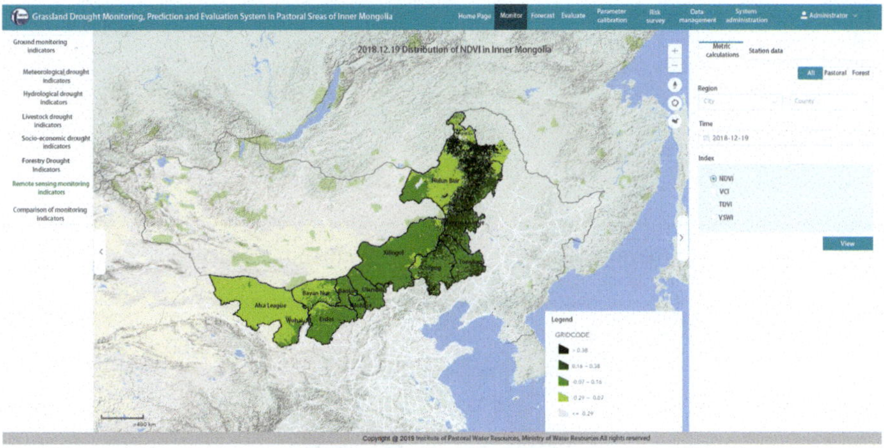

Fig. 23.3 Display of remote sensing monitoring index

23.5 Limitations and Prospects

Based on the research mentioned above and platform construction, some limitations still exist. First, it is difficult for a single drought index to accurately reflect the multi-type and multiscale characteristics of drought. Systematic research on a comprehensive monitoring and evaluation index system for grassland drought monitoring based on ground and remote sensing is required. Second, the evaluation of drought monitoring and other aspects is not perfect. Some key factors that are still lacking include temporal and spatial resource allocation, development and utilization conditions, and water conservancy project conditions. Third, the remote sensing drought monitoring

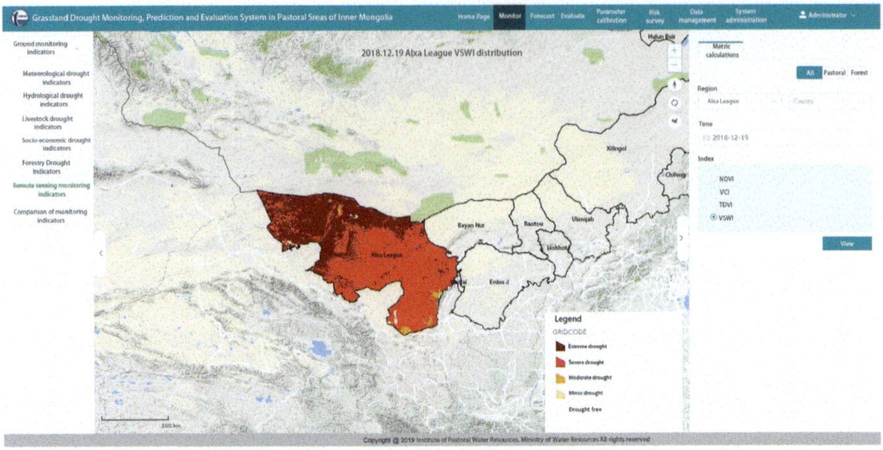

Fig. 23.4 Distribution of vegetation water supply index

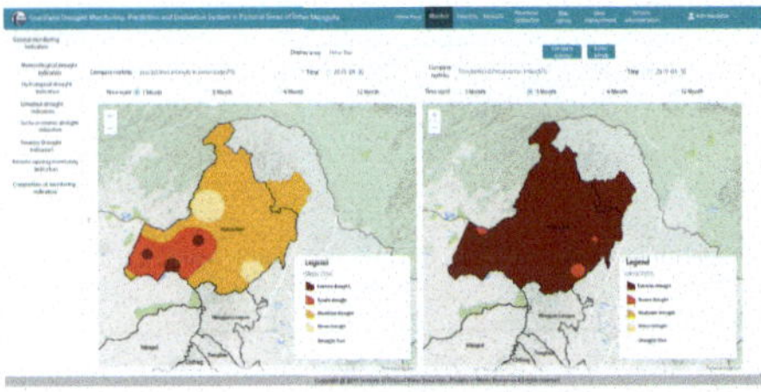

(a) different scales

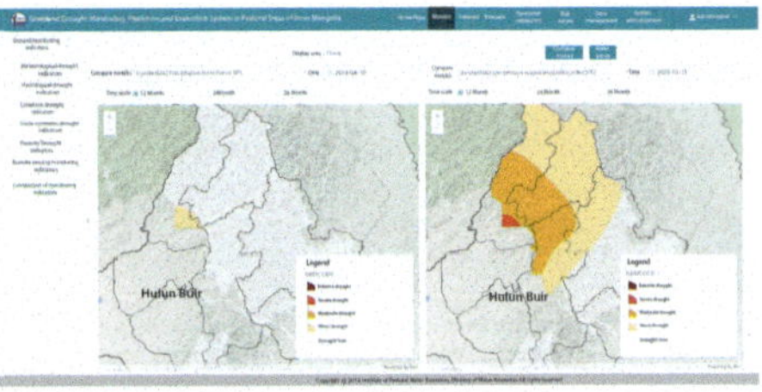

(b) different times

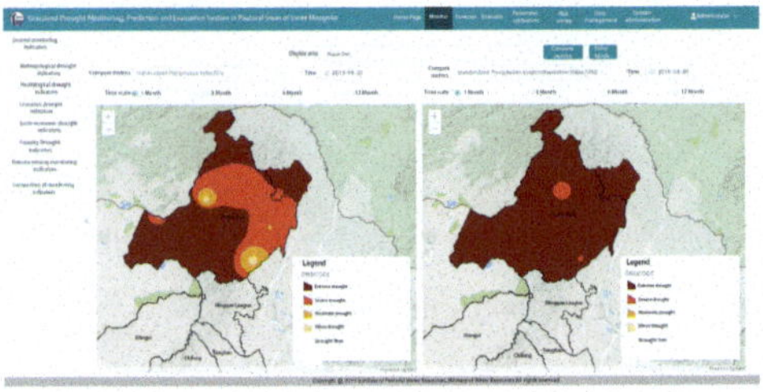

(c) different indicators

Fig. 23.5 Comparison of the same region using different scales (**a**), different times (**b**), and different indicators (**c**)

model still has a certain lag response. A variety of remote sensing drought monitoring methods have been considered to compensate for the shortcomings of various remote sensing drought monitoring models.

Despite these limitations, the prospects for drought monitoring and forecasting in the pastoral areas of Inner Mongolia are compelling. Future efforts will include

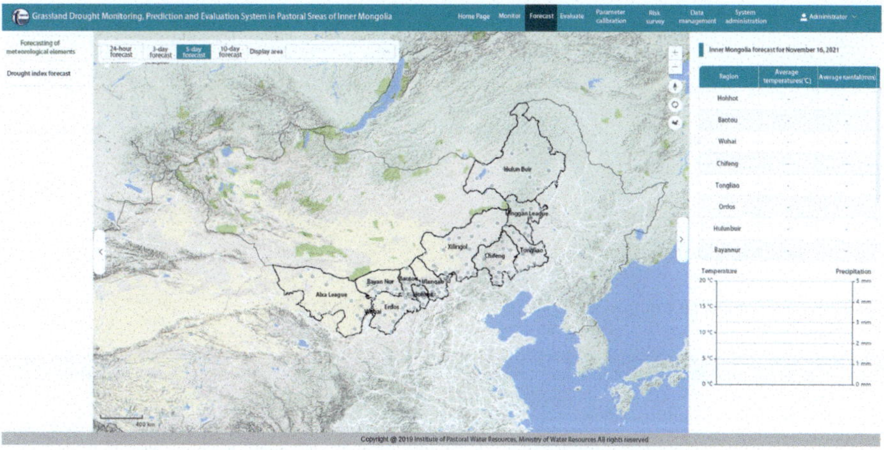

Fig. 23.6 Meteorological elements forecast page

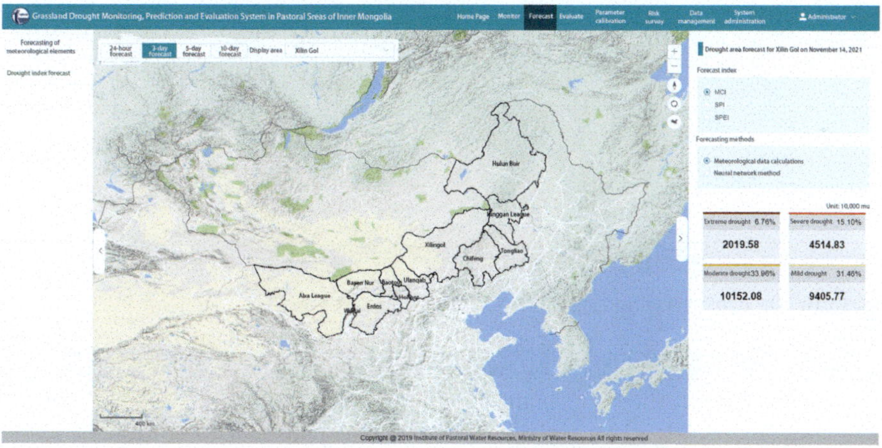

Fig. 23.7 Drought index forcast calculation page

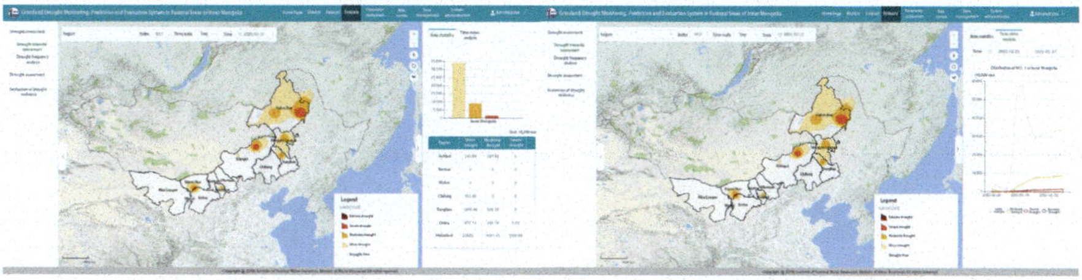

(a) area statistics (b) time series analysis

Fig. 23.8 Drought intensity assessment by **a** area statistics and **b** time series analysis

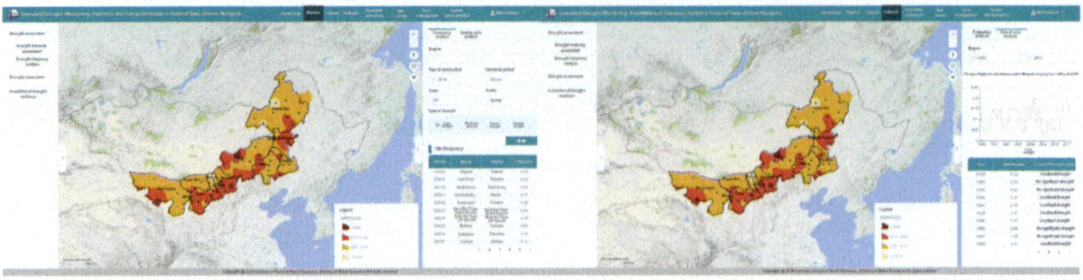

(a) Drought frequency analysis diagram (b) station ratio

Fig. 23.9 **a** Drought frequency analysis diagram and **b** station ratio

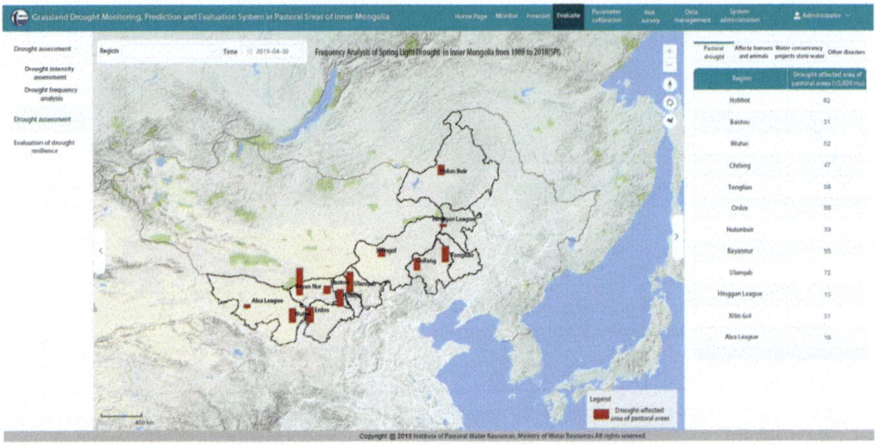

Fig. 23.10 Drought evaluation page

Fig. 23.11 Drought resistance evaluation results

the organic combination of surface drought and remote sensing drought indices to comprehensively evaluate the drought grade and temporal and spatial distribution characteristics of grassland in pastoral areas. In addition, a comprehensive evaluation index system for drought resistance in pastoral areas will be established by comprehensively considering the conditions of water conservancy projects and water resource distribution.

Based on the evaluation results, this study proposes measures for drought prevention and disaster reduction in the entire region and measures for drought prevention and disaster reduction. By focusing on theory and ignoring countermeasures and measures, this approach will bolster drought resistance initiatives.

Acknowledgements This research was funded by the Natural Science Foundation of Inner Mongolia Autonomous Region (No. 2021GG0050, MK0143B052021); the IWHR Research & Development Support Program (Nos. MK0145B02202).

References

Adnan RM, Mostafa RR, Islam ARMT, Gorgij AD, Kuriqi A, Kisi O (2021) Improving drought modeling using hybrid random vector functional link methods. Water 13(23):3379

Alley WM (1984) The Palmer drought severity index: limitations and assumptions. J Appl Meteorol Climatol 23(7):1100–1109

Barrett AB, Duivenvoorden S, Salakpi EE, Muthoka JM, Mwangi J, Oliver S, Rowhani P (2020) Forecasting vegetation condition for drought early warning systems in pastoral communities in Kenya. Remote Sens Environ 248:111886

Başakın EE, Ekmekcioğlu Ö, Özger M (2021) Drought prediction using hybrid soft-computing methods for semi-arid region. Model Earth Syst Environ 7:2363–2371

Beguería S, Vicente-Serrano SM, Reig F, Latorre B (2014) Standardized precipitation evapotranspiration index (SPEI) revisited: parameter fitting, evapotranspiration models, tools, datasets and drought monitoring. Int J Climatol 34(10):3001–3023

Belal AA, El-Ramady HR, Mohamed ES, Saleh AM (2014) Drought risk assessment using remote sensing and GIS techniques. Arab J Geosci 7:35–53

Changhan Z, Qiang Z, Hong W (1998) China's drought climate monitoring system. Drought Network News (1994–2001) 109

DarvishiSalookolaei D, Liu S, Babaei P (2017) Application of grey system theory in rainfall estimation. Control Optim Appl Math 2(2):15–32

Daryanto S, Wang L, Jacinthe PA (2015) Global synthesis of drought effects on food legume production. PLoS One 10(6):e0127401

Davis JM, Rappoport PN (1974) The use of time series analysis techniques in forecasting meteorological drought. Mon Weather Rev 102(2):176–180

Demisse GB, Tadesse T, Bayissa Y, Atnafu S, Argaw M, Nedaw D (2018) Vegetation condition prediction for drought monitoring in pastoralist areas: a case study in Ethiopia. Int J Remote Sens 39(14):4599–4615

Dong X, Hu C (2022) Remote sensing monitoring and evaluation of vegetation changes in Hulun Buir Grassland, Inner Mongolia autonomous region, China. Forests 13(12):2186

Feng A, Liu L, Wang G, Tang J, Zhang X, Chen Y, Liu P (2023) Drought monitoring from Fengyun satellite series: a comparative analysis with meteorological-drought composite index (MCI). Remote Sens 15(22):5410

Hao W, Zhao H, Bai Z, Han X (2009) Drought monitoring and evaluation of integrated services and early warning system development in inner Mongolia. Chin Agric Sci Bull 25(24):416–421
Hao L, Sun G, Liu Y, Zhou G, Wan J, Zhang L, He J (2016) Evapotranspiration and soil moisture dynamics in a temperate grassland ecosystem in Inner Mongolia, China. Trans ASABE 59(2):577–590
Jiang X, Chen S (2009) Application of weighted Markov SCGM (1, 1) C model to predict drought crop area. Syst Eng Theory Pract 29(9):179–185
Jingxuan L, Wei Q, Jun'e FU (2009) Review on drought monitoring using remote sensing. J China Inst Water Resour Hydropower Res 7(2):265–270
Kang Y, Guo E, Wang Y, Bao Y, Bao Y, Mandula N (2021) Monitoring vegetation change and its potential drivers in Inner Mongolia from 2000 to 2019. Remote Sens 13(17):3357
Krishnamurti TN, Bounoua L (2018) An introduction to numerical weather prediction techniques, 1st edn. CRC Press
Li C, Hu R, Yin S, Bao Y, Li Y (2015) Analysis of desertification feature of typical steppe grassland based on fractal theory: a case study of Xilin GOL League
Liu X, Wang Y, Peng J, Braimoh AK, Yin H (2013) Assessing vulnerability to drought based on exposure, sensitivity and adaptive capacity: a case study in middle Inner Mongolia of China. Chin Geogr Sci 23:13–25
Liu X, Zhu X, Pan Y, Li S, Liu Y, Ma Y (2016) Agricultural drought monitoring: progress, challenges, and prospects. J Geogr Sci 26:750–767
Liu L, Liu L, Liao J, Chen X, Chen X, Zhou G, Su Y, Xiang Z, Wang Z, Liu X, Li Y, Wu J, Xiong X, Shao H (2017) The Microwave Temperature Vegetation Drought Index (MTVDI) based on AMSR-E brightness temperatures for long-term drought assessment across China (2003–2010). Remote Sens Environ 199:302–320
Liu Y, You M, Zhu J, Wang F, Ran R (2019) Integrated risk assessment for agricultural drought and flood disasters based on entropy information diffusion theory in the middle and lower reaches of the Yangtze River, China. Int J Disaster Risk Reduct 38:101194
Luo D, Hu Y, Sun D (2022) Drought grade assessment method based on grey cloud incidence clustering model. Grey Syst Theory Appl 12(1):1–24
Luo D, Hao H, Zhu H (2023) Evaluation of regional drought disaster prevention ability and obstacle analysis. Arab J Geosci 16(3):150
Ma Q, Zhang J, Wang Y, Zhang F, Wu R (2016) Characteristics and prediction of drought in growing season in Inner Mongolia pastoral area. J Arid Land Resour Environ 30(07):157–163
Murthy CS, Singh J, Kumar P, Sesha Sai MVR (2016) Meteorological drought analysis over India using analytical framework on CPC rainfall time series. Nat Hazards 81(1):573–587
Naresh Kumar M, Murthy CS, Sesha Sai MVR, Roy PS (2009) On the use of Standardized Precipitation Index (SPI) for drought intensity assessment. Meteorol Appl J Forecast Pract Appl Train Tech Model 16(3):381–389
Pu Z, Kalnay E (2019) Numerical weather prediction basics: models, numerical methods, and data assimilation. In: Handbook of hydrometeorological ensemble forecasting, pp 67–97
Pulwarty RS, Sivakumar MVK (2014) Information systems in a changing climate: early warnings and drought risk management. Weather Clim Extrem 3:14–21
Rad AM, AghaKouchak A, Navari M, Sadegh M (2021) Progress, challenges, and opportunities in remote sensing of drought. Glob Drought Flood Obs Model Predict 1–28
Sheffield J, Wood EF (2008) Projected changes in drought occurrence under future global warming from multi-model, multi-scenario, IPCC AR4 simulations. Clim Dyn 31:79–105
Sheffield J, Wood EF (2012) Drought: past problems and future scenario Routledge. Taylor and Francis, United States
Shen R, Huang A, Li B, Guo J (2019) Construction of a drought monitoring model using deep learning based on multi-source remote sensing data. Int J Appl Earth Obs Geoinf 79:48–57
Svoboda M, LeComte D, Hayes M, Heim R, Gleason K, Angel J, Rippey B, Tinker R, Palecki M, Stooksbury D, Miskus D, Stephens S (2002) The drought monitor. Bull Am Meteor Soc 83(8):1181–1190

Weare B (2006) Monitoring and predicting agricultural drought: a global study
Wei W, Zhang J, Zhou L, Xie B, Zhou J, Li C (2021) Comparative evaluation of drought indices for monitoring drought based on remote sensing data. Environ Sci Pollut Res 28:20408–20425
Zarei A, Asadi E, Ebrahimi A, Jafari M, Malekian A, Nasrabadi HM, Chemura A, Maskell G (2020) Prediction of future grassland vegetation cover fluctuation under climate change scenarios. Ecol Ind 119:106858
Zargar A, Sadiq R, Naser B, Khan F (2011) A review of drought indices. Environ Rev 19(NA):333–349
Zhao SX, Wang WJ, Wu YJ, Li W, Quan Q (2021) Construction and application of comprehensive drought index in different steppe types.Trans Chinese Soc Agricult Eng 37(16):99–107
Zhao Q, Sun T, Zhang T, He L, Zhang Z, Shen Z, Xiong S (2021) High-precision potential evapotranspiration model using GNSS observation. Remote Sens 13(23):4848

Chapter 24
Application of Ion Chromatography Method on the Simultaneous Determination of Five Disinfection By-Products in Drinking Water

Liu Yang and Hui-li Gao

Abstract In order to meet the rapid analysis and detection requirements of environmental monitoring and improve the speed of analysis and detection, this paper establishes a method for the simultaneous determination of five types of by-products including chlorate, chlorite, dichloroacetic acid, trichloroacetic acid and bromate in drinking water through ion chromatography. This method takes potassium hydroxide as the mobile phase to detect the target analyte by a conductivity detector in the way of gradient elution, with the eluent flow rate of 1 ml/min. Experimental results: The linear correlation coefficient of the five disinfection by-products is greater than 0.999; the detection limit is 4.8 μg/L for $ClO_3{}^-$, 2.1 μg/L for $ClO_2{}^-$, 2.1 μg/L for DCAA, 3.4 μg/L for TCAA, 4.0 μg/L $BrO_3{}^-$; the relative standard deviation RSD is below 3.5%. The recovery rate of sample is 85.24 to 103%. The test results prove that this method has good accuracy, high sensitivity, simplicity and convenience, which can greatly improve the analysis efficiency of actual detection work.

Keywords Ion chromatography · Disinfection by-products · Drinking water

Drinking water disinfection is an important part of water supply projects. Its purpose is to kill the microbial pathogens in water and prevent the spread and prevalence of infectious diseases with water as a medium. Currently, commonly used methods for drinking water disinfection include chlorine disinfection, ozone disinfection, chlorine dioxide disinfection, chloramine disinfection and ultraviolet disinfection (Dong and Huang 2005). The disinfection of drinking water is accompanied with the chemical reactions between disinfectants and some natural organic matter and environmental organic pollutants in the water, thus producing a variety of disinfection by-products (DBPs), posing potential threats to human health (Hrudey 2009; Susan et al. 2007;

L. Yang · H. Gao (✉)
Shenyang Center, China Geological Survey, ShenyangLiaoning 110034, China
e-mail: 1621015293@qq.com

H. Xu (ed.), *The 6th International Symposium on Water Resource and Environmental Management*, Environmental Science and Engineering,
https://doi.org/10.1007/978-3-031-55989-1_24

Mills et al. 1998). Ozone disinfection will produce bromate (BrO_3^-); chlorine dioxide disinfection will produce chlorite (ClO_2^-) and chlorate (ClO_3^-) (Zhu et al. 2012); and chlorine disinfection will produce haloacetic acid. Studies have shown that the carcinogenic risk of haloacetic acids is far greater than the sum of other disinfection by-products. The contents of dichloroacetic acid (DCAA) and trichloroacetic acid (TCAA) are the largest in haloacetic acids, which have been defined as the potential human carcinogen by the U.S. Environmental Protection Agency (USEPA) (Zhu et al. 2017). China's Standards for drinking water quality GB5749-2022 (Standards for drinking water quality, 2022) clearly stipulates the types and limits of disinfection by-products that should be controlled by different water treatment and disinfection methods.

Ion chromatography is a method developed in the 1970s, which is mainly used to measure anions. This analytical method has high sensitivity, good stability, and simple sample pretreatment, which is the most commonly used method for analyzing anions (Lei et al. 2016; Huo 2014). In Part 10 of Standard examination methods for drinking water GB/T5750.10–2023 (Standard examination methods for drinking water, 2023), although there are the ion chromatographic determination methods for the five disinfection by-products mentioned in this paper (chlorate, chlorite, dichloroacetic acid, trichloroacetic acid and bromate), the test conditions are different, and it is troublesome to perform subentry tests on the same sample. Therefore, it is of great significance to establish a method for simultaneous measuring multiple analysis items with single sample injection.

In this study, potassium hydroxide was used as the mobile phase, a gradient elution method was applied, and the method was verified under optimal instrument conditions. The verification results show that the mass concentrations of these five disinfection by-products have a good linear relationship with the peak area, with the linear correlation coefficients all greater than 0.999. The detection limits are all smaller than the detection limits of each test element by the standard method, and the relative standard deviation RSD is 1.87 to 3.12%, and the recovery rate of actual samples is 85.24 to 103.00%. The analysis results prove that this method has good accuracy and precision, which can meet the requirements of actual sample analysis and detection.

24.1 Experiments

24.1.1 Principle

The ions to be measured in the water sample enter the anion exchange separation system (including guard column and analytical column) with the eluent-potassium hydroxide. According to the ion exchange separation mechanism, each ion is separated by their different affinities on the analytical column. Then, the detection sensitivity of the measured substance is improved by inhibiting the background of the suppressor. The conductivity values of various anionic components are measured by the conductivity detector, and the data are collected and processed by the chromatography workstation, so as to be qualitative by retention time and quantitative by peak area.

24.1.2 Main Instruments and Reagents

Ion chromatograph: Prin-Cen IC-2, Prin-Cen Technology;

Analytical column: Dionex IonPacTM AS19 (4*250 mm);

Guard column: Dionex IonPacTM AG19 (4*50 mm);

Mobile phase: online automatic electrolysis generates KOH;

Suppressor: AERS500 (4 mm);

Detector: conductivity detector;

1000 mg/L single-standard solution of five disinfection by-products (Northern Weiye Metrology Group Co., Ltd.);

Laboratory water: deionized water, resistivity of 18 MΩ·cm.

24.1.3 Experimental Methods

24.2 Instrument Conditions

Pump flow rate: 1 ml/min;

Injection volume: 500 μl;

Column temperature: 25 °C;

Pool temperature: 35 °C;

Detector temperature: 30 °C;

Suppressor current: 100 mA;

Collection time: 45 min.

Determination of chromatographic conditions. The flow rate and concentration of the eluent both affect the separation of the analyte. The flow rate affects the separation effect, and the concentration affects the peak time. After testing under different conditions, this paper selects KOH as the mobile phase and the gradient

Table 24.1 Gradient elution conditions

Gradient elution	Ramp	Lifting time (min)	Concentration (mmol/L)	Retention time (min)
	Init		4	1 8
	1	23	1 2	3 5
	2	3 5.1	4 0	4 0
	3	4 0.1	4	4 5

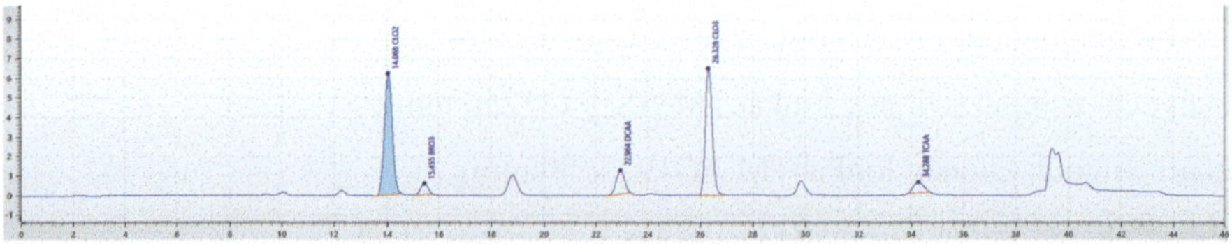

Fig. 24.1 Separation effect of five disinfection by-products

elution method (Table 24.1) with flow rate of 1 ml/min to simultaneously detect the five disinfection by-products. The separation effect is shown in Fig. 24.1.

24.3 Result and Discussion

24.3.1 Linear Relationship and Detection Limit

The mixed-standard solutions of different concentrations are accurately absorbed (Table 24.2) and tested according to the determined chromatographic conditions and instrumental analysis conditions. The standard curves of each ion are drawn with the concentration as the horizontal coordinate and the peak area as the vertical coordinate. The linear results are shown in Fig. 24.2, indicating that these five disinfection by-products all have good linear relationships under the test conditions. The detection limit of this method is calculated based on the standard concentration equivalent to 3 times the baseline noise. The detection limit of ClO_3^- is 4.8 μg/L, the detection limit of ClO_2^- is 2.1 μg/L, the detection limit of DCAA is 2.1 μg/L, the detection limit of TCAA is 3.4 μg/L, and the detection limit of BrO_3^- is 4.0 μg/L, which are all smaller than the detection limits in Standard examination methods for drinking water GB/T5750.10–2023.

Table 24.2 Mixed-standard curve concentration

Concentration (mg/L)	1	2	3	4	5	6
ClO_3^-	0.025	0.050	0.100	0.200	0.500	1.000
ClO_2^-	0.025	0.050	0.100	0.200	0.500	1.000
DCAA	0.125	0.025	0.050	0.100	0.250	0.500
TCAA	0.125	0.025	0.050	0.100	0.250	0.500
BrO_3^-	0.005	0.010	0.020	0.040	0.100	0.200

Fig. 24.2 Standard curve

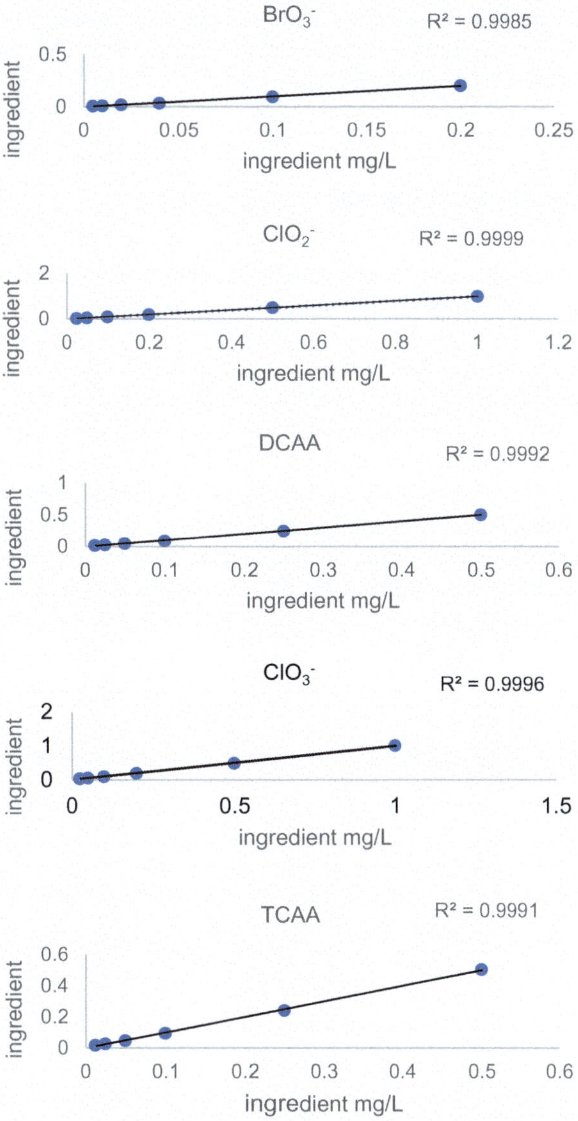

24.3.2 Verification of Precision

Determination of standard sample. Three standard samples of high, medium and low concentrations are prepared from the standard sample. Each sample is measured for 6 times in parallel according to the whole process, then calculating the average value, standard deviation, relative standard deviation and other parameters of each sample, as shown in Table 24.3.

Determination of actual sample. Three test samples sent by a certain company for inspection are used for matrix spike verification. The determination is carried out according to the experimental method in this paper, and the recovery rate of sample is calculated (Table 24.4).

24.4 Conclusion

This paper has experimentally verified that it is feasible to use ion chromatography to simultaneously determine the five disinfection by-products in drinking water. The experimental results show that the detection limits of the elements to be measured by this method meet the requirements in Standard examination methods for drinking water GB/T5750.10–2023 on the detection limits of these five detection elements. There is a good linear relationship between the mass concentrations of all elements to be measured and the peak area. The precision of test method and the recovery rate of sample all satisfy the quality requirements in the specification of testing quality management for geological laboratories (DZ/T0130-2006) on the water sample analysis. The application of this method to determine the five disinfection by-products in drinking water can greatly improve the efficiency of sample analysis and reduce the testing time.

Table 24.3 Determination of the standard samples with different concentrations

Number	Element	Concentration (mg/L)	Measured value (mg/L)						Average value (mg/L)	Standard deviation (mg/L)	Relative standard deviation (%)
Concentration 1	ClO_3^-	0.025	0.025	0.025	0.025	0.026	0.026	0.026	0.025	0.001	2.28
	ClO_2^-	0.025	0.025	0.025	0.025	0.026	0.026	0.026	0.025	0.001	3.12
	DCAA	0.125	0.123	0.122	0.127	0.130	0.128	0.126	0.126	0.003	2.36
	TCAA	0.125	0.123	0.123	0.127	0.131	0.129	0.127	0.127	0.003	2.37
	BrO_3^-	0.005	0.005	0.005	0.005	0.005	0.005	0.005	0.005	0.000	2.14
Concentration 2	ClO_3^-	0.100	0.099	0.099	0.102	0.104	0.103	0.103	0.102	0.002	2.15
	ClO_2^-	0.100	0.098	0.098	0.101	0.104	0.104	0.104	0.102	0.003	3.01
	DCAA	0.050	0.050	0.049	0.050	0.052	0.052	0.052	0.051	0.001	2.64
	TCAA	0.050	0.050	0.049	0.050	0.052	0.052	0.052	0.051	0.001	2.68
	BrO_3^-	0.005	0.005	0.005	0.005	0.005	0.005	0.005	0.005	0.000	2.61
Concentration 3	ClO_3^-	0.500	0.492	0.490	0.500	0.514	0.507	0.506	0.501	0.009	1.88
	ClO_2^-	0.500	0.493	0.487	0.499	0.517	0.508	0.505	0.502	0.011	2.17
	DCAA	0.250	0.248	0.247	0.252	0.259	0.256	0.254	0.253	0.005	1.87
	TCAA	0.250	0.247	0.246	0.253	0.259	0.255	0.253	0.252	0.005	1.89
	BrO_3^-	0.100	0.099	0.097	0.100	0.103	0.102	0.101	0.100	0.002	2.27

Table 24.4 The sample marking

Sample	Analyte	Sample concentration	Spiked solution concentration (mg/L)	Recovery rate (%)
No. 1	ClO_3^-	0.010	0.025	85.24
	ClO_2^-	0.010	0.025	90.23
	DCAA	0.051	0.125	94.51
	TCAA	0.052	0.125	90.11
	BrO_3^-	0.002	0.005	101.45
No. 2	ClO_3^-	0.040	0.100	96.46
	ClO_2^-	0.040	0.100	99.38
	DCAA	0.021	0.050	100.77
	TCAA	0.020	0.050	97.01
	BrO_3^-	0.002	0.020	94.11
No. 3	ClO_3^-	0.204	0.500	90.71
	ClO_2^-	0.205	0.500	98.12
	DCAA	0.104	0.250	103.00
	TCAA	0.101	0.250	94.11
	BrO_3^-	0.041	0.100	98.06

References

Dong LL, Huang JX (2005) Status and analytical techniques of disinfection by-products in drinking water. Prog Chem 02:350–358

Hrudey SE (2009) Chlorination disinfection by-products, public health risk tradeoffs and me. Water Res 43(8):2057–2092

Huo J (2014) Determination of seven types of anions in water by ion chromatography. Chin J Inorg Anal Chem 2:8–10

Lei K, Wang F, Wei Y et al (2016) Simultaneous determination of chlorite, chlorate, and bromate in drinking water by ion chromatography with large volume direct injection. Mod Prev Med 18:032

Mills CJ, Bull RJ, Cantor KP et al (1998) Workshop report. Health risks of drinking water chlorination by-products: report of an expert working group. Chronic Dis Can 19(3):91–102

Standard examination methods for drinking water (GB/T5750.10–2023)

Standards for drinking water quality (GB5749–2022)

Susan DR, Michael JP, Elizabeth DW, Rita S, David MD (2007) Occurrence, genotoxicity, and carcinogenicity of regulated and emerging disinfection by-products in drinking water: a review and roadmap for research. Mutat Research. Rev Mutat Res 636(1a3)

Zhu MX, Shen L, Zhu XM et al (2012) Formation of disinfection by-products by using chlorine dioxide in potable water. J Anhui Agric Sci 40(6):3508–3509

Zhu N, Zhang ZQ, Qin Y et al (2017) Determination of five disinfection by-products and six normal anions in water by ion chromatography. Chin J Public Health Eng 1(1):31–35

Chapter 25
Desalination of Seawater Using Solar Energy

Vladimir Dmitrienko, Mikhail Lemeshko, Stanislav Maslenikov, and Nikolai Semenov

Abstract The article is devoted to the method of desalination of salty or polluted water. The essence of the method is to use the energy of the sun to heat and evaporate the original, non-pure water. A feature of the method is the minimization of heat exchangers in the desalination process, which is provided by heating and evaporation of the source water directly in the solar collector. To reduce heat losses, the boiling point is lowered by vacuuming the space inside the collector. The separation of wet steam from the vapor-air mixture inside the collector housing is carried out by a vacuum pump. The removal of concentrated saline solution or concentrate of contaminated water from the boiling area is carried out by vacuum, which is created by a second vacuum pump or is carried out by one pump by switching the discharged pressure line to the concentrate discharge channel. A diagram of a desalination plant is presented, which shows the main elements of the desalination system. In the process of water evaporation, pressure constantly arises inside the collector, according to the sensor signal, the first vacuum pump is turned on and off, maintaining the pressure in the range of 10–30 kPa. To control the liquid in the collector, its sensor is used, along the way the salt water supply level valve opens and closes, at the same time the first vacuum pump is turned on to increase the pressure in the working volume.

Keywords Desalination of seawater · Solar energy · Heat exchange processes · Desalination method · Vacuuming

25.1 Introduction

For both humans and other representatives of flora and fauna, water is the most important natural resource, without which life is impossible. Almost 2/3 of the surface of our planet is occupied by ocean waters, but only about 3% of the total volume

V. Dmitrienko (✉) · M. Lemeshko · S. Maslenikov · N. Semenov
Institute of Service and Business (Branch) of DSTU in Shakhty, Rostov Region, 346500 Shakhty, Russia
e-mail: vadmitrienko@rambler.ru

H. Xu (ed.), *The 6th International Symposium on Water Resource and Environmental Management*, Environmental Science and Engineering,
https://doi.org/10.1007/978-3-031-55989-1_25

of water on the planet is fresh water, and not all of this water is suitable for human consumption (Savenko et al. 2012).

The constant growth of industrial and agricultural activities, an increase in the volume of human use of chemicals in everyday life contribute to an increase in the discharge of contaminated wastewater into surface reservoirs (Lashkov 2011). This is accompanied by a constant increase in pollution in drinking water sources.

Over the past twentieth century, the volume of freshwater use on the planet has increased 6 times, and in the next quarter century it may double. Currently, most of the world's population lives in conditions of limited consumption of fresh water (Madin and Charlotte 2016).

An increase in water consumption leads to the depletion of clean reservoirs, which is accompanied by an increase in the use of underground sources by the population (Lashkov 2011; Zakrutkin et al. 2016).

However, for a long time there has been a constant decrease in the quality of both surface and groundwater. The springs are being depleted and drying up. Due to changes in the chemical composition, the water in the wells becomes unsuitable for cooking. Small rivers and lakes, which often feed underground aquifers, are particularly adversely affected by human activity.

25.2 Reduction of Freshwater Quality and the Purpose of Research

According to the results of the analysis of the state of small rivers in the western part of the Rostov region, it can be noted that the mineralization of the Grushevka river exceeds the maximum permissible concentration (MPC) by almost three times, the content of iron by three times, and sulfates by almost 20 times (Drovovozova and Panenko 2019).

In the larger Tuzlov River, the total mineralization remains almost unchanged, but with a slight decrease in the content of sulfates, pollution with nitrate anions exceeding MPC by 15 times increases. This circumstance causes a constantly observed trend of significant increase in mineralization in the Don River.

If we talk about smaller rivers flowing into the above-mentioned ones, then their condition is more depressing, since the main discharge of pollution from industrial and agricultural enterprises, as well as small settlements, falls on their share.

In the Rostov region, one third of freshwater consumers are supplied from underground sources, which in most cases are confined to alluvial deposits and are near-surface (Nazarenko and Nazarenko 2007). That is, aquifers are hydrodynamically connected to river waters and, accordingly, are also subject to anthropogenic influences (Drovovozova and Marygin 2018).

Of course, industrial enterprises provide the largest volume of polluted water discharge, but despite the smaller number of residents in rural settlements there are no effective treatment facilities, which accordingly increases the proportion of pollution.

At the same time, livestock complexes and household farming are associated with high consumption of water, which is taken from wells and wells. This, in turn, leads to the depletion of sources of high-quality groundwater. Thus, in the coming years, there will be an increase in water consumption from sources that belong to the category of "dirty and very dirty".

In this case, the issue of providing drinking water will be especially acute, since the use of water with a high salt content leads to disorders of the gastrointestinal tract and the cardiovascular system. From the point of view of providing fresh water, the Rostov region has large reserves in the floodplains of large rivers. There are also about 250 lakes, but a significant part of them are represented by salt lakes. The situation is about the same with the main underground aquifers.

Prolonged use of hard water is also fraught with the occurrence of joint diseases (arthritis, polyarthritis), the formation of kidney stones and bile ducts.

But at the same time, the salt content should not be low, since soft water leaches salts from the body, causing the rapid development of rickets and thinning of bones, and can also provoke the risk of developing cardiovascular diseases. In this case, one of the options for ensuring the salt balance of groundwater is to reduce the concentration of mineralization by diluting them with distilled water.

Currently, a large number of effective methods and means of water purification from many pollutants have been developed, however, it is possible to achieve a significant reduction in salt content only with the use of very expensive distillation and reverse osmosis. While significant initial costs are quite acceptable for large consumers, they are not available for most small and individual farms. In this case, the task of desalination of small volumes of groundwater or seawater at minimal cost is very relevant.

The cheapest energy source on earth is the sun (Abdrakhmanov and Perevedentsev 1992; Drobyshev 2014; Popel et al. 2010). In recent years, solar energy has been widely used for water heating and heating (Popel et al. 2010; Kulikov 2015; Solovyova and Kondratenkov 2013). There are developments and prototypes of desalination plants using solar energy. The analysis shows that all of them have both advantages and disadvantages. Therefore, research on the development of effective methods of water desalination does not stop (Fried et al. 2012).

The main disadvantage of solar energy is its fluctuations both during the year and day, and within daylight hours due to clouds. This, in turn, affects the ability of installations to heat water to a temperature of intense evaporation, which significantly reduces their productivity.

Therefore, we have considered the possibility of using vacuuming to reduce the boiling point of water in the evaporator. This will allow to regulate the evaporation process to a certain extent by adjusting the pressure when the water temperature changes.

25.3 Materials and Methods

The object of the research was a specially made stand consisting of 10 rectangular pipes, which are interconnected by two perpendicular collectors.

Experimental studies were conducted in a thermally insulated chamber in order to determine the intensity of water evaporation when the stand was irradiated with solar radiation from a solar collector filled with water.

The air velocity during air conditioning was measured with an anemometer ATA-1038 with an accuracy of 0.1 m/s.

The solar collector was installed at an angle of 15 degrees to the horizon.

Water was supplied to the lower fitting for filling, and a tube was connected to the upper one, through which steam was discharged into the condenser.

A vessel for collecting distillate was connected at the outlet of the condenser.

Temperature control was carried out by thermocouples as part of the Aries complex. The ambient temperature was controlled by a psychrometer.

25.4 Discussion of Research Results

The first stage of the research was aimed at substantiating the design of the evaporator, since the pumping of air from its internal cavity can be accompanied by deformation of the walls. In the design, it was decided to use rectangular pipes to improve heat transfer. It was important to check the sufficiency of the wall thickness of the pipes.

The use of round pipes in the evaporator, similar to pipes in vacuum solar collectors, is the most advantageous option from the point of view of voltage distribution, but the manufacturing technology of the evaporator is more complicated and the heat reception of sunlight is worse.

To conduct experimental research, a stand was made consisting of ten aluminum pipes with a cross section of 40 $\times$ 60 mm and a length of 1.0 m, which are connected to the collectors at the ends by welding. Outlets with taps for water supply and steam extraction are connected to the collectors. The upper and side surfaces of the pipes are painted black. The volume of water in the pipes was 21.5 dm^3, and the area of the irradiated surface was 1.22 m^2.

When assessing the intensity of the disturbing heat flow to the surface of the test stand, reference data on the total solar radiation on a horizontal surface with a cloudless sky in Shakhty, Rostov region, located at 47.7o north latitude (Fig. 25.1) were used.

The analysis shows that the most optimal period of operation of the solar collector is April–September. In this case, the average monthly radiation will be 207.3 kW h/ m^2, and the daily radiation will be 6.9 kWh/m^2. Thus, with a stand area of 1.22 m^2, under ideal conditions, the thermal capacity of the stand will be 1.2 kW h/m^2.

When analyzing the operation of solar collectors for hot water supply, in most cases for the middle latitudes of our country, the temperature in the storage tank is

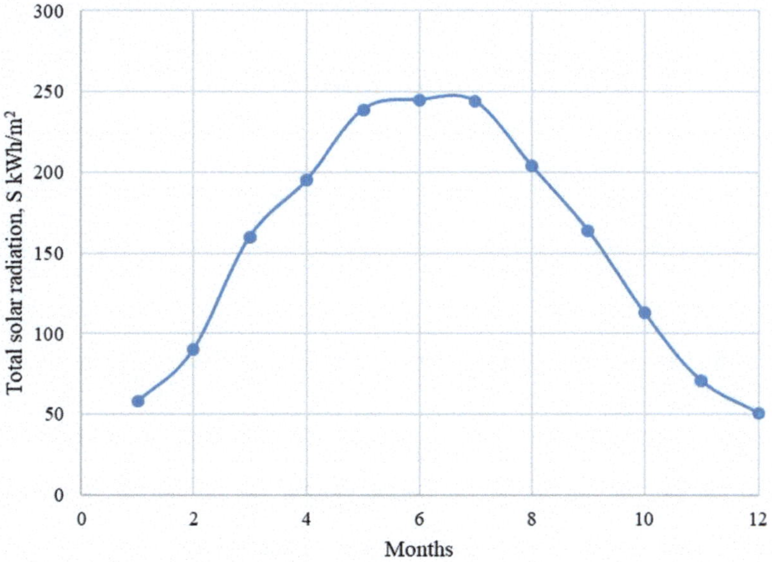

Fig. 25.1 Distribution of total solar radiation during the year in Shakhty

60–70 °C, since at higher temperatures the efficiency of the heater decreases and losses increase.

Estimating the dependence of the boiling point of water on the pressure shown in Fig. 25.2, it is possible to distinguish a pressure range of 20–30 kPa, in which the average boiling point will be 65 °C. Therefore, further research is aimed at evaluating the possibility of heating the water in the research stand to this temperature value.

Considering the results of a study of the illumination of a horizontal surface with a clear sky at various times of the day and year (Fig. 25.3), it is possible to identify a zone with minimal changes in illumination and, accordingly, solar radiation from 10.30 to 14 o'clock Moscow time. In this range, it was decided to conduct experimental studies of water heating in the stand.

The first studies conducted under real weather conditions showed that without thermal insulation of the stand, it is not possible to heat water even during the entire daylight hours above 47 °C at an ambient temperature of 27–31 °C. Therefore, the stand was placed in a frame, the lower part of which represented technonicol plates with a thickness of 0.1 m, on top of the walls it was covered with a transparent PVC film (Takaev et al. 2002). The air gap averaged 0.03 m.

Since July and August in our region were characterized by unstable weather with frequent rains and sharp temperature fluctuations, it was not possible to obtain a sufficiently objective sample to obtain average indicators. Figure 25.4 shows the results of measurements performed on August 28, 2023 from 11.10 to 15.10 Moscow time.

Measurements were carried out at various points of the stand and camera with thermocouples and the results were saved on a computer using the Aries complex.

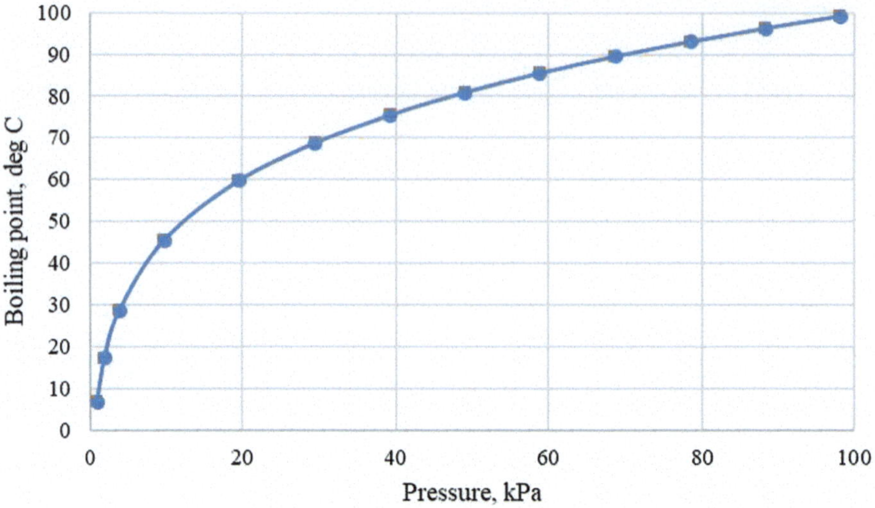

Fig. 25.2 Graph of the dependence of the boiling point of water on pressure

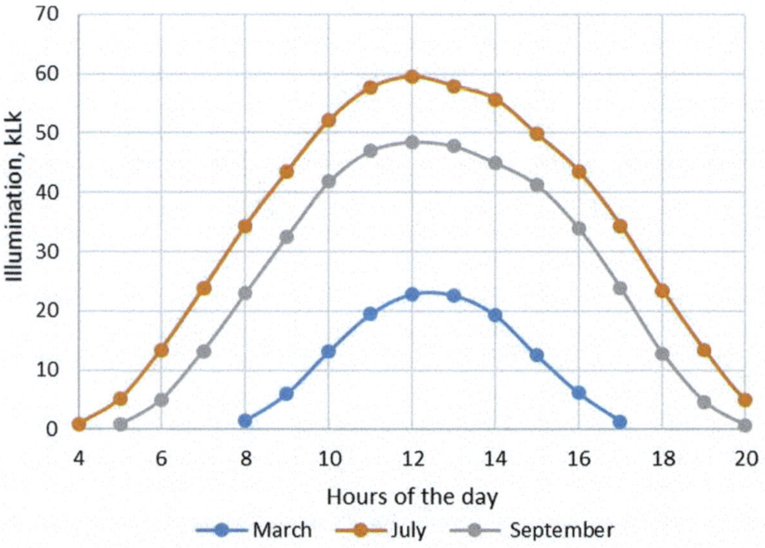

Fig. 25.3 Change of illumination during daylight hours

During four hours of measurements, the ambient temperature was in the range of 25.1–32.2 °C. The wind speed was 4.2–5.4 m/s. At an initial water temperature of 22.4 °C, the maximum value was observed after 2 h. 40 min. and it was 67.3 °C. This is the maximum value for the entire observation period, but it should be noted that in most cases clouds interfered.

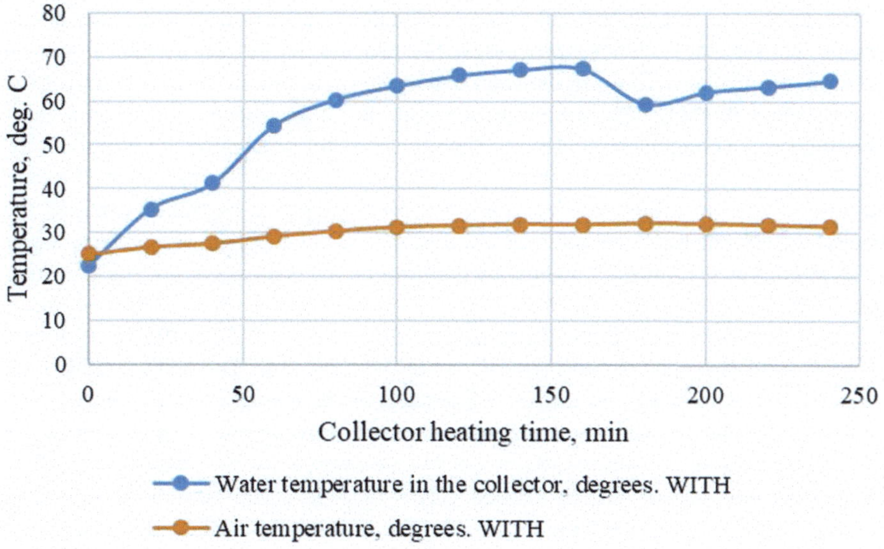

Fig. 25.4 The results of measuring the increase in water temperature in the test bench

The conducted studies allowed us to establish that the temperature increase begins to decrease after 90 min of irradiation, and when it reaches a value of about 65 °C, it is almost stable due to losses to the environment and an increase in heat for vaporization. Nevertheless, it has been established that even in the middle latitudes of Russia it is possible to use solar energy for desalination of mineralized water.

Therefore, a concept has been developed, which consists in the fact that it is advisable to carry out the evaporation of saline solution at a rationally low pressure using the thermal energy of the sun. In this case, the boiling process must be performed directly in the solar collector to minimize thermal and energy losses. Minimization of energy losses is ensured by minimizing the use of heat exchangers and reducing the length of heat-conducting channels.

The heating and evaporation process is performed at a pressure of the order of 20–30 Pa, which ensures the distillation process at a relatively low temperature, in the range of 50–65 0C. At the same time, both the collector surface and the heat-conducting channels of the desalination plant will have a relatively low thermal pressure with the environment, which will ensure lower heat losses.

The technical and constructive solutions developed in the course of research ensure a reduction in the cost of the desalination process, including seawater, due to the rational use of solar energy.

The technological features of the developed system include the operating pressure inside the solar collector, it is 0.1–0.3 atm. Reducing the pressure inside the solar collector reduces the boiling point and vaporization to values of 50–65 °C, and also reduces the time spent on the desalination process. Vacuuming inside the solar collector is supported by two vacuum pumps.

The desalination of seawater implemented in the desalination system under development includes the following processes: The process of heating seawater in a solar collector, the process of boiling seawater and its vaporization, steam removal from the solar collector housing, salt brine removal from the solar collector, steam condensation process in a condenser, heat exchange processes for preheating the initial desalinated water, processes for monitoring the level of salt solution in the solar collector, the process of controlling and regulating the pressure inside the solar collector, the process of replenishing the internal volume of the solar collector with the initial desalinated water. Figure 25.5 shows a diagram of a desalination system in a solar collector.

The system consists of the main element of the system, a solar collector of special design (7) with thermal insulation (6) of the non-working side. The first vacuum pump (11), which sucks water vapor into the condenser (9), the second vacuum pump, which removes salt solution from the collector without disturbing the evacuation mode, a container for the initial desalinated water (1), storage tanks for collecting distillate (8) and salt concentrate (3), controlled valves (4), (12). The system includes two thermally insulated heat exchangers (2), (9). The desalination system also includes a programmable controller (not shown in the figure), which programmatically controls the desalination process.

The desalination system works as follows. To fill the initial desalinated water volume of the collector (7), the water supply valve (12) is opened and the second vacuum pump (11) is turned on to take off steam. After the water is set to the required maximum level, the valve shuts off the water supply at a signal from the water level sensor. But the vacuum pump (11) continues to work to create the necessary discharge in the collector. At a signal from the pressure sensor, this pump is switched off. During the boiling and vaporization process, the pressure in the collector gradually increases and when the pressure increases above the set level, the vacuum pump (11) turns on

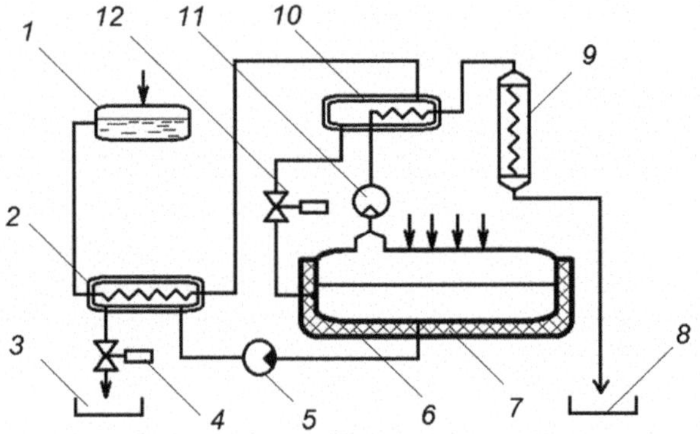

Fig. 25.5 Diagram of the desalination system in a solar collector: 1—a container with source water, 2—he first heat exchanger 3—salt concentrate storage, 4—the first controlled valve, 5—the first vacuum pump, 6—collector thermal insulation, 7—collector, 8—fresh water storage, 9—condenser, 10—the second heat exchange unit, 11—the second vacuum pump, 12—the second controlled valve

to pump out steam. To remove the concentrated saline solution, at a signal from the density sensor, the first vacuum pump (5) is turned on and the valve (4) opens, the saline solution accumulates in the container (3). Thus, the desalination system under the control of a programmable controller can operate in autonomous mode.

25.5 Conclusions

1. The developed desalination system uses the achievements of solar energy, which provides a significant reduction in energy costs for the desalination process and reduces the cost of desalinated water.
2. The main advantage of the proposed system is the minimization of losses in heat exchangers when using a renewable energy source.
3. The developed system of desalination of mineralized and seawater can operate in an autonomous mode.

References

Abdrakhmanov R, Perevedentsev Y (1992) Renewable energy sources. Kazan Publishing House, Kazan , Unita, p 134

Drobyshev A (2014) Solar and wind energy in the Krasnodar Territory, conditions for its utilization. Monograph. St. Petersburg: RGGMU, p 276

Drovovozova T, Marygin V (2018) Hydrochemical and geological assessment of groundwater in the Rostov region. Sciences of Europe 33–2(33):39–44

Drovovozova T, Panenko N (2019) The ecological state of small rivers of the Rostov region. Ecol Water Manag 1(01):1–17

Fried S, Kolomiets Y, Mordynsky A, Suleymanov M, Arsatov A, Oschepkov M (2012) The effectiveness of solar water heaters in climatic conditions of Russia. Izvestia of higher educational institutions. North Cauc Reg Ser: Tech Sci. 6, 21–26

Kulikov K (2015) Prospects for the use of solar collectors. Innov Sci 12–2:86–88

Lashkov G (2011) Reserves and problems of fresh water in the world. [Resources and problems of fresh water in the world]. All-Russian Scientific and Technical Conference [All-Russian scientific-technical conference]. Moscow, pp 306–309

Madin G, Charlotte E (2016) Water resource management in a vulnerable world: the hydro hazardscapes of climate change. Water Int 41(5):801–803

Nazarenko V, Nazarenko O (2007) The state of groundwater in the Rostov region. Environmental problems. A look into the future. Rostov N./D.: Publishing House of the Southern Federal University, pp 252–256

Popel O, Fried S, Kolomiets Y, Kiselyov S, Terekhova E (2010) Atlas of solar energy resources in Russia. United Institute of High Temperatures of the Russian Academy of Sciences, Moscow, p 54

Savenko V, Mikhaylov V, Zhuk V, Samokhin M, Zaslavskaya M, Frolova N (2012) Water reserves on earth. [Resources of water on the earth]. M.: Patterns of hydrological processes. pp 15–17

Solovyova E, Kondratenkov A (2013) The system of autonomous power supply of a building in the conditions of the II climatic zone. Bulletin of MG SU 10:208–215

Takaev B, Kazanjan B, Solodov A (2002) Air solar collector with transparent thermal insulation of capillary type. 1st All-Russian school-seminar of young scientists and specialists: collection of scientific tr. M. MEI. pp. 256–261

Zakrutkin V, Sklyarenko G, Bakaeva E, Reshetnyak O, Flexible E, Fomenko N (2016) Surface and groundwater within technogenically disturbed geosystems of the Donbas: formation of chemical composition and quality assessment. Rostov n/A: Publishing House of the Southern Federal University, 170 p

Chapter 26
Coal-Fired Power Plant Flue Gas Pollution and Treatment Technologies

Hailing Ma

Abstract This paper comprehensively reviews the flue gas pollution and its treatment technologies in coal-fired power plants. Initially, the emission characteristics of major pollutants in the flue gas of coal-fired power plants and their environmental hazards are discussed, with a focus on sulfur oxides (SO_x), nitrogen oxides (NO_x), and particulate matter (PM). The paper then delves into various technologies for reducing these pollutant emissions, including carbon dioxide capture, desulfurization, denitrification, and particulate matter removal techniques. The efficiency and practical application of these technologies are analyzed, such as Electrostatic Precipitators (ESP), Baghouse Filters, and Wet Electrostatic Precipitators (WESP), along with their potential environmental impacts. The paper also discusses the global application of these technologies in coal-fired power plants and compares the effectiveness of different technologies in pollutant control. Finally, the paper proposes suggestions for future research and technological improvements, particularly in the areas of particulate matter removal and final flue gas treatment.

Keywords Coal-fired power plants · Flue gas pollution · Carbon dioxide capture · Denitrification · Electrostatic precipitator (ESP) · Wet electrostatic precipitator (WESP)

26.1 Introduction

Coal-fired power generation is a significant method of energy supply (Zhang et al. 2017). Coal-fired power plants supply over 70% of China's electricity (Xu et al. 2016). However, the combustion of coal in boilers generates a large amount of pollutants (such as particulate matter (PM), nitrogen oxides (NO_x), and sulfur oxides (SO_x)). The resulting flue gas pollutes the atmosphere, causing acid rain and creating a

H. Ma (✉)
School of Engineering and Technology, The University of New South Wales, Canberra ACT 2600, Australia
e-mail: hailing.ma@unsw.edu.au

H. Xu (ed.), *The 6th International Symposium on Water Resource and Environmental Management*, Environmental Science and Engineering,
https://doi.org/10.1007/978-3-031-55989-1_26

greenhouse effect. To meet the "ultra-low emission" requirements of coal-fired power plants (Reddy et al. 2008): dust <10 mg/m^3, SO_2 <35 mg/m^3, NOX <50 mg/m^3, it is essential to install flue gas purification systems in the flues of coal-fired power plants to control pollutants. This paper will identify and summarize typical solutions to the air pollution problems caused by the combustion of fossil fuels in coal-fired power plants, thereby providing preventive measures against deadly respiratory diseases. The study will enhance public awareness of pollution control in areas preparing to construct coal-fired power plants.

26.2 Emission Characteristics and Environmental Pollution Hazards of Power Plant Flue Gas

The flue gas produced by the combustion of fossil fuels contains carbon dioxide and water vapor, along with other substances such as nitrogen, nitrous oxide, sulfur oxides, as well as fly ash (PM) and mercury from coal-fired power plants (China Environmental Protection Industry Association 2015). Studies have shown that emissions from coal combustion include: SO_2, NO_2 (sulfur and nitrous oxide), acidic gases as well as water vapor; micron-sized particles and toxic mercury; and greenhouse gases (CO, CO_2, and N_2O). The amounts of these emissions depend on the type of coal, its calorific value and source, and the amount burned. Power plant boilers burning coal produce flue gas. The main components of flue gas are nitrogen (N_2), CO_2, and water (H_2O). It carries PM and other pollutants. Depending on the combustion technology and fuel used, there may be traces of some oxides, such as sulfur oxides (SO_x) and nitrogen oxides (NO_x) (Dongyue 2015).

SO_2 is a significant air pollutant, primarily sourced from power plants burning high-sulfur coal (Li Qing 2002). Sulfur oxides and nitrogen oxides react with atmospheric water vapor and other chemicals in the presence of sunlight to form sulfuric and nitric acids. The acids are typically dissolved in suspended water droplets in clouds or fog. The acidity of these acid-rich droplets is comparable to that of lemon juice, which is washed from the air to the ground by rain or snow. This is what is known as acid rain. Acid rain or deposition leads to the acidification of groundwater and surface water, damage to life (especially forests and aquatic organisms), and the decay of buildings. This also proves that acid rain is caused by the reaction of nitrites and sulfur oxides with atmospheric water.

Fly ash is the waste left after coal combustion. Coal ash is toxic, and its production, especially when poorly managed, poses significant hazards (Lianqing and Zhe 2006). In 2008, a dyke at the Kingston coal-fired power plant in Tennessee, USA, failed, blocking a large coal ash treatment pond. This 40-acre pond leaked over a billion gallons of coal ash slurry into the adjacent valley, covering about 300 acres with thick toxic sludge, destroying three houses, damaging many others, and contaminating the Emory and Clinch Rivers. Water samples tested after the leak showed arsenic levels 149 times the permissible standard for drinking water. Levels of other toxic metals also rose, including lead, thallium, barium, cadmium, chromium, mercury, and nickel (Guoqiang et al. 2007). The emission of PM2.5 from fly ash is a major contributor to global mortality and morbidity rates.

PM in exhaust gases is considered harmful to human health because it penetrates deep into the lungs. Larger particles are less of a concern as they are effectively removed by the lung system. Fine particles are easily transported by air currents, so their harmful effects may be encountered far from the exhaust plume. In addition to health hazards, particulate matter in the form of smoke also negatively impacts human visibility.

26.3 Flue Gas Treatment Technologies in Coal-Fired Power Plants

26.3.1 Treatment of Carbon Dioxide in Coal-Fired Power Plants

Due to differences in power generation efficiency, fuel choices, and the age of power plants, there is a significant variation in carbon dioxide emissions per megawatt from fossil fuel power generation. The use of low-carbon fuels such as natural gas and improvements in energy conversion efficiency have been steadily reducing the average carbon dioxide emissions per megawatt (Bin et al. 2009; Shen et al. 2023). In the process of post-combustion carbon dioxide capture, the gas is separated from flue gases after combustion. Carbon dioxide sequestration in power plants is only effective after capturing the gas in a relatively pure form. Carbon dioxide sequestration refers to storing or using carbon dioxide in a way that does not enter the atmosphere. A small part of the captured carbon dioxide can be used to manufacture chemicals, while a large amount can enhance the production of oil and natural gas. Storage of carbon dioxide can be done in various ways, including underground storage or ocean sequestration. Methods to remove carbon dioxide from the gas stream include amine absorption, various adsorption techniques, and the use of membranes. In the capture process, 85 to 95% of the carbon dioxide in the fuel can be recovered. Carbon dioxide can also be captured using solid chemicals that react with the gas in solid form. These carbon dioxide capture processes require a lot of energy, leading to a reduction in power generation efficiency of up to 40%, and a decrease in net power generation of up to 40%. Post-combustion CO_2 capture is often considered one of the main strategies to reduce CO2 emissions from coal-fired power plants (Yongzheng et al. 2023; Shen et al. 2022).

Pre-combustion capture technology involves the reaction of fuel with oxygen or air, and possibly with steam, to produce syngas primarily consisting of carbon monoxide and hydrogen. Then, carbon monoxide reacts with steam in a catalytic reactor to produce carbon dioxide and more hydrogen. Carbon dioxide is then separated through physical or chemical absorption processes, producing hydrogen-rich fuel that can be used for many applications such as boilers, furnaces, gas turbines, engines, and fuel cells. Pre-combustion carbon dioxide capture must be part of the IGCC and will become the future of fossil fuel power generation (Changzheng et al. 2022). In oxy-fuel combustion technology, nearly pure oxygen is used instead of air for combustion, producing flue gas mainly consisting of CO_2 and H_2O. This simplifies the separation process, as water vapor can be easily condensed into liquid, leaving carbon dioxide for subsequent processing. Studies have found that retrofitting today's coal-fired power plants with post-combustion and oxy-fuel combustion is feasible. Both technologies are viable, safe, and have the potential to be cost-effective. The challenge lies in developing processes that can be economically deployed on a large scale (Zhenghui et al. 2014).

Samanta and Ghosh (2011) theoretically analyzed partial repowering of a 250 MW coal-fired power plant. It is suggested to burn a certain amount of coal in a pressurized combustion chamber (PPCC). Subsequently, the product gas expands in a gas turbine and the exhaust gas is sent to a boiler. The partial retrofit is expected to reduce the plant's carbon dioxide emissions by about 26.5%. Mborah and Agbomadzi (2010) proposed a method to remove carbon dioxide gas from the exhaust gas generated by the Takoradi thermal power station in Ghana, aiming to reduce the plant's carbon dioxide emissions to the atmosphere. The method is an improvement of the Fluor Daniel ECONAMINE FG process, eliminating the recovery of exhaust gas. This method works by using an amine solution that contacts the exhaust gas in a countercurrent manner to remove carbon dioxide from the exhaust gas. The work further reported that the method has been adopted by 23 large-scale carbon dioxide-producing companies and suggested a cost feasibility study for its application at the Takoradi plant. From another perspective, Martínez et al. (2011) analyzed a Ca cycle system that uses CaO as a renewable adsorbent to capture CO_2 from flue gas produced by power plants, as shown in Fig. 26.1.

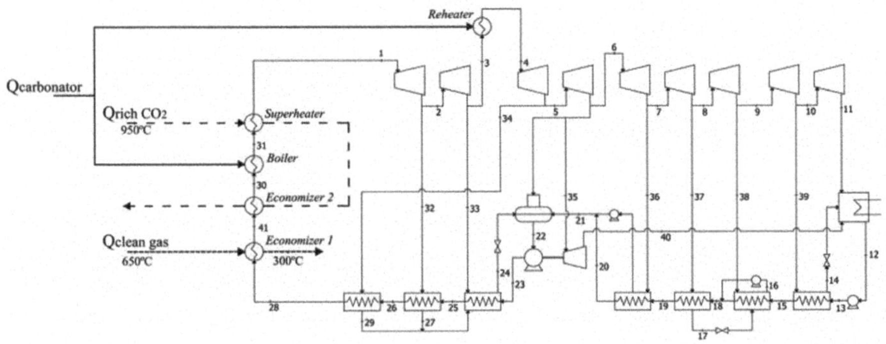

Fig. 26.1 Process flow diagram of the Ca cycle system for CO_2 capture from power plant flue gas (Martínez et al. 2011)

26.3.2 Treatment of SO_x and NOx in Coal-Fired Power Plants

For environmental considerations, many coal-fired power plants globally have installed DeNOx units to control nitrogen oxides (NO_x), and $DeSO_x$ units to remove sulfur oxides (SO_x) from the flue gas. Wet Flue Gas Desulfurization (WFGD) is the most common method used for SOx control. The primary measures in the combustion chamber to reduce nitrogen oxide emissions only achieve about 35% efficiency. This is insufficient to meet many environmental regulations. Selective Non-Catalytic Reduction (SNCR) and Selective Catalytic Reduction (SCR) are secondary measures for NOx removal. These technologies are more promising and are located after the boiler combustion zone. The secondary measures of SNCR and SCR can achieve efficiencies of about 50% and 95%, respectively (Min 2018; Jiayou and Qing 2011; Lixue 2011). Rasul et al. (Morgan et al. 2017) further pointed out that the nature of coal used in boilers and the plant's environmental management policies are two main factors determining the nature of flue gas treatment processes. In some countries, coal-fired power plants are required to have $DeNO_x$ and $DeSO_x$. However, this may vary from other countries. The study further elaborates that the coal used in Australia has a low sulfur content, hence the concentration of SO_x produced from burning Australian coal is relatively low. This factor has led to less stringent regulatory requirements. Therefore, Australian coal-fired power plants in the past did not need to be equipped with $DeNO_x$ or $DeSO_x$ units to clean or process nitrogen oxides and sulfur oxides in the flue gas stream.

Kikkawa et al. (2008) reported that Babcock-Hitachi K. K. is developing a technology to minimize the amount of nitrogen oxides, sulfur oxides, and other pollutants produced when burning coal in coal-fired power plants. Additionally, the company has developed a system using a unique amine solvent to absorb and recover carbon dioxide from flue gas. Babcock-Hitachi K. K. and Tokyo Electric Power Company confirmed that the system was installed in a pilot plant using flue gas from the Yokosuka coal-fired power station and achieved high carbon dioxide elimination performance. Kuroki et al. (2017) studied the removal of air pollutants containing nanoparticles using a wet plasma reactor, as shown in Fig. 26.2. Simulated exhaust gas was prepared for the study to assess the performance using polystyrene latex particles of diameters 29, 48, 100, 202, and 309 nm, as well as NO/SO_2 gas cylinders. The removal efficiencies of SO_2 and NO_x were over 99% and 77%, respectively. These results confirm that wet pulsed corona discharge plasma reactors can be used to simultaneously remove nanoparticles, NO_x, and SO_x. Morgan et al. (2017) also demonstrated the high efficiency of up to 74% for the removal of nitrogen oxides in atmospheric non-thermal plasma reactors, used for nitrogen oxide pollution control. The development and widespread application of such reactors will make the task of exhaust gas treatment more manageable.

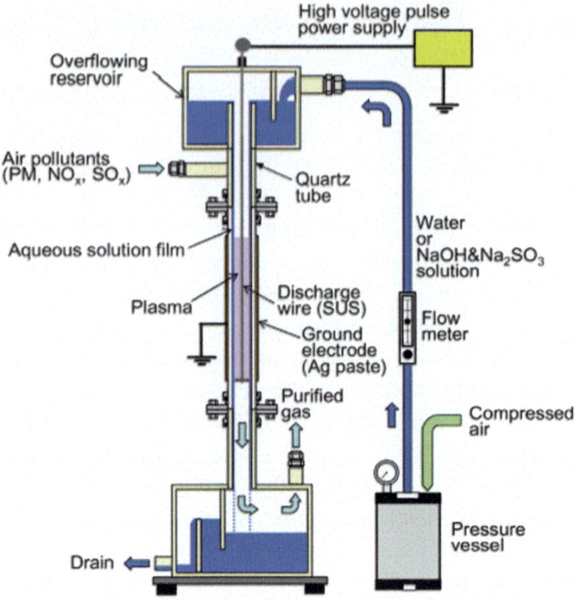

Fig. 26.2 Wet plasma reactor for the removal of air pollutants containing nanoparticles (Kuroki et al. 2017)

26.3.3 Treatment of Particulate Matter in Coal-Fired Power Plants

The emission of particulate matter from power plant exhaust gases is controlled using equipment such as cyclone separators, baghouse filters, and electrostatic precipitators (ESP). Cyclone separators are used for removing coarse particles, while electrostatic precipitators are used to remove a lower spectrum of coarse particles and fine particles. ESP is one of the most widely used devices for controlling particulate matter, especially fly ash. The efficiency of ESP depends on the resistivity of the particles, which can be significantly altered to the desired requirements by treating the flue gas with some known chemicals. However, baghouse filters are most suited for controlling fine particles and even partially removing ultrafine particles (Jiaxing et al. 2022; Jianwen et al. 2021). Increasing attention is being paid to the research on the formation and emission control of particulate matter. PM10 is one of the main atmospheric pollutants produced by coal combustion in China. Xiaolong et al. (2018) studied the Particulate Emission Control Device (PECD) and its classification and collection efficiency. The study experimentally determined the particulate emission efficiency and size distribution at the inlet and outlet of the PECD for four different coal-fired power plants using anthracite and bituminous coal. The mass concentrations of PM1, PM2.5, and PM10 within the aerodynamic diameter range of 0.03-10 μm were measured. The study found that the capture efficiency of ESP for PM and PM10 was 99.0–99.89% and 98.2–99.62%, respectively. Additionally, the collection efficiency of the baghouse filter for PM and PM10 was 99.94% and

99.76%, respectively. However, the study proved that for particle size ranges of 0.1–1 μm, the collection efficiency of ESP and baghouse filters was lower, at 90.8–98.6% and 99.54%, respectively (Table 26.1).

Wang et al. (2014) estimated the impact of particulate matter removal technologies on the total emissions in China, as shown in Fig. 26.3. It is suggested that replacing ESP with a mixed ESP/BAG would reduce the total emissions to 104,300 tons, of which PM2.5 emissions would be 47,480 tons. The study investigated the impact of dust collectors and wet flue gas desulfurization equipment on the PM2.5 emission characteristics of three coal-fired power plants. It was found that the PM2.5 removal efficiency of the mixed ESP/BAG was higher than other types of dust collectors, reaching 99%. The study noted that when the first electrostatic field is turned off, the efficiency of the mixed ESP/BAG decreases by 1.9%. The study also explored how operating conditions of the mixed ESP/BAG affect dust removal performance. On the other hand, Bandyopadhyay and Mandal in (2015) studied the operation of a Wet Electrostatic Precipitator (WESP) as an air pollution control device in Indian coal-fired thermal power plants to capture particulate matter. WESP was installed to reduce PM emissions to meet Indian chimney PM emission standards. The investigation showed that WESP could potentially reduce emissions of sulfur dioxide and carbon dioxide. The study suggests that operating WESP to purify flue gas from coal-fired thermal power plants can also be considered a greenhouse gas emission reduction option. However, the work pointed out that due to the air dilution of flue gas, nitrogen oxide emission levels could not be finally analyzed. Finally, it is recommended that plant operators use dilute alkaline reagents instead of water as the washing medium to achieve the multiple benefits of WESP cleaning all acidic gases and achieve better ash handling. Howard et al. (2019) compared the health benefits of tightening PM emission standards at a coal-fired power plant in Northeast Brazil with control costs to explore whether the current emission standard levels are strict enough under a simple cost–benefit framework. The study confirmed through modeling that, in all considered scenarios, the health benefits of tightening existing PM10 emission standards are more than 60 times higher than the emission control costs. The monetary value of avoiding hospitalization alone is at least four times the corresponding control costs. This work suggests that these results provide a

Table 26.1 Total collection efficiency for different PM sizes

Dust type	PM1	PM2.5	PM10	PM
Power plant 1—electrostatic precipitator	90.83	95.58	98.20	99.00
Power plant 2—electrostatic precipitator	98.59	99.16	99.62	99.89
Power plant 3—electrostatic precipitator	95.74	96.75	98.58	99.76
Power plant 4—baghouse filter	99.54	99.72	99.76	99.94

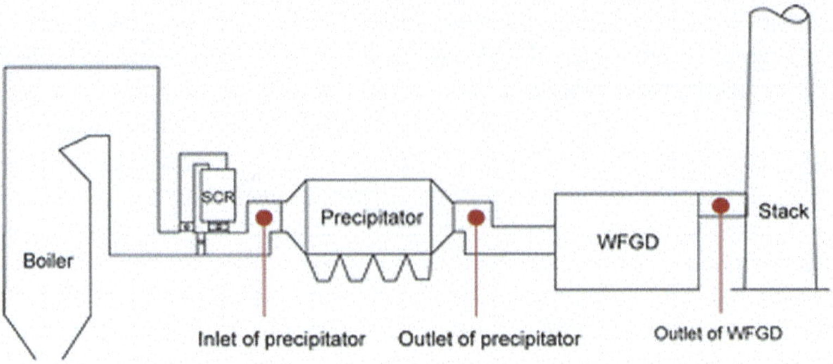

Fig. 26.3 Particulate matter removal technology in coal-fired power plants (Wang et al. 2014)

strong argument for considering tightening particulate matter emission standards in coal-fired power plants globally.

26.4 Conclusion

This paper has successfully conducted an extensive review of flue gas treatment in coal-fired power plants. It clearly pointed out the emission characteristics of these pollutants and the hazards they cause. An in-depth exploration was conducted on the control of these emissions, from carbon dioxide capture to desulfurization, denitrification, and particulate matter collection, and the best practices in practice were provided. Most of the studies considered are consistent with their research results and comply with standard procedures. The results of this review can serve as a platform for seeking ways to reduce atmospheric pollution and thereby prevent acid rain. In addition, carbon dioxide capture can alleviate the key issue of global warming, and providing quality air can act as a preventive measure against deadly respiratory diseases. However, more efforts are needed to improve the particulate matter collection system of baghouse filters and the final disposal methods.

References

Bandyopadhyay A, Mandal PK (2015) Investigation on multiple benefits of wet ESP for emission control in an Indian coal-fired thermal power plant. Clean Technol Environ Policy 17:421–431

Bin H, Shisen X, Shi G, Lianbo L, Jiye T, Hongwei N, Ming C (2009) technical and economic analysis of CO_2 capture systems in coal-fired power plants. Power Eng 5

Changzheng H, Yabo W, Shengchun L (2022) A comparison of the application of MEA solutions in biomass power plants and coal-fired power plants for CO2 Capture. Huadian Technol 44

Dongyue Y (2015) Research on comprehensive purification technology of flue gas in coal-fired power plants. North China Electric Power University (2015)

Electric Dust Removal Committee of China Environmental Protection Industry Association: Ultra-low Emission Technology of Flue Gas in Coal-Fired Power Plants. Ultra-low Emission Technology of Flue Gas in Coal-Fired Power Plants (2015)
Guoqiang W, Boshu H, Lili W (2007) Emission control technology of mercury in flue gas of coal-fired power plants. Environ Prot
Howard DB, Thé J, Soria R, Fann N, Schaeffer R, Saphores J-DM (2019) Health benefits and control costs of tightening particulate matter emissions standards for coal power plants-The case of Northeast Brazil. Environ Int 124:420–430
Jianwen C, Jingyun Z, Zhihong Z (2021) Exploration of the impact of injection pressure on the filtration performance of filter materials in baghouse dust collectors. China Environ Prot Ind 3
Jiaxing N, Shuguang Y, Zhenxin Z, Yuxuan Z, Gengchen C (2022) The impact of different air intake methods on filter bags in baghouse dust collectors. Coal Mine Mach 043
Jiayou G, Qing L (2011) The first domestic application of combined SNCR and SCR denitration process in coal-fired power plants. North China Electr Power Technol 000:31–34
Kikkawa H, Ishizaka H, Kai K, Nakamoto T (2008) DeNOx, DeSOx, and CO2 removal technology for power plant. Hitachi Rev 57:175
Kuroki T, Nishii S, Kuwahara T, Okubo M (2017) Nanoparticle removal and exhaust gas cleaning using gas-liquid interfacial nonthermal plasma. J Electrostat 87:86–92
Li Qing P (2002) Pollution and control of sulfur dioxide in coal-fired power plants. Shandong Electric Power Technology
Lianqing Y, Zhe H (2006) Study on radioactive pollution impact and control management technology of fly ash in coal-fired power plants. Energy Environ Prot
Lixue Z (2011) A brief analysis of boiler SNCR and SCR combined denitration methods. Sci Technol Innov Her 1
Martínez I, Murillo R, Grasa G, Abanades JC (2011) Integration of a Ca looping system for CO_2 capture in existing power plants. AIChE J 57:2599–2607
Mborah C, Agbomadzi S (2010) Removal of carbon dioxide gas from the exhaust gases generated at the Takoradi Thermal Power Station. Res J Environ Earth Sci 2:245–254
Min, H.: Study on deep low nitrogen combustion coupled with SNCR and SCR denitration in large power station boilers. Zhejiang University
Morgan N, Ibrahim D, Samirm A (2017) NOX emission reduction by non thermal plasma technique. J Energy Environ Chem Eng 2:25–31
Reddy S, Johnson D, Gilmartin J (2008) Fluor's Econamine FG PlusSM technology for CO_2 capture at coal-fired power plants. In: Power plant air pollutant control mega symposium 2008, pp 25–28
Samanta S, Ghosh S (2011) Efficiency improvement and CO_2 emission reduction of coal-fired power plant by repowering through pressurized pulverized coal combustion and waste heated feed water heating. J Power Technol 91
Shen M, Hu Z, Kong F, Tong L, Yin S, Liu C, Zhang P, Wang L, Ding Y (2023) Comprehensive technology and economic evaluation based on the promotion of large-scale carbon capture and storage demonstration projects. Rev Environ Sci Bio/Technol 22:823–885
Shen M, Tong L, Yin S, Liu C, Wang L, Feng W, Ding Y (2022) Cryogenic technology progress for CO_2 capture under carbon neutrality goals: a review. Sep Purif Technol 121734
Wang A, Song Q, Tu G, Wang H, Yue Y, Yao Q (2014) Influence of flue gas cleaning system on characteristics of PM 2.5 emission from coal-fired power plants. Int J Coal Sci & Technol 1:4–12
Xiaolong L, Daobin Z, Jiuxiang D, Wenjie Z, Junzhuang L, Junbo Y (2018) Research on particulate matter emission characteristics of coal-fired power plants under ultra-low emission conditions. China Environ Monit 34:6
Xu Y, Liu X, Cui J, Chen D, Xu M, Pan S, Zhang K, Gao X (2016) Field measurements on the emission and removal of PM2.5 from coal-fired power stations: 4. PM removal performance of wet electrostatic precipitators. Energy & Fuels 30:7465–7473
Yongzheng G, Tiankun W, Yan H, Rui Z, Dong X ((2023) Carbon dioxide capture, utilization, and storage technologies and engineering applications in coal-fired power plants. Clean Coal Technol

Zhang X, Liu J, Tang Y, Zhao X, Yang H, Gerbens-Leenes P, van Vliet MT, Yan J (2017) China's coal-fired power plants impose pressure on water resources. J Clean Prod 161:1171–1179
Zhenghui F, Wei L, Yulei X, Guanzhong S, Kui H (2014) Analysis of CO_2 emission reduction potential in thermal power plants based on coal washing processes. China Electric Power 47:125–131

The manufacturer's authorised representative in the EU is Springer Nature Customer Service Centre GmbH, Europaplatz 3, 69115 Heidelberg, Germany. If you have any concerns regarding our products, please contact ProductSafety@springernature.com

Printed and bound by CPI Group (UK) Ltd, Croydon, CR0 4YY
01/10/2026
02225935-0005